Lecture Notes in Mathematics   2095

For further volumes:
http://www.springer.com/series/304

Daniele Angella

Cohomological Aspects in Complex Non-Kähler Geometry

 Springer

Daniele Angella
Dipartimento di Matematica
Università di Pisa
Pisa, Italy

ISBN 978-3-319-02440-0   ISBN 978-3-319-02441-7 (eBook)
DOI 10.1007/978-3-319-02441-7
Springer Cham Heidelberg New York Dordrecht London

Lecture Notes in Mathematics ISSN print edition: 0075-8434
                              ISSN electronic edition: 1617-9692

Library of Congress Control Number: 2013952956

Mathematics Subject Classification (2010): 32Q99, 53C55, 32Q60, 32C35, 57T15, 32G05, 32G07,
                                            53D05, 53D18

Printed on acid-free paper

Springer is part of Springer Science+Business Media (www.springer.com)

In ricordo di mio nonno,
il Maestro Giovanni

Introduction

By a remarkable result by W. L. Chow, [Cho49, Theorem V], see also [Ser56], *projective manifolds* (that is, compact complex submanifolds of $\mathbb{CP}^n := \left(\mathbb{C}^{n+1} \setminus \{0\}\right) \big/ (\mathbb{C} \setminus \{0\})$, for $n \in \mathbb{N}$) are in fact *algebraic* (that is, they can be described as the zero set of finitely many homogeneous holomorphic polynomials). One is hence interested in relaxing the projective condition, looking for special properties on compact manifolds sharing a weaker structure than projective manifolds. For example, a large amount of developed analytic techniques allows to prove strong cohomological properties for compact *Kähler manifolds* (that is, compact complex manifolds endowed with a Kähler metric, namely, a Hermitian metric admitting a local potential function), [SvD30, Käh33], see also [Wei58], which are, in a certain sense, the "analytic-versus-algebraic", [Cho49, Theorem V], or the "$\mathbb{R}$-versus-$\mathbb{Q}$", [Kod54, Theorem 4], version of projective manifolds. On the one side, there is the class of Kähler manifolds, which are in fact endowed with three different structures, interacting each other: a *complex* structure, a *symplectic* structure, and a *metric* structure; it is the strong linking between them that allows to develop many analytic tools and hence to derive the very special properties of Kähler manifolds. On the other side (the Dark Side...), in order to further investigate any of such properties and to understand what of these three structures is actually involved and required, it is natural to look for manifolds not admitting any Kähler structure: a large amount of interesting non-Kähler manifolds has been provided since [Thu76]. In other words, one is led to study complex, symplectic, and metric contributions separately, possibly weakening either the interactions between them, or one of these structures. For example, by relaxing the metric condition, one could ask what properties of a compact complex manifold can be deduced by the existence of special Hermitian metrics defined by conditions similar to, but weaker than, the defining condition of the Kähler metrics (for example, metrics being *balanced* in the sense of M. L. Michelsohn [Mic82], *pluriclosed* [Bis89], *astheno-Kähler* [JY93, JY94], *Gauduchon* [Gau77], *strongly-Gauduchon* [Pop13]); by relaxing the complex structure, one is led to study properties of *almost-complex* manifold, possibly endowed with compatible symplectic structures.

In these notes, we are concerned with summarizing some recent results on the cohomological properties of compact complex manifolds endowed with no Kähler structure. Cohomological aspects of manifolds endowed with almost-complex structures, or with other special structures (such as, for example, symplectic, generalized-complex, . . .) are also considered.

We recall that a *complex* manifold X is endowed with a natural *almost-complex* structure, that is, an endomorphism $J \in \mathrm{End}(TX)$ of the tangent bundle of X such that $J^2 = -\mathrm{id}_{TX}$, which actually satisfies a further integrability condition, [NN57, Theorem 1.1]. By considering the decomposition into eigen-spaces, just the datum of the almost-complex structure yields a splitting of the complexified tangent bundle, namely,

$$TX \otimes \mathbb{C} = T^{1,0}X \oplus T^{0,1}X ,$$

and hence it induces also a splitting of the bundle of complex differential forms, namely,

$$\wedge^\bullet X \otimes_\mathbb{R} \mathbb{C} = \bigoplus_{p+q=\bullet} \wedge^{p,q} X .$$

Furthermore, on a complex manifold, the integrability condition of such an almost-complex structure yields a further structure on $\wedge^{\bullet,\bullet} X$, namely, a structure of double complex $\left(\wedge^{\bullet,\bullet} X, \partial, \bar\partial \right)$, where ∂ and $\bar\partial$ are the components of the $\mathbb{C}$-linear extension of the exterior differential d.

Hence, on a complex manifold X, one can consider both the de Rham cohomology

$$H^\bullet_{dR}(X;\mathbb{C}) := \frac{\ker \mathrm{d}}{\mathrm{im}\, \mathrm{d}}$$

and the Dolbeault cohomology

$$H^{\bullet,\bullet}_{\bar\partial}(X) := \frac{\ker \bar\partial}{\mathrm{im}\, \bar\partial} ;$$

whenever X is compact, the Hodge theory assures that they have finite dimension as $\mathbb{C}$-vector spaces. On a compact complex manifold, in general, no natural map between $H^{\bullet,\bullet}_{\bar\partial}(X)$ and $H^\bullet_{dR}(X;\mathbb{C})$ exists; on the other hand, the structure of double complex of $\left(\wedge^{\bullet,\bullet} X, \partial, \bar\partial \right)$ gives rise to a spectral sequence

$$E^{\bullet,\bullet}_1 \simeq H^{\bullet,\bullet}_{\bar\partial}(X) \Rightarrow H^\bullet_{dR}(X;\mathbb{C}) ,$$

from which one gets the *Frölicher inequality*, [Frö55, Theorem 2]: for every $k \in \mathbb{N}$,

$$\dim_{\mathbb{C}} H^k_{dR}(X;\mathbb{C}) \leq \sum_{p+q=k} \dim_{\mathbb{C}} H^{p,q}_{\bar{\partial}}(X) .$$

On a complex manifold, a "bridge" between the Dolbeault and the de Rham cohomology is provided, in a sense, by the *Bott-Chern cohomology*,

$$H^{\bullet,\bullet}_{BC}(X) := \frac{\ker \partial \cap \ker \bar{\partial}}{\operatorname{im} \partial\bar{\partial}} ,$$

and the *Aeppli cohomology*,

$$H^{\bullet,\bullet}_{A}(X) := \frac{\ker \partial\bar{\partial}}{\operatorname{im} \partial + \operatorname{im} \bar{\partial}} .$$

In fact, the identity induces the maps of (bi-)graded $\mathbb{C}$-vector spaces

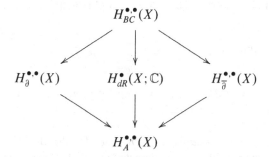

which, in general, are neither injective nor surjective.

These cohomology groups have been introduced by R. Bott and S. S. Chern in [BC65], and by A. Aeppli in [Aep65], and studied by several authors in different contexts: among others, B. Bigolin [Big69, Big70], A. Andreotti and F. Norguet [AN71], J. Varouchas [Var86], M. Abate [Aba88], L. Alessandrini and G. Bassanelli [AB96], S. Ofman [Ofm85a, Ofm85b, Ofm88], S. Boucksom [Bou04], J.-P. Demailly [Dem12], M. Schweitzer [Sch07], L. Lussardi [Lus10], R. Kooistra [Koo11], J.-M. Bismut [Bis11b, Bis11a], L.-S. Tseng, and S.-T. Yau [TY11].

We recall that whenever X is compact, the Hodge theory can be performed also for Bott-Chern and Aeppli cohomologies, [Sch07, §2], yielding their finite-dimensionality; more precisely, one has that, on a compact complex manifold X of complex dimension n endowed with a Hermitian metric,

$$H^{\bullet,\bullet}_{BC}(X) \simeq \tilde{\Delta}_{BC} \qquad \text{and} \qquad H^{\bullet,\bullet}_{A}(X) \simeq \tilde{\Delta}_{A} ,$$

where $\tilde{\Delta}_{BC}$ and $\tilde{\Delta}_A$ are 4th order self-adjoint elliptic differential operators; furthermore, the Hodge-$*$-operator associated to any Hermitian metric on X induces an isomorphism $H^{p,q}_{BC}(X) \simeq H^{n-q,n-p}_{A}(X)$, for every $p,q \in \mathbb{N}$.

By the definitions, the map $H_{BC}^{\bullet,\bullet}(X) \rightarrow H_{dR}^{\bullet}(X;\mathbb{C})$ is injective if and only if every ∂-closed $\overline{\partial}$-closed d-exact form is $\partial\overline{\partial}$-exact: a compact complex manifold fulfilling this property is said to satisfy the $\partial\overline{\partial}$-*Lemma*; see [DGMS75] by P. Deligne, Ph. A. Griffiths, J. Morgan, and D. P. Sullivan, where consequences of the validity of the $\partial\overline{\partial}$-Lemma on the real homotopy type of a compact complex manifold are investigated. When the $\partial\overline{\partial}$-Lemma holds, it turns out that actually all the above maps are isomorphisms, [DGMS75, Lemma 5.15, Remark 5.16, 5.21]: in particular, one gets a decomposition

$$H_{dR}^{\bullet}(X;\mathbb{C}) \simeq \bigoplus H_{\overline{\partial}}^{\bullet,\bullet}(X) \qquad \text{such that} \qquad H_{\overline{\partial}}^{\bullet_1,\bullet_2}(X) \simeq \overline{H_{\overline{\partial}}^{\bullet_2,\bullet_1}(X)} \,.$$

A very remarkable property of compact Kähler manifolds is that they satisfy the $\partial\overline{\partial}$-Lemma, [DGMS75, Lemma 5.11]: this follows from the Kähler identities, which can be proven as a consequence of the fact that the Kähler metrics osculate to order 2 the standard Hermitian metric of $\mathbb{C}^n$ at every point. Therefore, the above decomposition holds true, in particular, for compact Kähler manifolds, [Wei58, Théorème IV.3].

In particular, if X is a compact complex manifold satisfying the $\partial\overline{\partial}$-Lemma, then, for every $k \in \mathbb{N}$,

$$\dim_{\mathbb{C}} H_{dR}^k(X;\mathbb{C}) = \sum_{p+q=k} \dim_{\mathbb{C}} H_{BC}^{p,q}(X) \,.$$

In the first chapter, we study cohomological properties of compact complex manifolds, studying in particular the Bott-Chern and Aeppli cohomologies, and their relation with the $\partial\overline{\partial}$-Lemma.

In fact, we prove an *inequality à la Frölicher* for the Bott-Chern and Aeppli cohomologies, which provides also a characterization of the compact complex manifolds satisfying the $\partial\overline{\partial}$-Lemma just in terms of the dimensions of the Bott-Chern cohomology groups, [AT13b, Theorem A, Theorem B]; a key tool in the proof of the Frölicher-type inequality relies on exact sequences by J. Varouchas, [Var86]. More precisely, we state the following result.

Theorem (see Theorems 2.13 and 2.14). *Let X be a compact complex manifold. Then, for every $k \in \mathbb{N}$, the following inequality holds:*

$$\sum_{p+q=k} \left(\dim_{\mathbb{C}} H_{BC}^{p,q}(X) + \dim_{\mathbb{C}} H_A^{p,q}(X) \right) \geq 2 \dim_{\mathbb{C}} H_{dR}^k(X;\mathbb{C}) \,.$$

Furthermore, the equality

$$\sum_{p+q=k} \left(\dim_{\mathbb{C}} H_{BC}^{p,q}(X) + \dim_{\mathbb{C}} H_A^{p,q}(X) \right) = 2 \dim_{\mathbb{C}} H_{dR}^k(X;\mathbb{C})$$

holds for every $k \in \mathbb{N}$ if and only if X satisfies the $\partial\overline{\partial}$-Lemma.

Note that the equality $\sum_{p+q=k} \dim_{\mathbb{C}} H_{\overline{\partial}}^{p,q}(X) = \dim_{\mathbb{C}} H_{dR}^k(X;\mathbb{C})$ for every $k \in \mathbb{N}$ (which is equivalent to the degeneration of the Hodge and Frölicher spectral sequence at the first step, $E_1 \simeq E_\infty$) is not sufficient to let X satisfy the $\partial\bar\partial$-Lemma: in some sense, the above result states that the Bott-Chern cohomology, together with its dual, the Aeppli cohomology, encodes "more information" on the double complex $\left(\wedge^{\bullet,\bullet} X,\ \partial,\ \bar\partial\right)$ than just the Dolbeault cohomology.

As a straightforward consequence of the previous theorem, we obtain another proof, see [AT13b, Corollary 2.7], of the following stability result, see [Voi02, Proposition 9.21], [Wu06, Theorem 5.12], [Tom08, §B].

Corollary (see Corollary 2.2). *Satisfying the $\partial\bar\partial$-Lemma is a stable property under small deformations of the complex structure, that is, if $\{X_t\}_{t\in B}$ is a complex-analytic family of compact complex manifolds and X_{t_0} satisfies the $\partial\bar\partial$-Lemma for some $t_0 \in B$, then X_t satisfies the $\partial\bar\partial$-Lemma for every t in an open neighbourhood of t_0 in B.*

The previous results can be generalized to a more algebraic context, see [AT13a]. In particular, one gets applications concerning the cohomologies of symplectic manifolds and, more in general, of generalized complex manifolds, and to the study of the Hard Lefschetz Condition and of the $dd^{\mathcal{J}}$-Lemma, which we summarize in Appendix: Cohomological Properties of Generalized Complex Manifolds. (See Sects. 1.2 and 1.3 for preliminary results on symplectic and generalized complex structures and on their cohomologies.)

We recall that compact Kähler manifolds have special cohomological properties not only in the complex framework but also from the symplectic viewpoint: another important result, other than the Hodge decomposition theorem, [Wei58, Théorème IV.3], is the Lefschetz decomposition theorem, [Wei58, Théorème IV.5], which provides a decomposition of the de Rham cohomology in terms of primitive subgroups of the cohomology. Starting from J.-L. Koszul [Kos85] and J.-L. Brylinski [Bry88], several authors studied *symplectic geometry* from the cohomological point of view, see, e.g., [Mat95, Yan96, Cav05, TY12a, TY12b, Lin13]. More precisely, J.-L. Brylinski, aimed by drawing a parallel between the symplectic and the Riemannian cases, proposed in [Bry88] a Hodge theory for compact symplectic manifolds, introducing in particular the notion of symplectically harmonic form; O. Mathieu in [Mat95] and D. Yan in [Yan96] proved that any de Rham cohomology class admits a symplectically harmonic representative if and only if the so-called *Hard Lefschetz Condition* is satisfied. In [TY12a, TY12b], see also [TY11], L.-S. Tseng and S.-T. Yau introduced new cohomologies for symplectic manifolds: among them, in particular, they defined and studied a symplectic counterpart of the Bott-Chern and Aeppli cohomologies, further developing a Hodge theory.

By enlarging the study of the (co)tangent bundle to the study of the direct sum of tangent and cotangent bundles, complex structures and symplectic structures can be framed into a unified context, thanks to the notion of *generalized complex structure*, introduced by N. J. Hitchin in [Hit03] and developed, among others, by M. Gualtieri, [Gua04a, Gua11], and G. R. Cavalcanti, [Cav05], see also [Hit10, Cav07]. The notion of Bott-Chern cohomology, as well as the notion of $\partial\bar\partial$-Lemma, can be reformulated also in the generalized complex setting; in particular, one yields the so-called $dd^{\mathcal{J}}$-*Lemma*, see [Cav05]. Looking at symplectic structures as special cases of generalized complex structures, the $d d^{\mathcal{J}}$-Lemma turns out to be just the Hard Lefschetz Condition.

An inequality *à la* Frölicher, characterizing the validity of the Hard Lefschetz Condition, respectively, of the $d d^{\mathcal{J}}$-Lemma, holds also for symplectic manifolds, respectively for generalized complex manifolds, see Appendix: Cohomological Properties of Generalized Complex Manifolds.

In the second chapter, we consider *nilmanifolds* and, more in general, solv-manifolds. They are defined as compact quotients of connected simply-connected nilpotent, respectively solvable, Lie groups by co-compact discrete subgroups, and they constitute a fruitful and interesting source of examples in non-Kähler geometry. In fact, on the one hand, non-tori nilmanifolds admit no Kähler structure, [BG88, Theorem A], [Has89, Theorem 1, Corollary], and, on the other hand, focusing on *left-invariant* geometric structures on solvmanifolds, one can often reduce their study at the level of the associated Lie algebra; this turns out to hold true, in particular, for the de Rham cohomology of completely-solvable solvmanifolds, [Nom54, Hat60], and for the Dolbeault cohomology of nilmanifolds endowed with certain left-invariant complex structures, [Sak76, CFGU00, CF01, Rol09a, Rol11a], see, e.g., [Con06, Rol11a].

More precisely, on a nilmanifold $X = \Gamma \backslash G$, the inclusion of the subcom-plex composed of the G-left-invariant forms on X (which is isomorphic to the complex $(\wedge^{\bullet} \mathfrak{g}^*, d)$, where $\mathfrak{g}$ is the associated Lie algebra) turns out to be a quasi-isomorphism, [Nom54, Theorem 1], that is,

$$ i \colon H_{dR}^{\bullet}(\mathfrak{g}; \mathbb{R}) := H^{\bullet}\left(\wedge^{\bullet} \mathfrak{g}^*, d\right) \xrightarrow{\simeq} H_{dR}^{\bullet}(X; \mathbb{R}) \; ; $$

a similar result holds true also for completely-solvable solvmanifolds, [Hat60, Corollary 4.2], and for the Dolbeault cohomology of nilmanifolds endowed with left-invariant complex structures belonging to certain classes, [Sak76, Theorem 1], [CFGU00, Main Theorem], [CF01, Theorem 2, Remark 4], [Rol09a, Theorem 1.10], [Rol11a, Corollary 3.10].

As a matter of notation, denote by $H_{\sharp}^{\bullet,\bullet}(\mathfrak{g}_{\mathbb{C}})$, for $\sharp \in \left\{\bar{\partial}, \partial, BC, A\right\}$, the cohomology of the corresponding subcomplex of G-left-invariant forms on a solvmanifold $X = \Gamma \backslash G$, with Lie algebra $\mathfrak{g}$, endowed with a G-left-invariant complex structure. The following result states a *theorem à la Nomizu* also for the Bott-Chern and Aeppli cohomologies, [Ang11, Theorem 3.7, Theorem 3.8, Theorem 3.9].

Theorem (see Theorems 3.5, 3.6, Remark 3.10, and Theorem 3.7). *Let $X = \Gamma \backslash G$ be a solvmanifold endowed with a G-left-invariant complex structure J, and denote the Lie algebra naturally associated to G by $\mathfrak{g}$. Suppose that the inclusions of the subcomplexes of G-left-invariant forms on X into the corresponding complexes of differential forms on X yield the isomorphisms*

$$ i \colon H_{dR}^{\bullet}(\mathfrak{g}; \mathbb{C}) \xrightarrow{\simeq} H_{dR}^{\bullet}(X; \mathbb{C}) \qquad and \qquad i \colon H_{\bar{\partial}}^{\bullet,\bullet}(\mathfrak{g}_{\mathbb{C}}) \xrightarrow{\simeq} H_{\bar{\partial}}^{\bullet,\bullet}(X) \; ; $$

in particular, this holds true if one of the following conditions holds:

- *X is holomorphically parallelizable;*
- *J is an Abelian complex structure;*
- *J is a nilpotent complex structure;*
- *J is a rational complex structure;*
- $\mathfrak{g}$ *admits a torus-bundle series compatible with J and with the rational structure induced by* Γ*;*
- $\dim_{\mathbb{R}} \mathfrak{g} = 6$ *and* $\mathfrak{g}$ *is not isomorphic to* $\mathfrak{h}_7 := (0^3, 12, 13, 23)$*.*

Then also

$$i \colon H_{BC}^{\bullet,\bullet}(\mathfrak{g}_{\mathbb{C}}) \xrightarrow{\simeq} H_{BC}^{\bullet,\bullet}(X) \qquad and \qquad i \colon H_A^{\bullet,\bullet}(\mathfrak{g}_{\mathbb{C}}) \xrightarrow{\simeq} H_A^{\bullet,\bullet}(X)$$

are isomorphisms.

Furthermore, if $\mathcal{C}(\mathfrak{g})$ *denotes the set of G-left-invariant complex structures on* X*, then the set*

$$\mathcal{U} := \left\{ J \in \mathcal{C}(\mathfrak{g}) \ : \ i \colon H_{\sharp_J}^{\bullet,\bullet}(\mathfrak{g}_{\mathbb{C}}) \xrightarrow{\simeq} H_{\sharp_J}^{\bullet,\bullet}(X) \right\}$$

is open in $\mathcal{C}(\mathfrak{g})$*, for* $\sharp \in \{\partial, \bar{\partial}, BC, A\}$*.*

The above result allows to explicitly compute the Bott-Chern cohomology for the *Iwasawa manifold*

$$\mathbb{I}_3 := \mathbb{H}(3; \mathbb{Z}[i]) \backslash \mathbb{H}(3; \mathbb{C})$$

and for its *small deformations*, where

$$\mathbb{H}(3; \mathbb{C}) := \left\{ \begin{pmatrix} 1 & z^1 & z^3 \\ 0 & 1 & z^2 \\ 0 & 0 & 1 \end{pmatrix} \in GL(3; \mathbb{C}) \ : \ z^1, z^2, z^3 \in \mathbb{C} \right\}$$

$$\text{and} \quad \mathbb{H}(3; \mathbb{Z}[i]) := \mathbb{H}(3; \mathbb{C}) \cap GL(3; \mathbb{Z}[i]).$$

The Iwasawa manifold is one of the simplest example of compact non-Kähler complex manifold: as an example of a holomorphically parallelizable manifold, it has been studied by I. Nakamura, [Nak75], who computed its Kuranishi space and classified the small deformations of $\mathbb{I}_3$ by means of the dimensions of their Dolbeault cohomology groups.

In Sect. 3.2.4, [Ang11, §5.3], we explicitly compute the Bott-Chern cohomology of the small deformations of the Iwasawa manifold, showing that it makes possible to give a finer classification of the small deformations $\{X_t\}_{t \in \Delta(0,\varepsilon) \subset \mathbb{C}^6}$ of $\mathbb{I}_3$ than the Dolbeault cohomology: more precisely, classes *(ii)* and *(iii)* in I. Nakamura's classification [Nak75, §3] are further subdivided into subclasses *(ii.a)* and *(ii.b)*, respectively *(iii.a)* and *(iii.b)*, according to the value of $\dim_{\mathbb{C}} H_{BC}^{2,2}(X_t)$.

Left-invariant complex structures on *six-dimensional nilmanifolds* have been classified by M. Ceballos, A. Otal, L. Ugarte, and R. Villacampa in [COUV11]. Hence, in Sect. 3.3, we provide the dimensions of the Bott-Chern cohomology for each of these complex structures, as computed in [AFR12] jointly with M. G. Franzini and F. A. Rossi. In view of [AT13b, Theorem A, Theorem B], such dimensions measure the non-Kählerianity of six-dimensional nilmanifolds.

Enlarging the class of nilmanifolds to *solvmanifolds*, several results concerning de Rham cohomology have been studied by A. Hattori [Hat60], G. D. Mostow [Mos54], D. Guan [Gua07], S. Console and A. M. Fino [CF11], and by H. Kasuya [Kas13a, Kas12a, CFK13]; as for Dolbeault cohomology, results have been obtained by H. Kasuya [Kas13b, Kas12a, Kas11, Kas12d, Kas12b]; as for Bott-Chern coho- mology, we have obtained some results on Bott-Chern cohomology in joint work with H. Kasuya, [AK12, AK13a]. In Appendix: Cohomology of Solvmanifolds we summarize some of these results, providing the *Nakamura manifold* as an explicit example.

In the third chapter, we do not require the integrability of the almost-complex structure, and we study cohomological properties of *almost-complex manifolds*, that is, differentiable manifolds endowed with a (possibly non-integrable) almost- complex structure.[1] In this case, the Dolbeault cohomology is not defined. However, following T.-J. Li and W. Zhang, [LZ09], one can consider, for every $p, q \in \mathbb{N}$, the subgroup

$$H_J^{(p,q),(q,p)}(X; \mathbb{R}) := \left\{ [\alpha] \in H_{dR}^{p+q}(X; \mathbb{R}) : \alpha \in (\wedge^{p,q} X \oplus \wedge^{q,p} X) \cap \wedge^{p+q} X \right\}$$
$$\subseteq H_{dR}^{p+q}(X; \mathbb{R}),$$

and the complex counterpart

$$H_J^{(p,q)}(X; \mathbb{C}) := \left\{ [\alpha] \in H_{dR}^{p+q}(X; \mathbb{C}) : \alpha \in \wedge^{p,q} X \right\} \subseteq H_{dR}^{p+q}(X; \mathbb{C}) .$$

If X is a compact Kähler manifold, then $H_J^{(p,q)}(X; \mathbb{C}) \simeq H_{\bar{\partial}}^{p,q}(X)$ for every $p, q \in \mathbb{N}$, [DLZ10, Lemma 2.15, Theorem 2.16]; therefore these subgroups can be considered, in a sense, as a generalization of the Dolbeault cohomology groups to the non-Kähler, or to the non-integrable, case.

Two remarks need to be pointed out. Firstly, note that, in general, neither the equality in

$$\sum_{\substack{p+q=k \\ p \leq q}} H_J^{(p,q),(q,p)}(X; \mathbb{R}) \subseteq H_{dR}^{p+q}(X; \mathbb{R}), \quad \text{or} \quad \sum_{p+q=k} H_J^{(p,q)}(X; \mathbb{C}) \subseteq H_{dR}^{p+q}(X; \mathbb{C}),$$

---

[1]The theory developed by T.-J. Li and W. Zhang for almost-complex structures can actually be restated also in the symplectic, [AT12b], and in the **D**-complex settings, [AR12]. We recall that, in a sense, **D**-*complex geometry* is the "hyperbolic analogue" of complex geometry, and that many connections between it and other theory both in Mathematics and in Physics have been investigated in the last years, see, e.g., [HL83, AMT09, CMMS04, CMMS05, CM09, CFAG96, KMW10, ABDMO05, AS05, Kra10, Ros12a, Ros12b].

holds, nor the sum is direct, nor there are relations between the equality holding and the sum being direct, see, e.g., Proposition 4.1. Hence, one may be interested in studying compact almost-complex manifolds for which one of the above properties holds, at least for a fixed $k \in \mathbb{N}$, see [LZ09, DLZ10, DLZ11, FT10, AT11, AT12a, Zha13, ATZ12, DZ12, HMT11, LT12, DL13, DLZ12]. A remarkable result by T. Drăghici, T.-J. Li, and W. Zhang, [DLZ10, Theorem 2.3], states that every almost-complex structure J on a compact four-dimensional manifold X^4 satisfies the cohomological decomposition

$$H_{dR}^2\left(X^4; \mathbb{R}\right) = H_J^{(2,0),(0,2)}\left(X^4; \mathbb{R}\right) \oplus H_J^{(1,1)}\left(X^4; \mathbb{R}\right) .$$

Secondly, note that $J\lfloor_{\wedge^2 X}$ satisfies $\left(J\lfloor_{\wedge^2 X}\right)^2 = \mathrm{id}_{\wedge^2 X}$; therefore the above subgroups of $H_{dR}^2(X; \mathbb{R})$ can be interpreted as the subgroup represented by J-invariant forms,

$$H_J^+(X) := H_J^{(1,1)}(X; \mathbb{R}) = \left\{ [\alpha] \in H_{dR}^2(X; \mathbb{R}) : J\alpha = \alpha \right\} ,$$

and the subgroup represented by J-anti-invariant forms,

$$H_J^-(X) := H_J^{(2,0),(0,2)}(X; \mathbb{R}) = \left\{ [\alpha] \in H_{dR}^2(X; \mathbb{R}) : J\alpha = -\alpha \right\} .$$

Note also that if g is any Hermitian metric on X whose associated $(1, 1)$-form $\omega := g(J\cdot, \cdot\cdot) \in \wedge^{1,1} X \cap \wedge^2 X$ is d-closed (namely, g is an *almost-Kähler* metric on X), then $[\omega] \in H_J^+(X)$.

In fact, T.-J. Li and W. Zhang's interest in studying such subgroups and $\mathcal{C}^\infty$-*pure-and-full* almost-complex structures (that is, almost-complex structures for which the decomposition

$$H_{dR}^2(X; \mathbb{R}) = H_J^+(X) \oplus H_J^-(X)$$

holds, [LZ09, Definition 2.2, Definition 2.3, Lemma 2.2]) arises in investigating the symplectic cones of an almost-complex manifold, that is, the J-*tamed cone*

$$\mathcal{K}_J^t := \left\{ [\omega] \in H_{dR}^2(X; \mathbb{R}) : \omega_x\left(v_x, J_x v_x\right) > 0 \right.$$

$$\text{for every } v_x \in T_x X \setminus \{0\} \text{ and for every } x \in X \}$$

and the J-*compatible cone*

$$\mathcal{K}_J^c := \left\{ [\omega] \in H_{dR}^2(X; \mathbb{R}) : \omega_x\left(v_x, J_x v_x\right) > 0 \right.$$

$$\text{for every } v_x \in T_x X \setminus \{0\} \text{ and for every } x \in X, \text{ and } J\omega = \omega \} .$$

Indeed, they proved in [LZ09, Theorem 1.1] that, given a $\mathcal{C}^\infty$-pure-and-full almost-Kähler structure on a compact manifold X, the J-anti-invariant subgroup $H_J^-(X)$

of $H^2_{dR}(X;\mathbb{R})$ measures the quantitative difference between the J-tamed cone and the J-compatible cone, namely,

$$\mathcal{K}^t_J = \mathcal{K}^c_J \oplus H^-_J(X).$$

A natural question concerns the qualitative comparison between the tamed cone and the compatible cone: more precisely, one could ask whether, whenever an almost-complex structure J admits a J-tamed symplectic form, there exists also a J-compatible symplectic form. This turns out to be false, in general, for non-integrable almost-complex structures in dimension greater than 4, [MT00, Tom02], see also Theorem 4.6; on the other hand, it is not (yet) known whether, for almost-complex structures on compact four-dimensional manifolds, as asked by S. K. Donaldson, [Don06, Question 2], or for complex structures on compact manifolds of complex dimension greater than or equal to 3, as asked by T.-J. Li and W. Zhang, [LZ09, page 678], and by J. Streets and G. Tian, [ST10, Question 1.7], it holds that $\mathcal{K}^c_J$ is non-empty if and only if $\mathcal{K}^t_J$ is non-empty. We prove in Theorem 4.13 that no counterexample can be found among six-dimensional non-tori nilmanifolds endowed with left-invariant complex structures, [AT11, Theorem 3.3]; note that the same holds true, more in general, for higher dimensional nilmanifolds, as proven by N. Enrietti, A.M. Fino, and L. Vezzoni, [EFV12, Theorem 1.3].

Theorem (see Theorem 4.13). *Let $X = \Gamma \backslash G$ be a six-dimensional nilmanifold endowed with a G-left-invariant complex structure J. If X is not a torus, then there is no J-tamed symplectic structure on X.*

One can study further cones in cohomology, which are related to special metrics, other than Kähler metrics; a key tool is provided by the theory of cone structures on differentiable manifolds developed by D. P. Sullivan, [Sul76]. In order to compare, in particular, the cone associated to *balanced metrics* (that is, Hermitian metrics whose associated $(1,1)$-form is co-closed, [Mic82, Definition 1.4, Theorem 1.6]) and the cone associated to *strongly-Gauduchon metrics* (that is, Hermitian metrics whose associated $(1,1)$-form ω satisfies the condition that $\partial\left(\omega^{\dim_{\mathbb{C}} X-1}\right)$ is $\bar{\partial}$-exact [Pop09, Definition 3.1]), we give the following result, [AT12a, Theorem 2.9], which is the semi-Kähler counterpart of [LZ09, Theorem 1.1]. (We refer to Sect. 4.4.3 for the definitions of the cones $\mathcal{K}b^t_J$ and $\mathcal{K}b^c_J$ on a manifold X endowed with an almost-complex structure J.)

Theorem (see Theorem 4.19). *Let X be a compact $2n$-dimensional manifold endowed with an almost-complex structure J. Assume that $\mathcal{K}b^c_J \neq \varnothing$ (that is, there exists a semi-Kähler structure on X) and that $0 \notin \mathcal{K}b^t_J$. Then*

$$\mathcal{K}b^t_J \cap H^{(n-1,n-1)}_J(X;\mathbb{R}) = \mathcal{K}b^c_J$$

and

$$\mathcal{K}b^c_J + H^{(n,n-2),(n-2,n)}_J(X;\mathbb{R}) \subseteq \mathcal{K}b^t_J.$$

Moreover, if the equality $H_{dR}^{2n-2}(X;\mathbb{R}) = H_J^{(n,n-2),(n-2,n)}(X;\mathbb{R}) + H_J^{(n-1,n-1)}(X;\mathbb{R})$
holds, then

$$\mathcal{K}b_J^c + H_J^{(n,n-2),(n-2,n)}(X;\mathbb{R}) = \mathcal{K}b_J^t .$$

In order to better understand cohomological properties of compact almost-complex manifolds, and in view of the Hodge decomposition theorem for compact Kähler manifolds, it could be interesting to investigate the subgroups $H_J^{(p,q),(q,p)}(X;\mathbb{R})$ for almost-complex manifolds endowed with special structures. For example, we prove the following result, [ATZ12, Proposition 4.1], providing a strong difference between the Kähler case and the almost-Kähler case.

Proposition (see Proposition 4.8). *The differentiable manifold X underlying the Iwasawa manifold* $\mathbb{I}_3 := \mathbb{H}(3;\mathbb{Z}[\mathrm{i}]) \backslash \mathbb{H}(3;\mathbb{C})$ *admits a non-$\mathcal{C}^\infty$-pure-and-full almost-Kähler structure.*

A further study on almost-Kähler structures (J, ω, g) concerns the connections between $\mathcal{C}^\infty$-pure-and-fullness and the *Lefschetz-type property on 2-forms* firstly considered by W. Zhang, that is, the property that the Lefschetz operator

$$\omega^{n-2} \wedge \cdot : \wedge^2 X \to \wedge^{2n-2} X$$

takes g-harmonic 2-forms to g-harmonic $(2n-2)$-forms, see, e.g., Theorem 4.4; we refer to [ATZ12] for further results.

As a tool to study explicit examples, we provide a Nomizu-type theorem for the subgroups $H_J^{(p,q),(q,p)}(X;\mathbb{R})$ of a completely-solvable solvmanifold $X = \Gamma \backslash G$ endowed with a G-left-invariant almost-complex structure J, [ATZ12, Theorem 5.4], see Proposition 4.2, and Corollary 4.2.

A remarkable result by K. Kodaira and D. C. Spencer states that the Kähler property on compact complex manifolds is stable under *small deformations* of the complex structure, [KS60, Theorem 15]: more precisely, it states that, given a compact complex manifold admitting a Kähler structure, every small deformation still admits a Kähler structure; it can be proven as a consequence of the semi-continuity properties for the dimensions of the cohomology groups of a compact Kähler manifold. Hence, a natural question in non-Kähler geometry is to investigate the (in)stability of weaker properties than being Kähler. As a first result in this direction, L. Alessandrini and G. Bassanelli proved that, given a compact complex manifold, the property of admitting a *balanced metric* (that is, a Hermitian metric whose associated (1, 1)-form is co-closed) is not stable under small deformations of the complex structure, [AB90, Proposition 4.1]; on the other hand, they proved that the class of balanced manifolds is stable under modifications, [AB96, Corollary 5.7]. Another result in this context is the stability of the property of satisfying the $\partial\bar{\partial}$-Lemma under small deformations of the complex structure, as already recalled, see Corollary 2.2.

Therefore, it is natural to investigate stability properties for the cohomological decomposition by means of the subgroups $H_J^{(p,q),(q,p)}(X;\mathbb{R})$ on (almost-)complex manifolds (X, J). More precisely, we consider the Iwasawa manifold $\mathbb{I}_3 :=$ $\mathbb{H}(3;\mathbb{Z}[i])\backslash\mathbb{H}(3;\mathbb{C})$, showing that the subgroups $H_J^{(p,q),(q,p)}(X;\mathbb{R})$ provide a cohomological decomposition for $\mathbb{I}_3$ but not for some of its small deformations, Theorem 4.8. We prove the following result, [AT11, Theorem 3.2].

Theorem (see Theorem 4.7). *The properties of being C^∞-pure-and-full is not stable under small deformations of the complex structure.*

More in general, one could try to study directions along which the *curves of almost-complex structures* on a differentiable manifold preserve the property of being C^∞-pure-and-full. We use a procedure by J. Lee, [Lee04, §1], to construct curves of almost-complex structures through an almost-complex structure J, by means of J-anti-invariant real 2-forms, in order to provide examples, see, e.g., [AT11, Theorem 4.1], see Theorem 4.9.

Another problem in deformation theory is the study of *semi-continuity properties* for the dimensions of the subgroups $H_J^+(X)$ and $H_J^-(X)$. As a consequence of the Hodge theory for compact four-dimensional manifolds, T. Drăghici, T.-J. Li, and W. Zhang proved in [DLZ11, Theorem 2.6] that, given a curve $\{J_t\}_{t\in I\subseteq\mathbb{R}}$ of (C^∞-pure-and-full) almost-complex structures on a compact four-dimensional manifold X, the functions

$$I \ni t \mapsto \dim_\mathbb{R} H_{J_t}^-(X) \in \mathbb{N} \qquad \text{and} \qquad I \ni t \mapsto \dim_\mathbb{R} H_{J_t}^+(X) \in \mathbb{N}$$

are, respectively, upper-semi-continuous and lower-semi-continuous. In higher dimension this fails to be true, as we show in explicit examples, [AT12a, Proposition 4.1, Proposition 4.3], see Propositions 4.9 and 4.10. Motivated by such counterexamples, one can study a stronger semi-continuity property on almost-complex manifolds (namely, that, for every d-closed J-invariant real 2-form α, there exists a d-closed J_t-invariant real 2-form $\eta_t = \alpha + o(1)$, depending real-analytically in t, for $t \in (-\varepsilon, \varepsilon)$ with $\varepsilon > 0$ small enough): we give a formal characterization of the curves of almost-complex structures satisfying such a property, [AT12a, Proposition 4.5], see Proposition 4.11, and we provide also a counterexample to such a stronger semi-continuity property, [AT12a, Proposition 4.9], see Proposition 4.12.

The plan of these notes is as follows.

In Chap. 1, we collect the basic notions concerning (almost-)complex, symplectic, and generalized complex structures, we recall the main results on Hodge theory for Kähler manifolds, and we summarize the classical results on deformations of complex structures, on currents and de Rham homology, and on solvmanifolds.

In Chap. 2, we study cohomological aspects of compact complex manifolds, focusing in particular on the study of the Bott-Chern cohomology, [AT13b, AT13a]. By using exact sequences introduced by J. Varouchas, [Var86], we prove a Frölicher-type inequality for the Bott-Chern cohomology, Theorem 2.13, which also provides a characterization of the validity of the $\partial\bar{\partial}$-Lemma in terms of the dimensions of the Bott-Chern cohomology groups, Theorem 2.14. Finally, we collect some results concerning cohomological aspects of symplectic geometry and, more in general, of generalized complex geometry.

In Chap. 3, we study Bott-Chern cohomology of nilmanifolds, [Ang11, AK12], (see also [AK13a, AK13b]). We prove a result à la Nomizu for the Bott-Chern cohomology, showing that, for certain classes of complex structures on nilmanifolds, the Bott-Chern cohomology is completely determined by the associated Lie algebra endowed with the induced linear complex structure, Theorems 3.5, 3.6, and 3.7. As an application, in Sect. 3.2, we explicitly study the Bott-Chern and Aeppli cohomologies of the Iwasawa manifold and of its small deformations. Finally, we summarize some results concerning cohomologies of solvmanifolds.

In Chap. 4, we study cohomological properties of almost-complex manifolds, [AT11, AT12a, ATZ12]. Firstly, in Sect. 4.1, we recall the notion of C^∞-pure-and-full almost-complex structure, which has been introduced by T.-J. Li and W. Zhang in [LZ09] in order to investigate the relations between the compatible and the tamed symplectic cones on a compact almost-complex manifold and with the aim to throw light on a question by S. K. Donaldson, [Don06, Question 2]. In particular, we are interested in studying when certain subgroups, related to the almost-complex structure, let a splitting of the de Rham cohomology of an almost-complex manifold, and their relations with cones of metric structures. In Sect. 4.2, we focus on C^∞-pure-and-fullness on several classes of (almost-)complex manifolds, e.g., solvmanifolds endowed with left-invariant almost-complex structures, semi-Kähler manifolds, almost-Kähler manifolds. In Sect. 4.3, we study the behaviour of C^∞-pure-and-fullness under small deformations of the complex structure and along curves of almost-complex structures, investigating properties of stability, Theorems 4.7 and 4.9, and of semi-continuity for the dimensions of the invariant and anti-invariant subgroups of the de Rham cohomology with respect to the almost-complex structure, Propositions 4.9, 4.10, 4.11, and 4.12. In Sect. 4.4, we consider the cone of semi-Kähler structures on a compact almost-complex manifold and, in particular, by adapting the results by D. P. Sullivan on cone structures, [Sul76], we compare the cones of balanced metrics and of strongly-Gauduchon metrics on a compact complex manifold (see Theorem 4.19).

This work has been originated from the author's Ph.D. thesis at Dipartimento di Matematica of Università di Pisa, under the advice of prof. Adriano Tomassini, [Ang13b]. Part of the original results are contained in [AT11, AT12a, Ang11, AT13b, AR12, ATZ12, AC12, AT12b, Ang13a, AFR12, AK12, AT13a] (see also [AC13, AK13a, AK13b]).

Acknowledgments

The existence of these notes is mainly due to the advice, guidance, teachings, patience, suggestions, and support that my adviser Adriano Tomassini has given me during the last years: then my first thanks goes to him for making me grow up a lot.

I also wish to thank Jean-Pierre Demailly for his hospitality at Institut Fourier, for his explanations and his kind answers to my questions, and for his support and encouragement.

Many thanks to Marco Abate for all his kindness and for his very useful suggestions: his advices improved a lot the presentation of these notes.

A particular thanks goes to the director of the Ph.D. school "Galileo Galilei", Fabrizio Broglia, for his support and his help during my years in Pisa.

Thanks also to Ute McCrory, Friedhilde Meyer, and K. Vinodhini, for their kindly assistance in the preparation of the volume.

I think mathematics makes sense just when played in two, or more. Hence I would like to thank all my collaborators: Federico Alberto Rossi, Simone Calamai, Weiyi Zhang, Maria Giovanna Franzini, and Hisashi Kasuya.

My growth as a mathematician is due to very many useful conversations and discussions (on mathematics, and beyond), sometimes really brief, sometimes everlasting, but always very inspiring, especially with Lucia Alessandrini, Amedeo Altavilla, Claudio Arezzo, Paolo Baroni, Luca Battistella, Leonardo Biliotti, Junyan Cao, Carlo Collari, Laura Cremaschi, Alberto Della Vedova, Valentina Disarlo, Tian-Jun Li, Tedi Drăghici, Nicola Enrietti, Anna Fino, Alberto Gioia, Serena Guarino Lo Bianco, Greg Kuperberg, Minh Nguyet Mach, Maura Macrì, Gunnar Þór Magnússon, John Mandereau, Daniele Marconi, Costantino Medori, Samuele Mongodi, Isaia Nisoli, Marco Pasquali, Maria Rosaria Pati, David Petrecca, Massimiliano Pontecorvo, Maria Beatrice Pozzetti, Jasmin Raissy, Sönke Rollenske, Matteo Ruggiero, Matteo Serventi, Marco Spinaci, Cristiano Spotti, Herman Stel, Pietro Tortella, Luis Ugarte, Andrea Villa, and many others.

I wish to thank all the members and the staff of the three Departments of Mathematics where I spend part of my life: the Dipartimento di Matematica of the Università di Pisa; the Dipartimento di Matematica e Informatica of the Università di Parma; and the Institut Fourier in Grenoble. In particular, thanks to Daniele, Matteo, Alberto, Marco, Carlo, Jasmin, Cristiano, Paolo, Simone, Eridano, Pietro, Fabio, Amedeo, Martino, Giandomenico, Luigi, Francesca, Sara, Michele, Minh, Isaia, John, Andrea, David, Ana, Tiziano, Andrea, Maria Rosaria, Giovanni, Marco, Sara, Giuseppe, Matteo, Valentina, Laura, Paolo, Laura, Alessio, Flavia, Stefano, Cristina, Abramo, ...

Very special *gracias* to Serena, Maria Beatrice and Luca, Andrea, Simone, Paolo, Laura, Matteo, Chiara, Michele: morally, you know you should be considered as a sort of co-authors.

Pisa, Italy   Daniele Angella

Contents

1 Preliminaries on (Almost-)Complex Manifolds 1
  1.1 Almost-Complex Geometry and Complex Geometry 1
      1.1.1 Almost-Complex Structures 2
      1.1.2 Complex Structures, and Dolbeault Cohomology 4
  1.2 Symplectic Geometry 9
      1.2.1 Symplectic Structures 9
      1.2.2 Cohomological Aspects of Symplectic Geometry 12
  1.3 Generalized Geometry 20
      1.3.1 Generalized Complex Structures 20
      1.3.2 Cohomological Aspects of Generalized Complex
            Geometry 23
      1.3.3 Complex Structures and Symplectic Structures
            in Generalized Complex Geometry 24
  1.4 Kähler Geometry 27
      1.4.1 Kähler Metrics 27
      1.4.2 Hodge Theory for Kähler Manifolds 30
      1.4.3 $\partial\bar{\partial}$-Lemma and Formality for Compact Kähler Manifolds ... 31
  1.5 Deformations of Complex Structures 35
  1.6 Currents and de Rham Homology 42
  1.7 Solvmanifolds 44
      1.7.1 Lie Groups and Lie Algebras 44
      1.7.2 Nilmanifolds and Solvmanifolds 46
  Appendix: Low Dimensional Solvmanifolds and Special Structures 52
      A.1 Solvmanifolds up to Dimension 4 52
      A.2 Five-Dimensional Solvmanifolds 55
      A.3 Six-Dimensional Nilmanifolds 55
      A.4 Six-Dimensional Solvmanifolds 63

2   Cohomology of Complex Manifolds 65
   2.1   Cohomologies of Complex Manifolds 65
         2.1.1   The Bott-Chern Cohomology 66
         2.1.2   The Aeppli Cohomology 69
         2.1.3   The $\partial\bar{\partial}$-Lemma.. 73
   2.2   Cohomological Properties of Compact Complex
         Manifolds and the $\partial\bar{\partial}$-Lemma 78
         2.2.1   J. Varouchas' Exact Sequences............................... 79
         2.2.2   An Inequality $à$ la Frölicher for the Bott-Chern
                 Cohomology 81
         2.2.3   A Characterization of the $\partial\bar{\partial}$-Lemma in Terms
                 of the Bott-Chern Cohomology 84
   Appendix: Cohomological Properties of Generalized Complex
         Manifolds.. 88

3   Cohomology of Nilmanifolds.. 95
   3.1   Cohomology Computations for Special Nilmanifolds 95
         3.1.1   Left-Invariant Complex Structures on Solvmanifolds 96
         3.1.2   Classical Results on Computations
                 of the de Rham and Dolbeault Cohomologies................. 97
         3.1.3   The Bott-Chern Cohomology on Solvmanifolds 103
   3.2   The Cohomologies of the Iwasawa Manifold
         and of Its Small Deformations... 111
         3.2.1   The Iwasawa Manifold and Its Small Deformations 111
         3.2.2   The de Rham Cohomology of the Iwasawa
                 Manifold and of Its Small Deformations 120
         3.2.3   The Dolbeault Cohomology of the Iwasawa
                 Manifold and of Its Small Deformations 122
         3.2.4   The Bott-Chern and Aeppli Cohomologies
                 of the Iwasawa Manifold and of Its Small Deformations 126
   3.3   Cohomologies of Six-Dimensional Nilmanifolds 130
   Appendix: Cohomology of Solvmanifolds................................. 140

4   Cohomology of Almost-Complex Manifolds 151
   4.1   Subgroups of the de Rham (Co)Homology
         of an Almost-Complex Manifold..................................... 151
         4.1.1 C^∞-Pure-and-Full and Pure-and-Full
                 Almost-Complex Structures................................... 152
         4.1.2   Relations Between C^∞-Pure-and-Fullness
                 and Pure-and-Fullness 161
   4.2 C^∞-Pure-and-Fullness for Special Manifolds 165
         4.2.1   Special Classes of C^∞-Pure-and-Full
                 (Almost-)Complex Manifolds................................. 165
         4.2.2 C^∞-Pure-and-Full Solvmanifolds........................... 168
         4.2.3   Complex-C^∞-Pure-and-Fullness
                 for Four-Dimensional Manifolds.............................. 173

4.2.4   Almost-Complex Manifolds with Large
Anti-invariant Cohomology 175
4.2.5   Semi-Kähler Manifolds... 177
4.2.6   Almost-Kähler Manifolds and Lefschetz-Type Property 184
4.3 C^∞-Pure-and-Fullness and Deformations
of (Almost-)Complex Structures 195
4.3.1   Deformations of C^∞-Pure-and-Full
Almost-Complex Structures.................................. 196
4.3.2   The Semi-continuity Problem 212
4.4   Cones of Metric Structures 220
4.4.1   Sullivan's Results on Cone Structures 221
4.4.2   The Cones of Compatible, and Tamed Symplectic
Structures... 222
4.4.3   The Cones of Semi-Kähler,
and Strongly-Gauduchon Metrics............................ 228

References... 233

Index... 247

Chapter 1
Preliminaries on (Almost-)Complex Manifolds

Abstract In this preliminary chapter, we summarize some basic notions and some classical results in (almost-)complex and symplectic geometry. In particular, we start by setting some definitions and notation concerning (almost-)complex structures, Sect. 1.1, symplectic structures, Sect. 1.2, and generalized complex structures, Sect. 1.3; then we recall the main results in the Hodge theory for Kähler manifolds, Sect. 1.4, and in the Kodaira, Spencer, Nirenberg, and Kuranishi theory of deformations of complex structures, Sect. 1.5; furthermore, we summarize some basic definitions and some useful facts about currents and de Rham homology, Sect. 1.6, and about nilmanifolds and solvmanifolds, Sect. 1.7, in order to set the notation for the following chapters. (As a matter of notation, unless otherwise stated, by "manifold" we mean "connected differentiable manifold", and by "compact manifold" we mean "closed manifold".)

1.1 Almost-Complex Geometry and Complex Geometry

The tangent bundle of a complex manifold X is naturally endowed with an endomorphism $J \in \mathrm{End}(TX)$ such that $J^2 = -\mathrm{id}_{TX}$, satisfying a further integrability property. It is hence natural to study differentiable manifolds endowed with such an endomorphism, the so-called *almost-complex* manifolds. It turns out that the vanishing of the *Nijenhuis tensor* Nij_J characterizes the almost-complex structures J on X naturally induced by a structure of complex manifold [NN57, Theorem 1.1].

In this section, we recall the notions of almost-complex structure, complex manifold, and Dolbeault cohomology, and some of their properties.

D. Angella, *Cohomological Aspects in Complex Non-Kähler Geometry*,
Lecture Notes in Mathematics 2095, DOI 10.1007/978-3-319-02441-7_1,
© Springer International Publishing Switzerland 2014

1.1.1 Almost-Complex Structures

Let X be a (differentiable) manifold. We start by recalling the notion of almost-complex structure.

Definition 1.1 ([Ehr49]). Let X be a differentiable manifold. An *almost-complex structure* J on X is an endomorphism $J \in \mathrm{End}(TX)$ such that $J^2 = -\mathrm{id}_{TX}$.

Let J be an almost-complex structure on X. Extending J by $\mathbb{C}$-linearity to $TX \otimes \mathbb{C}$, we get the decomposition

$$TX \otimes \mathbb{C} = T^{1,0}X \oplus T^{0,1}X ,$$

where $T^{1,0}X$ (respectively, $T^{0,1}X$) is the sub-bundle of $TX \otimes \mathbb{C}$ given by the i-eigen-spaces (respectively, the $(-\mathrm{i})$-eigen-spaces) of $J \in \mathrm{End}\,(TX \otimes \mathbb{C})$: that is, for every $x \in X$,

$$\left(T^{1,0}X\right)_x = \{v_x - \mathrm{i}\, J_x v_x \,:\, v_x \in T_x X\} ,$$
$$\left(T^{0,1}X\right)_x = \{v_x + \mathrm{i}\, J_x v_x \,:\, v_x \in T_x X\} .$$

Considering the dual of J, again denoted by $J \in \mathrm{End}\,(T^*X)$, we get analogously a decomposition at the level of the cotangent bundle:

$$T^*X \otimes \mathbb{C} = \left(T^{1,0}X\right)^* \oplus \left(T^{0,1}X\right)^* ,$$

where $\left(T^{1,0}X\right)^*$ (respectively, $\left(T^{0,1}X\right)^*$) is the sub-bundle of $T^*X \otimes \mathbb{C}$ given by the i-eigen-spaces (respectively, the $(-\mathrm{i})$-eigen-spaces) of the $\mathbb{C}$-linear extension $J \in \mathrm{End}\,(T^*X \otimes \mathbb{C})$. Extending the endomorphism J to the bundle $\wedge^\bullet\,(T^*X) \otimes \mathbb{C}$ of complex-valued differential forms, we get, for every $k \in \mathbb{N}$, the bundle decomposition

$$\wedge^k\,(T^*X) \otimes \mathbb{C} = \bigoplus_{p+q=k} \wedge^p\left(T^{1,0}X\right)^* \otimes \wedge^q\left(T^{0,1}X\right)^* .$$

As a matter of notation, we will denote by $\mathcal{C}^\infty\,(X;F)$ the space of smooth sections of a vector bundle F over X, and, for every $k \in \mathbb{N}$ and $(p,q) \in \mathbb{N}^2$, we will denote by $\wedge^k X := \mathcal{C}^\infty\left(X; \wedge^k\,(T^*X)\right)$ the space of smooth sections of $\wedge^k\,(T^*X)$ over X and by $\wedge^{p,q} X := \wedge_J^{p,q} X := \mathcal{C}^\infty\left(X; \wedge^p\left(T^{1,0}X\right)^* \otimes \wedge^q\left(T^{0,1}X\right)^*\right)$ the space of smooth sections of $\wedge^p\left(T^{1,0}X\right)^* \otimes \wedge^q\left(T^{0,1}X\right)^*$ over X.

Since $\mathrm{d}\left(\wedge^0 X \otimes_{\mathbb{R}} \mathbb{C}\right) \subseteq \wedge^{1,0}X \oplus \wedge^{0,1}X$ and $\mathrm{d}\left(\wedge^1 X \otimes_{\mathbb{R}} \mathbb{C}\right) \subseteq \wedge^{2,0}X \oplus \wedge^{1,1}X \oplus \wedge^{0,2}X$, since every differential form is locally a finite sum of decomposable differential forms, and by the Leibniz rule, the $\mathbb{C}$-linear extension of the exterior differential, $\mathrm{d} \colon \wedge^\bullet X \otimes \mathbb{C} \to \wedge^{\bullet+1}X \otimes \mathbb{C}$, splits into four components:

$$\mathrm{d} = A + \partial + \bar{\partial} + \bar{A}$$

where

$$A: \wedge^{\bullet,\bullet}X \to \wedge^{\bullet+2,\bullet-1}X \,, \qquad \partial: \wedge^{\bullet,\bullet}X \to \wedge^{\bullet+1,\bullet}X \,, \qquad \bar{\partial}: \wedge^{\bullet,\bullet}X \to \wedge^{\bullet,\bullet+1}X \,,$$

$$\bar{A}: \wedge^{\bullet,\bullet}X \to \wedge^{\bullet-1,\bullet+2}X \,;$$

in terms of these components, the condition $\mathrm{d}^2 = 0$ is written as

$$
\begin{cases}
A^2 = 0 \\
A\partial + \partial A = 0 \\
A\bar{\partial} + \partial^2 + \bar{\partial}A = 0 \\
A\bar{A} + \partial\bar{\partial} + \bar{\partial}\partial + A\bar{A} = 0 \\
\partial\bar{A} + \bar{\partial}^2 + \bar{A}\partial = 0 \\
\bar{A}\bar{\partial} + \bar{\partial}\bar{A} = 0 \\
\bar{A}^2 = 0
\end{cases}
\,.
$$

Remark 1.1. More in general, given a $2n$-dimensional manifold X, an *almost-c.p.s. structure*, [Vai07, Sect. 1], is the datum of an endomorphism $K \in \mathrm{End}(TX)$ such that $K^2 = \lambda \, \mathrm{id}_{TX}$ where $\lambda \in \{-1, 0, 1\}$: more precisely, the case $\lambda = -1$ corresponds to *almost-complex structures*, the case $\lambda = 0$ corresponds to *almost-subtangent structures*, and the case $\lambda = 1$ corresponds to *almost-product structures*.

Among almost-product structures, **D**-complex structures deserve special interest. By definition, an *almost-**D**-complex structure* (also called *almost-para-complex structure*) on a manifold X is an endomorphism $K \in \mathrm{End}(TX)$ such that $K^2 = \mathrm{id}_{TX}$ and $\mathrm{rk}\, T^+X = \mathrm{rk}\, T^-X = \frac{1}{2} \dim X$, where, for $\pm \in \{+, -\}$, the sub-bundle $T^\pm X$ of TX is the (± 1)-eigen-bundle of $K \in \mathrm{End}(TX)$. Recall also that a **D**-*holomorphic map* between two almost-**D**-complex manifolds (X_1, K_1) and (X_2, K_2) is a smooth map $f: X_1 \to X_2$ such that $\mathrm{d}\, f \circ K_1 = K_2 \circ \mathrm{d}\, f$. Recall that an almost-**D**-complex structure is said to be *integrable* (and hence it is called a **D**-*complex structure*, or also a *para-complex structure*) if $\left[T^+X, T^+X \right] \subseteq T^+X$ and $[T^-X, T^-X] \subseteq T^-X$. Equivalently, the integrability condition can be straightforwardly expressed as the vanishing of the *Nijenhuis tensor* of K, defined as $\mathrm{Nij}_K(\cdot, \cdot\cdot) := [\cdot, \cdot\cdot] + [K\cdot, K\cdot\cdot] - K[K\cdot, \cdot\cdot] - K[\cdot, K\cdot\cdot]$. Furthermore, for an almost-**D**-complex structure on an n-dimensional manifold X, the integrability is also equivalent to being naturally associated to a structure on X defined in terms of local homeomorphisms with open sets in $\mathbf{D}^n$ and **D**-holomorphic changes of coordinates, see, e.g., [CMMS04, Proposition 3], where $\mathbf{D}^n := \mathbb{R}^n + \tau \mathbb{R}^n$, with $\tau^2 = 1$, is the algebra of *double numbers*.

Finally, we recall that, given a $2n$-dimensional manifold endowed with an almost-**D**-complex structure K, a **D**-*Hermitian metric* on X is a pseudo-Riemannian metric of signature (n, n) such that $g(K\cdot, K\cdot\cdot) = -g(\cdot, \cdot\cdot)$. A **D**-*Kähler structure* on a manifold X is the datum of an integrable **D**-complex structure K and a **D**-Hermitian metric g such that its associated K-anti-invariant form $\omega := g(K\cdot, \cdot\cdot)$ is d-closed, equivalently, the datum of a K-*compatible* (that is, a K-anti-invariant) *symplectic form* on X, see, e.g., [AMT09, Sect. 5.1], [CMMS04, Theorem 1].

We refer, e.g., to [HL83, AMT09, CMMS04, CMMS05, CM09, CFAG96, KMW10, ABDMO05, AS05, Kra10, Ros12a, Ros12b, AR12, Ros13] and the references therein for more notions and results in **D**-complex geometry and for motivations for its study.

1.1.2   Complex Structures, and Dolbeault Cohomology

If X is a complex manifold, then there is a natural almost-complex structure on X: locally, in a holomorphic coordinate chart $\left(U, \{z^\alpha =: x^{2\alpha-1} + i\, x^{2\alpha}\}_{\alpha \in \{1,\dots,\dim_{\mathbb{C}} X\}}\right)$, with $\left(U, \{x^\alpha\}_{\alpha \in \{1,\dots,2\dim_{\mathbb{C}} X\}}\right)$ a (differential) coordinate chart, one defines, for every $\alpha \in \{1, \dots, \dim_{\mathbb{C}} X\}$,

$$J\left(\frac{\partial}{\partial x^{2\alpha-1}}\right) \overset{\text{loc}}{:=} \frac{\partial}{\partial x^{2\alpha}}, \qquad J\left(\frac{\partial}{\partial x^{2\alpha}}\right) \overset{\text{loc}}{:=} -\frac{\partial}{\partial x^{2\alpha-1}};$$

note that this local definition does not depend on the coordinate chart, by the Cauchy and Riemann equations.

Conversely, we recall that an almost-complex structure J on a differentiable manifold X is called *integrable* if it is the natural almost-complex structure induced by a structure of complex manifold on X.

The following theorem by A. Newlander and L. Nirenberg characterizes the integrable almost-complex structures on a manifold X in terms of the *Nijenhuis tensor* Nij_J, defined as

$$\mathrm{Nij}_J(\cdot, \cdot\cdot) := [\cdot, \cdot\cdot] + J[J\cdot, \cdot\cdot] + J[\cdot, J\cdot\cdot] - [J\cdot, J\cdot\cdot].$$

(See also [KN96, Appendix 8] for the integrability of real-analytic almost-complex structures.)

Theorem 1.1 ([NN57, Theorem 1.1]). *Let X be a differentiable manifold. An almost-complex structure J on X is integrable if and only if $\mathrm{Nij}_J = 0$.*

By a straightforward computation, the integrability of an almost-complex structure J turns out to be equivalent to the vanishing of the components A and $\bar{A}$ of the exterior differential, equivalently, to $\left(\wedge^{\bullet,\bullet} X, \partial, \bar{\partial}\right)$ being a double complex of $\mathcal{C}^\infty(X; \mathbb{C})$-modules (see, e.g., [Wel08, Sect. 2.6], [Mor07, Proposition 8.2]).

Therefore, for a complex manifold X, one can consider, for every $p \in \mathbb{N}$, the differential complex $\left(\wedge^{p, \bullet} X, \bar{\partial} \right)$ and its cohomology, defining the *Dolbeault cohomology*, as the bi-graded $\mathbb{C}$-vector space

$$H_{\bar{\partial}}^{\bullet, \bullet}(X) := \frac{\ker \bar{\partial}}{\operatorname{im} \bar{\partial}} \,.$$

For every $(p, q) \in \mathbb{N}^2$, denote by $\mathcal{A}_X^{p,q}$ the (fine) sheaf of germs of (p, q)-forms on X. For every $p \in \mathbb{N}$, denote by Ω_X^p the sheaf of germs of *holomorphic p-forms* on X, that is, the kernel sheaf of the map $\bar{\partial} \colon \mathcal{A}_X^{p,0} \to \mathcal{A}_X^{p,1}$. By the Dolbeault and Grothendieck Lemma, see, e.g., [Dem12, I.3.29], one has that

$$0 \to \Omega_X^p \to \mathcal{A}_X^{p, \bullet}$$

is a fine resolution of Ω_X^p; hence, one gets the following result.

Theorem 1.2 (Dolbeault Theorem, [Dol53]). *Let X be a complex manifold. For every $p, q \in \mathbb{N}$,*

$$H_{\bar{\partial}}^{p,q}(X) \simeq \check{H}^q(X; \Omega^p) \,.$$

This gives a sheaf-theoretic interpretation of the Dolbeault cohomology. On the other hand, also an analytic interpretation can be provided.

Suppose X is a compact complex manifold of complex dimension n, and fix g a Hermitian metric on X and vol the induced volume form on X (recall that every complex manifold is orientable, see, e.g., [GH94, pp. 17–18]); denote by $\omega := g(J \cdot, \cdot \cdot) \in \wedge^{1,1} X \cap \wedge^2 X$ the associated $(1, 1)$-form to g. Recall that g induces a Hermitian inner product $\langle \cdot, \cdot \cdot \rangle$ on the space $\wedge^{\bullet, \bullet} X$ of global differential forms on X, and that the *Hodge-$*$-operator* associated to g is the $\mathbb{C}$-linear map

$$* \lfloor_{\wedge^{p,q} X} \colon \wedge^{p,q} X \to \wedge^{n-q, n-p} X$$

defined requiring that, for every $\alpha, \beta \in \wedge^{p,q} X$,

$$\alpha \wedge *\bar{\beta} = \langle \alpha, \beta \rangle \operatorname{vol} \,.$$

Define

$$\bar{\partial}^* := - * \partial * \colon \wedge^{\bullet, \bullet} X \to \wedge^{\bullet, \bullet-1} X \,;$$

the operator $\bar{\partial}^* \colon \wedge^{\bullet, \bullet} X \to \wedge^{\bullet, \bullet-1} X$ is the adjoint[1] of $\bar{\partial} \colon \wedge^{\bullet, \bullet} X \to \wedge^{\bullet, \bullet+1} X$ with respect to $\langle \cdot, \cdot \cdot \rangle$. Define

---

[1] This fact can be generalized for a *bi-differential bi-graded algebra of* PD-*type*, [Kas12b, Sect. 2], see also [AK12, Sect. 2.1].

$$\overline{\square} := \left[\overline{\partial}, \overline{\partial}^* \right] := \overline{\partial}\,\overline{\partial}^* + \overline{\partial}^*\,\overline{\partial} \colon \wedge^{\bullet,\bullet} X \to \wedge^{\bullet,\bullet} X \ ;$$

$\overline{\square}$ being a 2nd order self-adjoint elliptic differential operator (see, e.g., [Kod05, Theorem 3.16]), one gets the following result.

Theorem 1.3 (Hodge Theorem, [Hod89]). *Let X be a compact complex manifold endowed with a Hermitian metric. There is an orthogonal decomposition*

$$\wedge^{\bullet,\bullet} X = \ker \overline{\square} \ \overset{\perp}{\oplus} \ \overline{\partial} \wedge^{\bullet,\bullet-1} X \ \overset{\perp}{\oplus} \ \overline{\partial}^* \wedge^{\bullet,\bullet+1} X \ ,$$

and hence an isomorphism

$$H_{\overline{\partial}}^{\bullet,\bullet}(X) \simeq \ker \overline{\square} \ .$$

In particular, $\dim_{\mathbb{C}} H_{\overline{\partial}}^{\bullet,\bullet}(X) < +\infty$.

Note that, for any $(p,q) \in \mathbb{N}^2$, the Hodge-$*$-operator $* \colon \wedge^{p,q} X \to \wedge^{n-q,n-p} X$ sends a $\overline{\square}$-harmonic (p,q)-form ψ (that is, $\psi \in \wedge^{p,q} X$ is such that $\overline{\square}\psi = 0$) to a $\square$-harmonic $(n-q,n-p)$-form $*\psi$, where $\square := [\partial, \partial^*] := \partial\partial^* + \partial^*\partial \in \mathrm{End}\,(\wedge^{\bullet,\bullet} X)$ is the conjugate operator to $\overline{\square}$, and hence, by conjugating, one gets a $\overline{\square}$-harmonic $(n-p,n-q)$-form $\overline{*\psi}$. Hence, one gets the following result.

Theorem 1.4 (Serre Duality, [Ser55, Théorème 4]). *Let X be a compact complex manifold of complex dimension n, endowed with a Hermitian metric. For every $p, q \in \mathbb{N}$, the Hodge-$*$-operator induces an isomorphism*

$$* \colon H_{\overline{\partial}}^{p,q}(X) \overset{\simeq}{\to} \overline{H_{\overline{\partial}}^{n-p,n-q}(X)} \ .$$

Since a $\overline{\partial}$-closed form is not necessarily d-closed, Dolbeault cohomology classes do not define, in general, de Rham cohomology classes, that is, in general, on a compact complex manifold, there is no natural map between the Dolbeault cohomology and the de Rham cohomology (as we will see, in the special case of compact Kähler manifolds, or more in general of compact complex manifolds satisfying the $\partial\overline{\partial}$-Lemma, the de Rham cohomology actually can be decomposed by means of the Dolbeault cohomology groups, [Wei58, Théorème IV.3], [DGMS75, Lemma 5.15, Remark 5.16, 5.21]). Nevertheless, the Frölicher inequality provides a relation between the dimension of the Dolbeault cohomology and the dimension of the de Rham cohomology; it follows by considering the Hodge and Frölicher spectral sequence, which we recall here.

The structure of double complex of $\left(\wedge^{\bullet,\bullet} X, \partial, \overline{\partial} \right)$ gives rise to two natural filtrations of $\wedge^{\bullet} X \otimes \mathbb{C}$, namely (for $(p,q) \in \mathbb{N}^2$ and for $k \in \mathbb{N}$,)

$$'F^p\left(\wedge^k X \otimes \mathbb{C} \right) := \bigoplus_{\substack{r+s=k \\ r \geq p}} \wedge^{r,s} X \quad \text{and} \quad ''F^q\left(\wedge^k X \otimes \mathbb{C} \right) := \bigoplus_{\substack{r+s=k \\ s \geq q}} \wedge^{r,s} X \ ;$$

these filtrations induce two spectral sequences (see, e.g., [McC01, Sect. 2.4], [GH94, Sect. 3.5]),

$$\left\{ \left(E_r^{\bullet,\bullet}, d_r \right) = \left({}'E_r^{\bullet,\bullet}, {}'d_r \right) \right\}_{r \in \mathbb{N}} \quad \text{and, respectively,} \quad \left\{ \left({}''E_r^{\bullet,\bullet}, {}''d_r \right) \right\}_{r \in \mathbb{N}} ,$$

called *Hodge and Frölicher spectral sequences* (or *Hodge to de Rham spectral sequences*): one has

$${}'E_1^{\bullet,\bullet} \simeq H_{\bar\partial}^{\bullet,\bullet}(X) \;\Rightarrow\; H_{dR}^{\bullet}(X;\mathbb{C}) \quad \text{and} \quad {}''E_1^{\bullet,\bullet} \simeq H_{\partial}^{\bullet,\bullet}(X) \;\Rightarrow\; H_{dR}^{\bullet}(X;\mathbb{C}) .$$

An explicit description of $\{(E_r, d_r)\}_{r \in \mathbb{N}}$ is given in [CFUG97]: for any $p, q \in \mathbb{N}$ and $r \in \mathbb{N}$, its terms are

$$E_r^{p,q} \simeq \frac{\mathcal{X}_r^{p,q}}{\mathcal{Y}_r^{p,q}} ,$$

where, for $r = 1$,

$$\mathcal{X}_1^{p,q} := \left\{ \alpha \in \wedge^{p,q} X \;:\; \bar\partial \alpha = 0 \right\} , \qquad \mathcal{Y}_1^{p,q} := \bar\partial \wedge^{p,q-1} X ,$$

and, for $r \geq 2$,

$$\mathcal{X}_r^{p,q} := \left\{ \alpha^{p,q} \in \wedge^{p,q} X \;:\; \bar\partial \alpha^{p,q} = 0 \text{ and, for any } i \in \{1, \dots, r-1\} , \right.$$

$$\text{there exists } \alpha^{p+i,q-i} \in \wedge^{p+i,q-i} X$$

$$\left. \text{such that } \partial \alpha^{p+i-1,q-i+1} + \bar\partial \alpha^{p+i,q-i} = 0 \right\} ,$$

$$\mathcal{Y}_r^{p,q} := \left\{ \partial \beta^{p-1,q} + \bar\partial \beta^{p,q-1} \in \wedge^{p,q} X \;:\; \text{for any } i \in \{2, \dots, r-1\} , \right.$$

$$\text{there exists } \beta^{p-i,q+i-1} \in \wedge^{p-i,q+i-1} X$$

$$\left. \text{such that } \partial \beta^{p-i,q+i-1} + \bar\partial \beta^{p-i+1,q+i-2} = 0 \text{ and } \bar\partial \beta^{p-r+1,q+r-2} = 0 \right\} ,$$

see [CFUG97, Theorem 1], and, for any $r \geq 1$, the map $d_r \colon E_r^{\bullet,\bullet} \to E_r^{\bullet+r,\bullet-r+1}$ is given by

$$d_r \colon \{[\alpha^{p,q}] \in E_r^{p,q}\}_{(p,q) \in \mathbb{N}^2} \mapsto \left\{\left[\partial \alpha^{p+r-1,q-r+1}\right] \in E_r^{p+r,q-r+1}\right\}_{(p,q) \in \mathbb{N}^2} ,$$

see [CFUG97, Theorem 3].

As a consequence of ${}'E_1^{\bullet,\bullet} \simeq H_{\bar\partial}^{\bullet,\bullet}(X) \;\Rightarrow\; H_{dR}^{\bullet}(X;\mathbb{C})$, one gets the following inequality by A. Frölicher.

Theorem 1.5 (Frölicher Inequality, [Frö55, Theorem 2]). *Let X be a compact complex manifold. Then, for every $k \in \mathbb{N}$,*

$$\dim_{\mathbb{C}} H^k_{dR}(X; \mathbb{C}) \leq \sum_{p+q=k} \dim_{\mathbb{C}} H^{p,q}_{\bar{\partial}}(X) .$$

As a matter of notation, for $k \in \mathbb{N}$ and $(p, q) \in \mathbb{N}^2$, we will denote by $b_k := \dim_{\mathbb{R}} H^k_{dR}(X; \mathbb{R})$, respectively $h^{p,q}_{\bar{\partial}} := \dim_{\mathbb{C}} H^{p,q}_{\bar{\partial}}(X)$, the kth *Betti number*, respectively the $(p, q)^{th}$ *Hodge number* of X.

Remark 1.2. Other than the Dolbeault cohomology, other cohomologies can be defined for a complex manifold X; more precisely, since, for every $(p, q) \in \mathbb{N}^2$,

$$\wedge^{p-1,q-1} X \xrightarrow{\partial\bar{\partial}} \wedge^{p,q} X \xrightarrow{\partial+\bar{\partial}} \wedge^{p+1,q} X \oplus \wedge^{p,q+1} X \qquad \text{and}$$

$$\wedge^{p-1,q} X \oplus \wedge^{p,q-1} X \xrightarrow{(\partial, \bar{\partial})} \wedge^{p,q} X \xrightarrow{\partial\bar{\partial}} \wedge^{p+1,q+1} X$$

are complexes, one can define the *Bott-Chern cohomology* $H^{\bullet,\bullet}_{BC}(X)$ and the *Aeppli cohomology* $H^{\bullet,\bullet}_A(X)$ of X as

$$H^{\bullet,\bullet}_{BC}(X) := \frac{\ker \partial \cap \ker \bar{\partial}}{\operatorname{im} \partial\bar{\partial}} \qquad \text{and} \qquad H^{\bullet,\bullet}_A(X) := \frac{\ker \partial\bar{\partial}}{\operatorname{im} \partial + \operatorname{im} \bar{\partial}} ;$$

we refer to Sect. 2.1 for further details.

In the next chapter, we will provide a Frölicher-type inequality also for the Bott-Chern cohomology, Theorem 2.13, showing that it allows to characterize the compact complex manifolds satisfying the $\partial\bar{\partial}$-Lemma just in terms of the dimensions of the Bott-Chern cohomology and of the de Rham cohomology, Theorem 2.14.

(We refer also to [AT13a], where Dolbeault, Bott-Chern, and Aeppli cohomologies are considered in a more general context.)

Remark 1.3. In Complex Analysis, properties concerning the existence of exhaustion functions with convexity properties may have consequences on the vanishing of the cohomology. In fact, the Hörmander theorem, [Hör65, Theorems 2.2.4 and 2.2.5], [Hör90, Theorem 4.2.2, Corollary 4.2.6], states that the Dolbeault cohomology groups $H^{p,q}_{\bar{\partial}}(D)$ of a *strictly pseudo-convex* domain D in $\mathbb{C}^n$ (that is, a domain admitting a smooth proper strictly pluri-sub-harmonic exhaustion function) vanish for $q \geq 1$, for any $p \in \mathbb{N}$. More in general, A. Andreotti and H. Grauert proved in [AG62, Proposition 27] that the Dolbeault cohomology groups $H^{r,s}_{\bar{\partial}}(D)$ of a *q-complete* domain in $\mathbb{C}^n$ (that is, a domain in $\mathbb{C}^n$ admitting a smooth proper exhaustion function whose Levi form has at least $n - q + 1$ positive eigen-values, [AG62, Rot55]) vanish for $s \geq q$, for any $r \in \mathbb{N}$, see also [AV65a, AV65b, Theorem 5], see also [Ves67]. (Furthermore, given a domain D of a Stein manifold with boundary of class $\mathcal{C}^2$, if $H_{\bar{\partial}}^{r,s}(D) = \{0\}$ for $s \geq q$ and for any $r \in \mathbb{N}$, then D is q-complete, [EVS80, Theorem 3.8].)

See [AC12, AC13] for other results connecting convexity properties (see [HL12, HL11]) and vanishing of the cohomology, see also [Sha86, Theorem 1], [Wu87, Theorem 1], see also [HL11, Proposition 5.7].

1.2 Symplectic Geometry

In this section, we recall some definitions and results concerning symplectic manifolds, that is, differentiable manifolds endowed with a non-degenerate d-closed 2-form. An interesting class of examples of symplectic manifolds is provided by the Kähler manifolds. Moreover, given a differentiable manifold X, its cotangent bundle T^*X is endowed with a natural symplectic structure (see, e.g., [CdS01, Sect. 2]): in fact, symplectic geometry has applications and motivations in the study of Hamiltonian Mechanics, see, e.g., [CdS01, Part VII].

1.2.1 Symplectic Structures

Let X be a compact $2n$-dimensional manifold. We start by recalling the notion of symplectic structure.

Definition 1.2 ([Wey97, Sect. VI]). Let X be a differentiable manifold. A *symplectic form* on X is a non-degenerate d-closed 2-form $\omega \in \wedge^2 X$.

The main difference between symplectic geometry and Riemannian geometry is provided by G. Darboux's theorem.

Theorem 1.6 (Darboux Theorem, [Dar82]). *Let X be a $2n$-dimensional manifold endowed with a symplectic form ω. Then, for every $x \in X$, there exists a coordinate chart $\left(U, \{x^j\}_{j \in \{1,\dots,2n\}}\right)$, with $x \in U$, such that*

$$\omega \stackrel{loc}{=} \sum_{j=1}^{n} d x^{2j-1} \wedge d x^{2j} .$$

As for (almost-)complex manifolds, on a symplectic manifold X endowed with a symplectic form ω, one has a decomposition of differential forms in symplectic-type components, the so-called Lefschetz decomposition; it is a consequence of an $\mathfrak{sl}(2; \mathbb{R})$-representation on $\wedge^2 X$ by means of operators related to the symplectic structure.

More precisely, define the operators L, Λ, $H \in \mathrm{End}^\bullet \left(\wedge^\bullet X \right)$ as

$$L : \wedge^\bullet X \to \wedge^{\bullet+2} X , \qquad \alpha \mapsto \omega \wedge \alpha ,$$

$$\Lambda : \wedge^\bullet X \to \wedge^{\bullet-2} X , \qquad \alpha \mapsto -\iota_\Pi \alpha ,$$

$$H : \wedge^\bullet X \to \wedge^\bullet X , \qquad \alpha \mapsto \sum_k (n - k) \, \pi_{\wedge^k X} \alpha$$

(where $\iota_\xi \colon \wedge^\bullet X \to \wedge^{\bullet-2} X$ denotes the interior product with $\xi \in \wedge^2 (TX)$, and, for $k \in \mathbb{N}$, the map $\pi_{\wedge^k X} \colon \wedge^\bullet X \to \wedge^k X$ denotes the natural projection onto $\wedge^k X$). Note that, using the symplectic-$\star$-operator $\star_\omega$, one can write, [Yan96, Lemma 1.5],

$$\Lambda = -\star_\omega L \star_\omega .$$

The following result holds.[2]

Theorem 1.7 ([Yan96, Corollary 1.6]). *Let X be a manifold endowed with a symplectic structure. Then*

$$[L, H] = 2 L , \qquad [\Lambda, H] = -2 \Lambda , \qquad [L, \Lambda] = H ,$$

and hence

$$\mathfrak{sl}(2; \mathbb{R}) \simeq \langle L, \Lambda, H \rangle \to \mathrm{End}^\bullet \left(\wedge^\bullet X \right)$$

gives an $\mathfrak{sl}(2; \mathbb{R})$-representation on $\wedge^\bullet X$.

(See also [Huy05, Proposition 1.2.2] for a proof.)

The above $\mathfrak{sl}(2; \mathbb{R})$-representation, having finite H-spectrum,[3] induces a decomposition of the space of the differential forms.

Theorem 1.8 ([Yan96, Corollary 2.6]). *Let X be a manifold endowed with a symplectic structure. Then one has the* Lefschetz decomposition *on differential forms,*

$$\wedge^\bullet X = \bigoplus_{r \in \mathbb{N}} L^r \, \mathrm{P} \wedge^{\bullet-2r} X ,$$

---

[2] See, e.g., [Hum78, Sect. 7] for general results concerning $\mathfrak{sl}(2; \mathbb{K})$-representations.

[3] Recall that an $\mathfrak{sl}(2; \mathbb{R})$-representation on a (possibly non-finite dimensional) $\mathbb{R}$-vector space V is called of *finite H-spectrum* if V can be decomposed into the direct sum of eigen-spaces of H and H has only finitely-many distinct eigen-values, [Yan96, Definition 2.2].

where

$$P\wedge^\bullet X := \ker \Lambda$$

is the space of primitive forms.

Remark 1.4. Note that, for every $k \in \mathbb{N}$,

$$P\wedge^k X = \ker L^{n-k+1}\big\lfloor_{\wedge^k X} ,$$

see, e.g., [Huy05, Proposition 1.2.30(v)].

In general, see, e.g., [TY12b, p. 422], the Lefschetz decomposition of $A^{(k)} \in \wedge^k X$ reads as

$$A^{(k)} = \sum_{r \geq \max\{k-n,0\}} \frac{1}{r!} L^r B^{(k-2r)}$$

where, for $r \geq \max\{k-n, 0\}$,

$$B^{(k-2r)} := \left(\sum_{\ell \in \mathbb{N}} a_{r,\ell,(n,k)} \frac{1}{\ell!} L^\ell \Lambda^{r+\ell} \right) A^{(k)} \in P\wedge^{k-2r} X$$

and, for $r \geq \max\{k-n, 0\}$ and $\ell \in \mathbb{N}$,

$$a_{r,\ell,(n,k)} := (-1)^\ell \cdot (n-k+2r+1)^2 \cdot \prod_{i=0}^{r} \frac{1}{n-k+2r+1-i}$$

$$\cdot \prod_{j=0}^{\ell} \frac{1}{n-k+2r+1+j} \in \mathbb{Q} .$$

We recall that

$$L\big\lfloor_{\bigoplus_{k=-1}^{n-2} \wedge^{n-k-2}X} \colon \bigoplus_{k=-1}^{n-2} \wedge^{n-k-2} X \to \wedge^{n-k} X$$

is injective, [Yan96, Corollary 2.8], and that, for every $k \in \mathbb{N}$,

$$L^k \colon \wedge^{n-k} X \to \wedge^{n+k} X$$

is an isomorphism, [Yan96, Corollary 2.7].

1.2.1.1 Primitive Currents

Let X be a $2n$-dimensional compact manifold endowed with a symplectic structure ω. Denote by $\mathcal{D}_\bullet X := \mathcal{D}^{2n-\bullet} X$ the space of currents, and consider the de Rham homology $H^{dR}_\bullet(X; \mathbb{R}) := H^\bullet(\mathcal{D}_\bullet X, d)$. (See Sect. 1.6 for definitions and results concerning currents and de Rham homology.)

Following [Lin13, Definition 4.1], set, by duality,

$$L: \mathcal{D}_\bullet X \to \mathcal{D}_{\bullet-2} X , \qquad S \mapsto S(L \cdot) ,$$

$$\Lambda: \mathcal{D}_\bullet X \to \mathcal{D}_{\bullet+2} X , \qquad S \mapsto S(\Lambda \cdot) ,$$

$$H: \mathcal{D}_\bullet X \to \mathcal{D}_\bullet X , \qquad S \mapsto S(-H \cdot) ;$$

note that

$$[L, H] = 2L , \qquad [\Lambda, H] = -2\Lambda , \qquad [L, \Lambda] = H .$$

A current $S \in \mathcal{D}^k X$ is said *primitive* if $\Lambda S = 0$, equivalently, if $L^{n-k+1} S = 0$, see, e.g., [Lin13, Proposition 4.3]; denote by $PD^\bullet X := PD_{2n-\bullet} X$ the space of primitive currents on X.

In [Lin13], Y. Lin proved the following result.

Theorem 1.9 ([Lin13, Lemma 4.2, Proposition 4.3]). *Let X be a compact manifold endowed with a symplectic structure ω. Then $\langle L, \Lambda, H \rangle$ gives an $\mathfrak{sl}(2; \mathbb{R})$-module structure on $\mathcal{D}^\bullet X$. In particular, one has the Lefschetz decomposition on the space of currents,*

$$\mathcal{D}^\bullet X = \bigoplus_{r \in \mathbb{N}} L^r PD^{\bullet-2r} X = \bigoplus_{r \in \mathbb{N}} L^r PD_{2n-\bullet+2r} X .$$

Finally, if $j: Y \hookrightarrow X$ is a compact oriented submanifold of X of codimension k (possibly with non-empty boundary), then the *dual current* $[Y] \in \mathcal{D}_k X$ associated with Y is defined, by setting, for every $\varphi \in \wedge^k X$,

$$[Y](\varphi) := \int_Y j^*(\varphi) .$$

If Y is a closed oriented submanifold, then the dual current $[Y]$ is d-closed. According to [TY12a, Lemma 4.1], the dual current $[Y]$ is primitive if and only if Y is co-isotropic.

1.2.2 Cohomological Aspects of Symplectic Geometry

Cohomological properties of symplectic manifolds have been studied starting from the works by J.-L. Koszul, [Kos85], and by J.-L. Brylinski, [Bry88]. Drawing a parallel between the symplectic and the Riemannian cases, J.-L. Brylinski proposed in [Bry88] a Hodge theory for compact symplectic manifolds (X, ω), introducing a symplectic Hodge-$\star$-operator $\star_\omega$ and the notion of ω-symplectically-harmonic form (i.e., a form being both d-closed and d^Λ-closed, where the symplectic co-differential is defined as $d^\Lambda\lfloor_{\wedge^k X} := (-1)^{k+1} \star_\omega d \star_\omega$ for every $k \in \mathbb{N}$): in this context, O. Mathieu in [Mat95], and D. Yan in [Yan96], proved that any de Rham cohomology class admits an ω-symplectically-harmonic representative if and only if the Hard Lefschetz Condition is satisfied. In [TY12a, TY12b], see also [TY11], L.-S. Tseng and S.-T. Yau introduced new cohomologies for symplectic manifolds (X, ω): among them, in particular, they defined and studied

$$H^\bullet_{d+d^\Lambda}(X, \mathbb{R}) \; :- \; \frac{\ker\left(d + d^\Lambda\right)}{\operatorname{im} d\, d^\Lambda} \, ,$$

developing a Hodge theory for this cohomology; furthermore, they studied the dual currents of Lagrangian and co-isotropic submanifolds, and they defined a homology theory on co-isotropic chains, which turns out to be naturally dual to a primitive cohomology. In the context of generalized geometry, [Gua04a, Gua11, Cav05, Cav07], the cohomology $H^\bullet_{d+d^\Lambda}(X; \mathbb{R})$ can be interpreted as the symplectic counterpart of the Bott-Chern cohomology of a complex manifold, see [TY11]. Inspired also by their works, Y. Lin developed in [Lin13] a geometric measure theoretic approach to symplectic Hodge theory, proving in particular that, on any compact symplectic manifold, every primitive cohomology class of positive degree admits a symplectically-harmonic representative not supported on the entire manifold, [Lin13, Theorems 1.1 and 5.3].

In this section, we recall some notions and results concerning Hodge theory for compact symplectic manifolds; we refer to [Bry88, Mat95, Yan96, Cav05, TY12a, TY12b, Lin13] for further details.

1.2.2.1 Symplectic Hodge Theory

Let X be a compact $2n$-dimensional manifold endowed with a symplectic form ω. By exploiting the parallelism with Riemannian geometry, one can try to develop a Hodge theory also for compact symplectic manifolds, [Bry88]. The first tool that can be introduced is an analogue of the Hodge-$*$-operator; hence one can define a symplectic counterpart of the co-differential operator, and define the notion of symplectically-harmonic form, investigating the existence of symplectically-harmonic representatives in any de Rham cohomology class.

Note that every symplectic manifold is orientable, $\frac{\omega^n}{n!}$ giving a canonical orientation.

Denote by I the natural isomorphism of vector bundles induced by ω, namely, for every $x \in X$,

$$T_x X \ni v \mapsto I(v)(\cdot) := \omega(v, \cdot) \in \operatorname{Hom}(T_x X; \mathbb{R}) \, .$$

Then, for every $k \in \mathbb{N}$, the form ω gives rise to a bi-$\mathcal{C}^\infty(X; \mathbb{R})$-linear form on $\wedge^k X$ denoted by $(\omega^{-1})^k$, which is skew-symmetric, respectively symmetric, according that k is odd, respectively even, and defined on the simple elements $\alpha^1 \wedge \ldots \wedge \alpha^k \in \wedge^k X$ and $\beta^1 \wedge \ldots \wedge \beta^k \in \wedge^k X$ as

$$\left(\omega^{-1}\right)^k \left(\alpha^1 \wedge \ldots \wedge \alpha^k, \beta^1 \wedge \ldots \wedge \beta^k\right) := \det\left(\omega^{-1}\left(\alpha^\ell, \beta^m\right)\right)_{\ell, m \in \{1, \ldots, k\}},$$

where

$$\omega^{-1}\left(\alpha^\ell, \beta^m\right) := \omega\left(I^{-1}\left(\alpha^\ell\right), I^{-1}\left(\beta^m\right)\right)$$

for every $\ell, m \in \{1, \ldots, k\}$. In a Darboux coordinate chart $\left(U, \{x^j\}_{j \in \{1, \ldots, 2n\}}\right)$, the canonical *Poisson bi-vector* $\Pi := \omega^{-1} \in \wedge^2 TX$ associated to ω is written as $\omega^{-1} \overset{\text{loc}}{=} \sum_{j=1}^n \frac{\partial}{\partial x^{2j-1}} \wedge \frac{\partial}{\partial x^{2j}}$.

The *symplectic-$\star$-operator*

$$\star_\omega \colon \wedge^\bullet X \to \wedge^{2n-\bullet} X,$$

introduced by J.-L. Brylinski, [Bry88, Sect. 2], is defined requiring that, for every $k \in \mathbb{N}$, and for every $\alpha, \beta \in \wedge^k X$,

$$\alpha \wedge \star_\omega \beta = \left(\omega^{-1}\right)^k (\alpha, \beta) \frac{\omega^n}{n!}.$$

By continuing in the parallelism between Riemannian geometry and symplectic geometry, one can introduce the d$^\Lambda$ operator with respect to a symplectic structure ω as

$$\mathrm{d}^\Lambda\big\lfloor_{\wedge^k X} := (-1)^{k+1} \star_\omega \mathrm{d} \star_\omega$$

for any $k \in \mathbb{N}$, and interpret it as the symplectic counterpart of the Riemannian d* operator with respect to a Riemannian metric. In light of this, J.-L. Brylinski proposed in [Bry88] a Hodge theory for compact symplectic manifolds, conjecturing that, on a compact manifold endowed with a symplectic structure ω, every de Rham cohomology class admits a (possibly non-unique) ω-*symplectically-harmonic representative*, namely, a d-closed d$^\Lambda$-closed representative, [Bry88, Conjecture 2.2.7]. (Note that $\mathrm{d}\,\mathrm{d}^\Lambda + \mathrm{d}^\Lambda \mathrm{d} = 0$, [Bry88, Theorem 1.3.1], [Kos85, p. 265], provides a strong difference in the parallelism between symplectic geometry and Riemannian geometry; in particular, it follows that a ω-symplectically-harmonic representative, whenever it exists, is not unique.)

Remark 1.5. For an almost-Kähler structure (J, ω, g) on a compact manifold X (that is, $\omega \in \wedge^2 X$ is a symplectic form on X, and $J \in \mathrm{End}\,(TX)$ is an almost-complex structure on X, and g is a J-Hermitian metric on X such that ω is the associated $(1, 1)$-form to g), the symplectic-$\star$-operator $\star_\omega$ and the Hodge-$*$-operator $*_g$ are related by

$$\star_\omega \ = \ J *_g \, ,$$

and hence

$$\mathrm{d}^\Lambda \ = \ -(\mathrm{d}^c)^{*_g}$$

where $\mathrm{d}^c := J^{-1} \, \mathrm{d} \, J$ and $(\mathrm{d}^c)^{*_g} \lfloor_{\wedge^k X} := (-1)^{k+1} *_g \mathrm{d} *_g$ for every $k \in \mathbb{N}$ (note that, when J is integrable, then $\mathrm{d}^c = -\mathrm{i} \left(\partial - \bar{\partial} \right)$). Moreover, on a compact manifold X endowed with a Kähler structure (J, ω, g), by the Hodge decomposition theorem, [Wei58, Théorème IV.3], the pure-type components with respect to J of the harmonic representatives of the de Rham cohomology classes are themselves harmonic. Hence, it follows that Brylinski's conjecture holds true for compact Kähler manifolds, [Bry88, Corollary 2.4.3].

Remark 1.6. In the framework of generalized complex geometry, see Sect. 1.3, the d^Λ operator associated to a symplectic structure should be interpreted as the symplectic counterpart of the operator $\mathrm{d}^c := -\mathrm{i} \left(\partial - \bar{\partial} \right)$ associated to a complex structure, [Cav05].

Consider now, for $k \in \mathbb{N}$, the map $L^k \colon \wedge^{n-k} X \to \wedge^{n+k} X$. Since $[L, \mathrm{d}] = 0$, it induces a map $L^k \colon H_{dR}^{n-k}(X; \mathbb{R}) \to H_{dR}^{n+k}(X; \mathbb{R})$ in cohomology. One says that X satisfies the *Hard Lefschetz Condition*, shortly HLC, if

$$\text{for every } k \in \mathbb{N}, \qquad L^k \colon H_{dR}^{n-k}(X; \mathbb{R}) \xrightarrow{\simeq} H_{dR}^{n+k}(X; \mathbb{R}) \, . \qquad \text{(HLC)}$$

O. Mathieu in [Mat95], and D. Yan in [Yan96], provided counterexamples to Brylinski's conjecture, characterizing the compact symplectic manifolds satisfying Brylinski's conjecture in terms of the validity of the Hard Lefschetz Condition. Furthermore, S.A. Merkulov in [Mer98], see also [Cav05], and V. Guillemin in [Gui01], proved that the Hard Lefschetz Condition on compact symplectic manifolds is equivalent to satisfying the $\mathrm{d}\,\mathrm{d}^\Lambda$-Lemma, namely, to every d-exact d^Λ-closed form being $\mathrm{d}\,\mathrm{d}^\Lambda$-exact. See Theorem 1.15 for a summary of several equivalent statements to Hard Lefschetz Condition.

Note that, by the Lefschetz decomposition theorem, [Wei58, Théorème IV.5] (see Sect. 1.4), compact Kähler manifolds satisfy the Hard Lefschetz Condition.

1.2.2.2 Symplectic Cohomologies

Let X be a compact $2n$-dimensional manifold endowed with a symplectic structure ω.

In [TY12a], L.-S. Tseng and S.-T. Yau introduced also the $\left(d + d^A\right)$-*cohomology*, [TY12a, Sect. 3.2],

$$H^\bullet_{d+d^A}(X;\mathbb{R}) := \frac{\ker\left(d + d^A\right)}{\operatorname{im} d\, d^A},$$

and the $\left(d\, d^A\right)$-*cohomology*, [TY12a, Sect. 3.3],

$$H^\bullet_{d\,d^A}(X;\mathbb{R}) := \frac{\ker d\, d^A}{\operatorname{im} d + \operatorname{im} d^A};$$

such cohomologies are, in a sense, the symplectic counterpart of the Bott-Chern and Aeppli cohomologies of complex manifolds, see [TY12a, Sect. 5] and [TY11] for further discussions.

Furthermore, they provided a Hodge theory for such cohomologies, proving the following result.

Theorem 1.10 ([TY12a, Theorem 3.5, Corollary 3.6]). *Let X be a compact manifold endowed with a symplectic structure ω. Let (J, ω, g) be an almost-Kähler structure on X. For a fixed $\lambda > 0$, the 4th order self-adjoint differential operator*

$$D_{d+d^A} := \left(dd^A\right)\left(dd^A\right)^* + \left(dd^A\right)^*\left(dd^A\right) + \left(d^* d^A\right)\left(d^* d^A\right)^* + \left(d^* d^A\right)^*\left(d^* d^A\right)$$

$$+ \lambda \left(d^* d + \left(d^A\right)^* d^A\right).$$

is elliptic, with $\ker D_{d+d^A} = \ker d \cap \ker d^A \cap \ker \left(d\, d^A\right)^*$.
Furthermore, there exist an orthogonal decomposition

$$\wedge^\bullet X = \ker D_{d+d^A} \oplus d\, d^A \wedge^\bullet X \oplus \left(d^* \wedge^{\bullet+1} X + \left(d^A\right)^* \wedge^{\bullet-1} X\right)$$

and an isomorphism

$$H^\bullet_{d+d^A}(X;\mathbb{R}) \simeq \ker D_{d+d^A}.$$

In particular, $\dim_\mathbb{R} H^\bullet_{d+d^A}(X;\mathbb{R}) < +\infty.$

An analogous statement holds for the $\left(d\, d^A\right)$-cohomology.

Theorem 1.11 ([TY12a, Theorem 3.16, Corollary 3.17]). *Let X be a compact manifold endowed with a symplectic structure ω. Let (J, ω, g) be an almost-Kähler structure on X. For a fixed $\lambda > 0$, the 4th order self-adjoint differential operator*

$$D_{d\,d^A} := \left(d\,d^A\right)\left(d\,d^A\right)^* + \left(d\,d^A\right)^*\left(d\,d^A\right) + \left(d\left(d^A\right)^*\right)\left(d\left(d^A\right)^*\right)^*$$

$$+ \left(d\left(d^A\right)^*\right)^*\left(d\left(d^A\right)^*\right) + \lambda \left(dd^* + d^A\left(d^A\right)^*\right).$$

is elliptic, with $\ker D_{d\,d^A} = \ker\left(d\,d^A\right) \cap \ker d^* \cap \ker\left(d^A\right)^*.$

Furthermore, there exist an orthogonal decomposition

$$\wedge^\bullet X = \ker D_{dd^\Lambda} \oplus \left(d \wedge^{\bullet-1} X + d^\Lambda \wedge^{\bullet+1} X \right) \oplus \left(dd^\Lambda \right)^* \wedge^\bullet X$$

and an isomorphism

$$H_{dd^\Lambda}^\bullet(X;\mathbb{R}) \simeq \ker D_{dd^\Lambda} .$$

In particular, $\dim_\mathbb{R} H_{dd^\Lambda}^\bullet(X;\mathbb{R}) < +\infty.$

As for the Bott-Chern and the Aeppli cohomologies, the $\left(d + d^\Lambda \right)$-cohomology and the $\left(dd^\Lambda \right)$-cohomology groups turn out to be isomorphic by means of the Hodge-$*$-operator associated to any Riemannian metric being compatible with ω.

Theorem 1.12 ([TY12a, Lemma 3.23, Proposition 3.24, Corollary 3.25]). *Let X be a 2n-dimensional compact manifold endowed with a symplectic structure ω. Let (J, ω, g) be an almost-Kähler structure on X. The operators D_{d+d^Λ} and D_{dd^Λ} satisfy*

$$*_g D_{d+d^\Lambda} = D_{dd^\Lambda} *_g ,$$

and hence $*_g$ induces an isomorphism

$$*_g \colon H_{d+d^\Lambda}^\bullet(X;\mathbb{R}) \xrightarrow{\simeq} H_{dd^\Lambda}^{2n-\bullet}(X;\mathbb{R}) .$$

Moreover, the cohomology $H_{d+d^\Lambda}^\bullet(X;\mathbb{R})$ is invariant under symplectomorphisms and Hamiltonian isotopies, [TY12a, Proposition 2.8].

One has the following commutation relations between the differential operators d, d^Λ, and dd^Λ, and the elements L, Λ, and H of the $\mathfrak{sl}(2;\mathbb{R})$-triple, see, e.g., [TY12a, Lemma 2.3]:

$$[d, L] = 0 , \quad [d^\Lambda, L] = -d , \quad [dd^\Lambda, L] = 0 ,$$
$$[d, \Lambda] = d^\Lambda , \quad [d^\Lambda, \Lambda] = 0 , \quad [dd^\Lambda, \Lambda] = 0 ,$$
$$[d, H] = d , \quad [d^\Lambda, H] = -d^\Lambda , \quad [dd^\Lambda, H] = 0 .$$

Hence, by defining the *primitive* $\left(d + d^\Lambda \right)$-*cohomology* as

$$PH_{d+d^\Lambda}^\bullet(X;\mathbb{R}) := \frac{\ker d \cap \ker d^\Lambda \cap P \wedge^\bullet X}{\operatorname{im} dd^\Lambda \cap P \wedge^\bullet X} = \frac{\ker d \cap P \wedge^\bullet X}{\operatorname{im} dd^\Lambda \lfloor_{P \wedge^\bullet X}}$$

(where the second equality follows from [TY12a, Lemma 3.9]), one gets the following result.

Theorem 1.13 ([TY12a, Theorem 3.11]). *Let X be a 2n-dimensional compact manifold endowed with a symplectic structure ω. Then there exist a decomposition*

$$H^{\bullet}_{d+d^{\Lambda}}(X;\mathbb{R}) \;=\; \bigoplus_{r\in\mathbb{N}} L^r \, PH^{\bullet-2r}_{d+d^{\Lambda}}(X;\mathbb{R})$$

and, for every $k \in \mathbb{N}$, an isomorphism

$$L^k \colon H^{n-k}_{d+d^{\Lambda}}(X;\mathbb{R}) \xrightarrow{\simeq} H^{n+k}_{d+d^{\Lambda}}(X;\mathbb{R}) \,,$$

Analogously, by defining the *primitive $\left(d\,d^{\Lambda}\right)$-cohomology* as

$$PH^{\bullet}_{d\,d^{\Lambda}}(X;\mathbb{R}) := \frac{\ker d\,d^{\Lambda} \cap P\wedge^{\bullet} X}{\left(\operatorname{im}d + \operatorname{im}d^{\Lambda}\right) \cap P\wedge^{\bullet} X}$$

$$= \frac{\ker d\,d^{\Lambda} \cap P\wedge^{\bullet} X}{\operatorname{im}\left(d + LH^{-1}\,d^{\Lambda}\right)\big|_{P\wedge^{\bullet-1}X} + \operatorname{im}d^{\Lambda}\big|_{P\wedge^{\bullet+1}X}}$$

(where the second equality follows from [TY12a, Lemma 3.20]), one gets the following result.

Theorem 1.14 ([TY12a, Theorem 3.21]). *Let X be a 2n-dimensional compact manifold endowed with a symplectic structure ω. Then there exist a decomposition*

$$H^{\bullet}_{d\,d^{\Lambda}}(X;\mathbb{R}) \;=\; \bigoplus_{r\in\mathbb{N}} L^r \, PH^{\bullet-2r}_{d\,d^{\Lambda}}(X;\mathbb{R})$$

and, for every $k \in \mathbb{N}$, an isomorphism

$$L^k \colon H^{n-k}_{d\,d^{\Lambda}}(X;\mathbb{R}) \xrightarrow{\simeq} H^{n+k}_{d\,d^{\Lambda}}(X;\mathbb{R}) \,,$$

The identity map induces the following natural maps in cohomology:

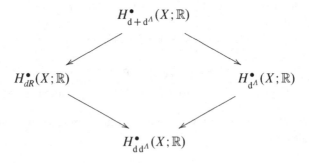

Recall that a symplectic manifold is said to satisfy the $d d^\Lambda$-*Lemma* if every d-exact d^Λ-closed form is $d d^\Lambda$-exact, [DGMS75], namely, if $H^\bullet_{d + d^\Lambda}(X; \mathbb{R}) \to H^\bullet_{dR}(X; \mathbb{R})$ is injective.

Remark 1.7. (In view of [DGMS75, Lemma 5.15] and [AT13a, Lemma 1.4]) note that

$$\ker d^\Lambda \cap \operatorname{im} d = \operatorname{im} d d^\Lambda \qquad \text{if and only if} \qquad \ker d \cap \operatorname{im} d^\Lambda = \operatorname{im} d d^\Lambda .$$

Indeed, since $\star^2_\omega = \operatorname{id}_{\wedge^\bullet X}$, [Bry88, Lemma 2.1.2], and $d d^\Lambda + d^\Lambda d = 0$, [Bry88, Theorem 1.3.1], one has

$$\star_\omega \ker d = \ker d^\Lambda , \qquad \star_\omega \operatorname{im} d = \operatorname{im} d^\Lambda , \qquad \star_\omega \operatorname{im} d d^\Lambda = \operatorname{im} d d^\Lambda .$$

Recalling that a $2n$-dimensional compact manifold X endowed with a symplectic form ω is said to satisfy the Hard Lefschetz Condition if and only if

$$\text{for every } k \in \mathbb{N} , \qquad L^k \colon H^{n-k}_{dR}(X; \mathbb{R}) \xrightarrow{\simeq} H^{n+k}_{dR}(X; \mathbb{R}) , \qquad \text{(HLC)}$$

we summarize in the following result the connection between the $d d^\Lambda$-Lemma, the Hard Lefschetz Condition, and the existence of ω-symplectically harmonic representatives in any de Rham cohomology class.

Theorem 1.15 ([Mat95, Corollary 2], [Yan96, Theorem 0.1], [Mer98, Proposition 1.4], [Gui01], [DGMS75, Lemma 5.15, Remarks 5.16, 5.21], [TY12a, Proposition 3.13], [Cav05, Theorem 5.4], [AT13a, Theorem 4.4], [AT12b, Remark 2.3]). *Let X be a compact manifold endowed with a symplectic structure ω. The following conditions are equivalent:*

  (i) *every de Rham cohomology class admits a representative being both d-closed and d^Λ-closed (i.e., Brylinski's conjecture [Bry88, Conjecture 2.2.7] holds on X);*
 (ii) *the Hard Lefschetz Condition holds on X;*
(iii) *the natural map $H^\bullet_{d + d^\Lambda}(X; \mathbb{R}) \to H^\bullet_{dR}(X; \mathbb{R})$ induced by the identity is injective;*
 (iv) *the natural map $H^\bullet_{d + d^\Lambda}(X; \mathbb{R}) \to H^\bullet_{dR}(X; \mathbb{R})$ induced by the identity is actually an isomorphism;*
  (v) *X satisfies the $d d^\Lambda$-Lemma;*
 (vi) *for every $k \in \mathbb{Z}$, it holds $\dim_\mathbb{R} H^k_{d + d^\Lambda}(X; \mathbb{R}) + \dim_\mathbb{R} H^k_{d d^\Lambda}(X; \mathbb{R}) = 2 \dim_\mathbb{R} H^k_{dR}(X; \mathbb{R})$;*
(vii) *the Lefschetz decomposition $\wedge^\bullet X = \bigoplus_{r \in \mathbb{N}} L^r P \wedge^{\bullet - 2r} X$ on forms induces a decomposition in cohomology, namely, $H^\bullet_{dR}(X; \mathbb{R}) = \bigoplus_{r \in \mathbb{N}} L^r H^{(0, \bullet - 2r)}_\omega (X; \mathbb{R})$, where $H^{(r,s)}_\omega(X; \mathbb{R}) := \{[L^r \beta^{(s)}] \in H^{2r+s}_{dR}(X; \mathbb{R}) : \beta^{(s)} \in P \wedge^s X\}$.*

Note that, by the Lefschetz decomposition theorem, the compact Kähler manifolds satisfy the Hard Lefschetz Condition; in other terms, note that, given a Kähler structure (J, ω, g) on a compact manifold X, one has $\star_\omega = J *_g$, [Bry88, Theorem 2.4.1], and hence every de Rham cohomology class admits an ω-symplectically-harmonic representative.

1.3 Generalized Geometry

Generalized complex geometry, introduced by N.J. Hitchin in [Hit03] and developed, among others, by M. Gualtieri, [Gua04a, Gua11], and G.R. Cavalcanti, [Cav05], see also [Hit10, Cav07], allows to frame symplectic structures and complex structures in the same context. (In a sense, this add more significance to the term "symplectic", which was invented by H. Weyl, [Wey97, Sect. VI], substituting the Greek root in the term "complex" with the corresponding Latin root.)

We recall here the notion of generalized complex structure, referring to [Cav07, Gua04a, Gua11, Cav05] for more details.

1.3.1 Generalized Complex Structures

Let X be a compact differentiable manifold of dimension $2n$.

Note that a complex structure J on X can be seen as an isomorphism $J: TX \overset{\simeq}{\to} TX$ satisfying the algebraic condition $J^2 = -\mathrm{id}_{TX}$ and the integrability condition $\mathrm{Nij}_J = 0$. Analogously, a symplectic structure ω on X can be seen as an isomorphism $\omega(\cdot)(\cdot\cdot) := \omega(\cdot, \cdot\cdot): TX \overset{\simeq}{\to} T^*X$ satisfying the algebraic condition $\omega(\cdot)(\cdot\cdot) = -\omega(\cdot\cdot)(\cdot)$ and the integrability condition $\mathrm{d}\,\omega = 0$.

Therefore, consider the bundle $TX \oplus T^*X$. It can be endowed with the natural symmetric pairing

$$\langle \cdot | \cdot\cdot \rangle : (TX \oplus T^*X) \times (TX \oplus T^*X) \to \mathbb{R}, \quad \langle X + \xi | Y + \eta \rangle := \frac{1}{2}\left(\xi(Y) + \eta(X)\right).$$

Fixed a d-closed 3-form H on X (possibly, the zero form $H = 0$), define a suitable bracket for an integrability condition on $TX \oplus T^*X$ as follows. On the space $\mathcal{C}^\infty (X; TX \oplus T^*X)$ of smooth sections of $TX \oplus T^*X$ over X, define the *H-twisted Courant bracket* as

$$[\cdot, \cdot\cdot]_H : \mathcal{C}^\infty (X; TX \oplus T^*X) \times \mathcal{C}^\infty (X; TX \oplus T^*X) \to \mathcal{C}^\infty (X; TX \oplus T^*X) ,$$

$$[X + \xi, Y + \eta]_H := [X, Y] + \mathcal{L}_X\eta - \mathcal{L}_Y\xi - \frac{1}{2}\,\mathrm{d}\left(\iota_X\eta - \iota_Y\xi\right) + \iota_Y\iota_X H$$

(where $\iota_X \in \mathrm{End}^{-1} (\wedge^\bullet X)$ denotes the interior product with $X \in \mathcal{C}^\infty(X; TX)$ and $\mathcal{L}_X := [\iota_X, \mathrm{d}] \in \mathrm{End}^0 (\wedge^\bullet X)$ denotes the Lie derivative along $X \in \mathcal{C}^\infty(X; TX)$);

the H-twisted Courant bracket can be seen also as a derived bracket induced by the H-twisted differential $d_H := d + H \wedge \cdot$, see [Gua04a, Sect. 3.2], [Gua11, Sect. 2].

Consider also the *Clifford action*[4] of $TX \oplus T^*X$ on the space of differential forms with respect to the natural pairing $\langle \cdot \mid \cdot \cdot \rangle$,

$$\text{Cliff}\left(TX \oplus T^*X, \langle \cdot \mid \cdot \cdot \rangle\right) \times \wedge^\bullet X \to \wedge^{\bullet-1} X \oplus \wedge^{\bullet+1} X , \quad (X+\xi) \cdot \varphi = \iota_X \varphi + \xi \wedge \varphi ,$$

and its bi-$\mathbb{C}$-linear extension $\text{Cliff}((TX \oplus T^*X) \otimes_{\mathbb{R}} \mathbb{C}) \times (\wedge^\bullet X \otimes_{\mathbb{R}} \mathbb{C}) \to (\wedge^{\bullet-1} X \otimes_{\mathbb{R}} \mathbb{C}) \oplus (\wedge^{\bullet+1} X \otimes_{\mathbb{R}} \mathbb{C})$.

We give now several equivalent definitions of generalized complex structure.

Definition 1.3 ([Gua04a, Definitions 4.14, 4.18], [Gua11, Definition 3.1]). Let X be a differentiable manifold and H be a d closed 3-form on X. An *H-twisted generalized complex structure* is an endomorphism $\mathcal{J} \in \text{End}(TX \oplus T^*X)$ such that

- $\mathcal{J}^2 = -\text{id}_{TX \oplus T^*X}$, and
- $\mathcal{J}$ is orthogonal with respect to $\langle \cdot \mid \cdot \cdot \rangle$, and
- the Nijenhuis tensor

$$\text{Nij}_{\mathcal{J},H} := -[\mathcal{J} \cdot, \mathcal{J} \cdot \cdot]_H + \mathcal{J}[\mathcal{J} \cdot, \cdot \cdot]_H + \mathcal{J}[\cdot, \mathcal{J} \cdot \cdot]_H + \mathcal{J}[\cdot, \cdot \cdot]_H$$

$$\in \left(TX \oplus T^*X\right) \otimes_{\mathbb{R}} \left(TX \oplus T^*X\right) \otimes_{\mathbb{R}} \left(TX \oplus T^*X\right)^*$$

of $\mathcal{J}$ with respect to the H-twisted Courant bracket vanishes identically.

Equivalently, by setting $L := L_{\mathcal{J}}$ the i-eigen-bundle of the $\mathbb{C}$-linear extension of $\mathcal{J}$ to $(TX \oplus T^*X) \otimes_{\mathbb{R}} \mathbb{C}$, one gets the following definition.

Proposition 1.1 ([Gua04a, Proposition 4.3]). *Let X be a differentiable manifold and H be a d-closed 3-form on X. A generalized complex structure on X is identified by a sub-bundle L of $(TX \oplus T^*X) \otimes_{\mathbb{R}} \mathbb{C}$ such that*

- *L is maximal isotropic with respect to $\langle \cdot \mid \cdot \cdot \rangle$, and*
- *L is involutive with respect to the H-twisted Courant bracket, and*
- *$L \cap \bar{L} = \{0\}$.*

Equivalently, by choosing a complex form ρ whose Clifford annihilator

$$L_\rho := \left\{v \in \left(TX \oplus T^*X\right) \otimes_{\mathbb{R}} \mathbb{C} : v \cdot \rho = 0\right\}$$

is the i-eigen-bundle $L_{\mathcal{J}}$ of the $\mathbb{C}$-linear extension of $\mathcal{J}$ to $(TX \oplus T^*X) \otimes_{\mathbb{R}} \mathbb{C}$), one gets the following definition.

---

[4]We recall that the Clifford algebra associated to $TX \oplus T^*X$ and $\langle \cdot \mid \cdot \cdot \rangle$ is

$$\text{Cliff}(TX \oplus T^*X, \langle \cdot \mid \cdot \cdot \rangle) = \left(\bigoplus_{k \in \mathbb{Z}} \bigotimes_{j=1}^{k} (TX \oplus T^*X)\right) \Big/ \left\{v \otimes_{\mathbb{R}} v - \langle v \mid v \rangle : v \in TX \oplus T^*X\right\} .$$

Proposition 1.2 ([Gua04a, Theorem 4.8]). *Let X be a differentiable manifold and H be a d-closed 3-form on X. A generalized complex structure on X is identified by a sub-bundle $U := U_{\mathcal{J}}$ of complex rank 1 of $\wedge^{\bullet} X \otimes_{\mathbb{R}} \mathbb{C}$ being locally generated by a form $\rho = \exp(B + i\omega) \wedge \Omega$, where $B \in \wedge^2 X$, and $\omega \in \wedge^2 X$, and $\Omega = \theta^1 \wedge \cdots \wedge \theta^k \in \wedge^k X \otimes_{\mathbb{R}} \mathbb{C}$ with $\{\theta^1, \ldots, \theta^k\}$ a set of linearly independent complex 1-forms, such that*

- $\Omega \wedge \bar{\Omega} \wedge \omega^{n-k} \neq 0$, and
- *there exists $v \in (TX \oplus T^*X) \otimes_{\mathbb{R}} \mathbb{C}$ such that $d_H \rho = v \cdot \rho$, where $d_H := d + H \wedge \cdot$.*

According to [Gua04a, Sect. 4.1], [Gua11, Definition 3.7], the bundle $U_{\mathcal{J}}$ as in Proposition 1.2 is called the *canonical bundle* of the generalized complex structure $\mathcal{J}$.

By definition, the *type* of a generalized complex structure $\mathcal{J}$ on X, [Gua04a, Sect. 4.3], [Gua11, Definition 3.5], is the upper-semi-continuous function

$$\text{type}(\mathcal{J}) := \frac{1}{2} \dim_{\mathbb{R}} \left(T^*X \cap \mathcal{J} T^*X \right)$$

on X, equivalently, [Gua11, Definition 1.1], the degree of the form Ω.

A point x of a generalized complex manifold is called a *regular point* if the type of the generalized complex structure is locally constant at x.

(See Remark 1.10 for the notion of H-twisted generalized Kähler structure.)

A generalized complex structure $\mathcal{J}$ on X induces a $\mathbb{Z}$-graduation on the space of complex differential forms on X, [Gua04a, Sect. 4.4], [Gua11, Proposition 3.8]. Namely, define, for $k \in \mathbb{Z}$,

$$U^k := U^k_{\mathcal{J}} := \wedge^{n-k} \bar{L}_{\mathcal{J}} \cdot U_{\mathcal{J}} \subseteq \wedge^{\bullet} X \otimes_{\mathbb{R}} \mathbb{C},$$

where $L_{\mathcal{J}}$ is the i-eigenspace of the $\mathbb{C}$-linear extension of $\mathcal{J}$ to $(TX \oplus T^*X) \otimes_{\mathbb{R}} \mathbb{C}$ and $U^n_{\mathcal{J}} := U_{\mathcal{J}}$ is the canonical bundle of $\mathcal{J}$.

For a $\langle \cdot \mid \cdot \rangle$-orthogonal endomorphism $\mathcal{J} \in \text{End}(TX \oplus T^*X)$ satisfying $\mathcal{J}^2 = -\text{id}_{TX \oplus T^*X}$, the $\mathbb{Z}$-graduation $U^{\bullet}_{\mathcal{J}}$ still makes sense, and the condition $\text{Nij}_{\mathcal{J},H} = 0$ turns out to be equivalent, [Gua04a, Theorem 4.3], [Gua11, Theorem 3.14], to

$$d_H \colon U^{\bullet}_{\mathcal{J}} \to U^{\bullet+1}_{\mathcal{J}} \oplus U^{\bullet-1}_{\mathcal{J}}.$$

Therefore, on a compact differentiable manifold endowed with a generalized complex structure $\mathcal{J}$, one has, [Gua04a, Sect. 4.4], [Gua11, Sect. 3],

$$d_H = \partial_{\mathcal{J},H} + \bar{\partial}_{\mathcal{J},H} \quad \text{where} \quad \partial_{\mathcal{J},H} \colon U^{\bullet}_{\mathcal{J}} \to U^{\bullet+1}_{\mathcal{J}} \quad \text{and} \quad \bar{\partial}_{\mathcal{J},H} \colon U^{\bullet}_{\mathcal{J}} \to U^{\bullet-1}_{\mathcal{J}}.$$

Define also, [Gua04a, p. 52], [Gua11, Remark at page 97],

$$d^{\mathcal{J}}_H := -i \left(\partial_{\mathcal{J},H} - \bar{\partial}_{\mathcal{J},H} \right) \colon U^{\bullet}_{\mathcal{J}} \to U^{\bullet+1}_{\mathcal{J}} \oplus U^{\bullet-1}_{\mathcal{J}}.$$

1.3.2 Cohomological Aspects of Generalized Complex Geometry

Let X be a compact complex manifold endowed with an H-twisted generalized complex structure.

By considering the decomposition $\wedge^\bullet X \otimes_\mathbb{R} \mathbb{C} = \bigoplus_{k \in \mathbb{Z}} U_\mathcal{J}^k$ and the operators $d_H : U_\mathcal{J}^\bullet \to U_\mathcal{J}^{\bullet+1} \oplus U_\mathcal{J}^{\bullet-1}$, and $\partial_{\mathcal{J},H} : U_\mathcal{J}^\bullet \to U_\mathcal{J}^{\bullet+1}$ and $\bar\partial_{\mathcal{J},H} : U_\mathcal{J}^\bullet \to U_\mathcal{J}^{\bullet-1}$, one can study the following cohomologies:

$$GH_{dR_H}(X) := \frac{\ker d_H}{\operatorname{im} d_H} ,$$

and

$$GH^\bullet_{\bar\partial_{\mathcal{J},H}}(X) := \frac{\ker \bar\partial_{\mathcal{J},H}}{\operatorname{im} \bar\partial_{\mathcal{J},H}} \qquad \text{and} \qquad GH^\bullet_{\partial_{\mathcal{J},H}}(X) := \frac{\partial_{\mathcal{J},H}}{\partial_{\mathcal{J},H}} ,$$

and

$$GH^\bullet_{BC_{\mathcal{J},H}}(X) := \frac{\ker \partial_{\mathcal{J},H} \cap \ker \bar\partial_{\mathcal{J},H}}{\operatorname{im} \partial_{\mathcal{J},H}\bar\partial_{\mathcal{J},H}} \qquad \text{and}$$

$$GH^\bullet_{A_{\mathcal{J},H}}(X) := \frac{\ker \partial_{\mathcal{J},H}\bar\partial_{\mathcal{J},H}}{\operatorname{im} \partial_{\mathcal{J},H} + \operatorname{im} \bar\partial_{\mathcal{J},H}} .$$

Note that, for $H = 0$, one has $GH_{dR_0}(X) = \bigoplus_{k \in \mathbb{Z}} H^k_{dR}(X; \mathbb{C})$.

By [Gua04a, Proposition 5.1], [Gua11, Proposition 3.15], it follows that $\dim_\mathbb{C} GH^\bullet_{\partial_{\mathcal{J},H}}(X) < +\infty$ and $\dim_\mathbb{C} GH^\bullet_{\bar\partial_{\mathcal{J},H}}(X) < +\infty$.

By abuse of notation, one says that X satisfies the $\partial_{\mathcal{J},H}\bar\partial_{\mathcal{J},H}$-Lemma if $\left(U_\mathcal{J}^\bullet, \partial_{\mathcal{J},H}, \bar\partial_{\mathcal{J},H} \right)$ satisfies the $\partial_{\mathcal{J},H}\bar\partial_{\mathcal{J},H}$-Lemma, and one says that X satisfies the $d_H d_H^\mathcal{J}$-Lemma if $(U_\mathcal{J}^\bullet, d_H, d_H^\mathcal{J})$ satisfies the $d_H d_H^\mathcal{J}$-Lemma. Actually, it turns out that X satisfies the $d_H d_H^\mathcal{J}$-Lemma if and only if X satisfies the $\partial_{\mathcal{J},H}\bar\partial_{\mathcal{J},H}$-Lemma, [Cav06, Remark at page 129]: indeed, note that $\ker \partial_{\mathcal{J},H}\bar\partial_{\mathcal{J},H} = \ker d_H d_H^\mathcal{J}$, and $\ker \partial_{\mathcal{J},H} \cap \ker \bar\partial_{\mathcal{J},H} = \ker d_H \cap \ker d_H^\mathcal{J}$, and $\operatorname{im} \partial_{\mathcal{J},H} + \operatorname{im} \bar\partial_{\mathcal{J},H} = \operatorname{im} d_H + \operatorname{im} d_H^\mathcal{J}$.

Moreover, the following result by G.R. Cavalcanti holds.

Theorem 1.16 ([Cav05, Theorem 4.2], [Cav06, Theorem 4.1, Corollary 2]).
A manifold X endowed with an H-twisted generalized complex structure $\mathcal{J}$ satisfies the $d_H d_H^\mathcal{J}$-Lemma if and only if $(\ker d_H^\mathcal{J}, d_H) \hookrightarrow (U_\mathcal{J}^\bullet, d_H)$ is a quasi-isomorphism of differential $\mathbb{Z}$-graded $\mathbb{C}$-vector spaces. In this case, it follows that the splitting $\wedge^\bullet X \otimes_\mathbb{R} \mathbb{C} = \bigoplus_{k \in \mathbb{Z}} U_\mathcal{J}^k$ gives rise to a decomposition in cohomology.

An application of [DGMS75, Propositions 5.17, 5.21] gives the following result.

Theorem 1.17 ([Cav05, Theorem 4.4], [Cav06, Theorem 5.1]). *A manifold* X *endowed with an* H-*twisted generalized complex structure* $\mathcal{J}$ *satisfies the* $d_H \, d_H^{\mathcal{J}}$-*lemma if and only if the canonical spectral sequence[5] degenerates at the first level and the decomposition of complex forms into sub-bundles* $U_{\mathcal{J}}^k$, *varying* $k \in \mathbb{Z}$, *induces a decomposition in cohomology.*

(See Remark 1.10 for $d_H \, d_H^{\mathcal{J}}$-Lemma on H-twisted generalized Kähler manifolds.)

1.3.3 Complex Structures and Symplectic Structures in Generalized Complex Geometry

In view of the following generalized Darboux theorem by M. Gualtieri, symplectic structures and complex structures provide the basic examples of generalized complex structures.

Theorem 1.18 ([Gua04a, Theorem 4.35], [Gua11, Theorem 3.6]). *For any regular point of a* $2n$-*dimensional generalized complex manifold with type equal to* k, *there is an open neighbourhood endowed with a set of local coordinates such that the generalized complex structure is a* B-*field transform of the standard generalized complex structure of* $\mathbb{C}^k \times \mathbb{R}^{2n-2k}$.

Hence, in the following examples, we recall the standard generalized complex structure of constant type n (that is, locally equivalent to the standard complex structure of $\mathbb{C}^n$), the generalized complex structure of constant type 0 (that is, locally equivalent to the standard symplectic structure of $\mathbb{R}^{2n}$), and the B-field transform of a generalized complex structure, on a $2n$-dimensional manifold. See also [Gua04a, Example 4.12].

Example 1.1 (Generalized Complex Structures of Type n, [Gua04a, Examples 4.11 and 4.25]). Let X be a compact $2n$-dimensional manifold endowed with a complex structure $J \in \mathrm{End}(TX)$. Consider the (0-twisted) generalized complex structure

$$\mathcal{J}_J := \left(\begin{array}{c|c} -J & 0 \\ \hline 0 & J^* \end{array} \right) \in \mathrm{End}\left(TX \oplus T^*X\right) ,$$

where $J^* \in \mathrm{End}(T^*X)$ denotes the dual endomorphism of $J \in \mathrm{End}(TX)$.

---

[5]We recall that, given a manifold X endowed with a generalized complex structure $\mathcal{J}$, the *canonical spectral sequence* is the spectral sequence being naturally associated to the double complex obtained from $\left(U_{\mathcal{J}}^\bullet, \partial_{\mathcal{J},H}, \overline{\partial}_{\mathcal{J},H}\right)$: more precisely, to the double complex $\left(U_{\mathcal{J}_J}^{\bullet_1 - \bullet_2} \otimes_\mathbb{C} \mathbb{C}\beta^{\bullet_2}, \, \partial_{\mathcal{J}_J} \otimes_\mathbb{R} \mathrm{id}, \, \overline{\partial}_{\mathcal{J}_J} \otimes_\mathbb{R} \beta\right)$, where $\{\beta^m : m \in \mathbb{Z}\}$ is an infinite cyclic multiplicative group generated by some β, [Cav05, Sect. 4.2], see [Bry88, Sect. 1.3], [Goo85, Sect. II.2], [Con85, Sect. II]; (compare also [AT13a, Sect. 1]).

The i-eigenspace of the $\mathbb{C}$-linear extension of $\mathcal{J}_J$ to $(TX \oplus T^*X) \otimes_{\mathbb{C}} \mathbb{C}$ is

$$L_{\mathcal{J}_J} = T_J^{0,1}X \oplus \left(T_J^{1,0}X \right)^* .$$

The canonical bundle is

$$U_{\mathcal{J}_J}^n = \wedge_J^{n,0}X .$$

Hence, one gets that, [Gua04a, Example 4.25],

$$U_{\mathcal{J}_J}^\bullet = \bigoplus_{p-q=\bullet} \wedge_J^{p,q}X , \qquad \text{and } \partial_{\mathcal{J}_J} = \partial_J \quad \text{and} \quad \overline{\partial}_{\mathcal{J}_J} = \overline{\partial}_J ;$$

note that $\mathrm{d}^{\mathcal{J}_J}$ is the operator $\mathrm{d}_J^c := -\mathrm{i}(\partial - \overline{\partial})$, [Gua04a, Remark 4.26].

Take an infinite cyclic multiplicative group $\{\beta^m : m \in \mathbb{Z}\}$ generated by some β. The Hodge and Frölicher spectral sequence associated to the canonical double complex $\left(U_{\mathcal{J}_J}^{\bullet_1 - \bullet_2} \otimes_{\mathbb{C}} \mathbb{C}\beta^{\bullet_2}, \partial_{\mathcal{J}_J} \otimes_{\mathbb{R}} \mathrm{id}, \overline{\partial}_{\mathcal{J}_J} \otimes_{\mathbb{R}} \beta \right)$ degenerates at the first level if and only if the Hodge and Frölicher spectral sequence associated to the double complex $\left(\wedge_J^{\bullet,\bullet}X, \partial_J, \overline{\partial}_J \right)$ does, [Cav05, Remark at page 76]. X satisfies the $\mathrm{d}\,\mathrm{d}^{\mathcal{J}_J}$-Lemma if and only if X satisfies the $\mathrm{d}\,\mathrm{d}_J^c$-Lemma.

For $\sharp \in \left\{ \overline{\partial}, \partial, BC, A \right\}$,

$$GH_{\sharp_{\mathcal{J}_J}}^\bullet (X) = \mathrm{Tot}^\bullet H_{\sharp_J}^{\bullet,-\bullet}(X) = \bigoplus_{p-q=\bullet} H_{\sharp_J}^{p,q}(X) .$$

Example 1.2 (Generalized Complex Structures of Type 0, [Gua04a, Example 4.10]). Let X be a compact $2n$-dimensional manifold endowed with a symplectic structure $\omega \in \wedge^2 X \simeq \mathrm{Hom}\,(TX; T^*X)$. Consider the (0-twisted) generalized complex structure

$$\mathcal{J}_\omega := \left(\begin{array}{c|c} 0 & -\omega^{-1} \\ \hline \omega & 0 \end{array} \right) ,$$

where $\omega^{-1} \in \mathrm{Hom}\,(T^*X; TX)$ denotes the inverse of $\omega \in \mathrm{Hom}\,(TX; T^*X)$.

The i-eigenspace of the $\mathbb{C}$-linear extension of $\mathcal{J}_\omega$ to $(TX \otimes_{\mathbb{R}} \mathbb{C}) \oplus (T^*X \otimes_{\mathbb{R}} \mathbb{C})$ is

$$L_{\mathcal{J}_\omega} = \{X - \mathrm{i}\,\omega\,(X, \cdot) : X \in TX \otimes_{\mathbb{R}} \mathbb{C}\} ,$$

which has Clifford annihilator $\exp(\mathrm{i}\,\omega)$. The canonical bundle is

$$U_{\mathcal{J}_\omega}^n = \mathbb{C}\,\langle \exp(\mathrm{i}\,\omega) \rangle .$$

In particular, one gets that, [Cav06, Theorem 2.2],

$$U_{\mathcal{J}_\omega}^{n-\bullet} = \exp(i\,\omega)\left(\exp\left(\frac{\Lambda}{2i}\right)(\wedge^\bullet X \otimes_\mathbb{R} \mathbb{C})\right) .$$

Note that, [Cav06, Sect. 2.2],

$$d^{\mathcal{J}_\omega} = d^\Lambda .$$

By considering the natural isomorphism

$$\varphi: \bigoplus_{k\in\mathbb{Z}} \wedge^k X \otimes_\mathbb{R} \mathbb{C} \ni \alpha \mapsto \exp(i\,\omega)\left(\exp\left(\frac{\Lambda}{2i}\right)\alpha\right) \in \bigoplus_{k\in\mathbb{Z}} \wedge^k X \otimes_\mathbb{R} \mathbb{C} ,$$

one gets that, [Cav06, Corollary 1],

$$\varphi\left(\wedge^\bullet X \otimes_\mathbb{R} \mathbb{C}\right) \simeq U^{n-\bullet} ,$$

and

$$\varphi\, d = \bar{\partial}_{\mathcal{J}_\omega}\varphi \quad \text{and} \quad \varphi\, d^{\mathcal{J}_\omega} = -2\,i\,\partial_{\mathcal{J}_\omega}\varphi ;$$

in particular,

$$GH_{\bar{\partial}_{\mathcal{J}_\omega}}^\bullet(X) \simeq H_{dR}^{n-\bullet}(X;\mathbb{C}) .$$

Example 1.3 (B-Transform, [Gua04a, Sect. 3.3]). Let X be a compact $2n$-dimensional manifold endowed with an H-twisted generalized complex structure $\mathcal{J}$, and let B be a d-closed 2-form. Consider the H-twisted generalized complex structure

$$\mathcal{J}^B := \exp(-B)\,\mathcal{J}\,\exp B \qquad \text{where} \qquad \exp B = \left(\begin{array}{c|c} \mathrm{id}_{TX} & 0 \\ \hline B & \mathrm{id}_{T^*X} \end{array}\right) .$$

The i-eigenspace of the $\mathbb{C}$-linear extension of $\mathcal{J}$ to $(TX \oplus T^*X)\otimes_\mathbb{R}\mathbb{C}$ is, [Cav05, Example 2.3],

$$L_{\mathcal{J}^B} = \{X + \xi - \iota_X B \;:\; X + \xi \in L_{\mathcal{J}}\} ,$$

and the canonical bundle is, [Cav05, Example 2.6],

$$U_{\mathcal{J}^B}^n = \exp B \wedge U_{\mathcal{J}}^n .$$

Hence one gets that, [Cav06, Sect. 2.3],

$$U^\bullet_{\mathcal{J}^B} = \exp B \wedge U^\bullet_{\mathcal{J}}, \qquad \text{and that} \qquad \partial_{\mathcal{J}^B} = \exp(-B)\,\partial_{\mathcal{J}}\exp B \qquad \text{and}$$

$$\bar{\partial}_{\mathcal{J}^B} = \exp(-B)\,\bar{\partial}_{\mathcal{J}}\exp B\;.$$

In particular, $\mathcal{J}$ satisfies the $\partial_{\mathcal{J}}\bar{\partial}_{\mathcal{J}}$-Lemma if and only if $\mathcal{J}^B$ satisfies the $\partial_{\mathcal{J}^B}\bar{\partial}_{\mathcal{J}^B}$-Lemma.

1.4 Kähler Geometry

Note that, given a manifold X endowed with a symplectic form ω, there is always a (possibly non-integrable) almost-complex structure J on X such that $g := \omega(\cdot,\,J\cdot)$ is a Hermitian metric on X with ω as the associated $(1, 1)$-form, see, e.g., [CdS01, Corollary 12.7] (in fact, the set of such almost-complex structures is contractible, see, e.g., [AL94, Corollary II.1.1.7], [CdS01, Proposition 13.1]; see also [Gro85, Corollary 2.3.C$'_2$], which proves that the space of almost-complex structures on X tamed by a given 2-form on X is contractible). Instead, the datum of an integrable almost-complex structure with the above property yields a Kähler structure on X. The notion of Kähler manifold has been studied for the first time by J.A. Schouten and D. van Dantzig [SvD30], see also [Sch29], and by E. Kähler [Käh33], and the terminology has been fixed by A. Weil [Wei58].

Kähler structures can be defined in different ways, according to the point of view which is stressed, Sect. 1.4.1. The presence of three different structures (complex, symplectic, and Riemannian) allows to make use of the tools available for any of them; in addition, the relations between such structures make available further tools, which yield many interesting results on Hodge theory, Sect. 1.4.2. Finally, we will study a cohomological property of compact Kähler manifolds, namely, the $\partial\bar{\partial}$-Lemma, Sect. 1.4.3: other than being a very useful tool in Kähler geometry (compare, e.g., its role in S.-T. Yau's proof [Yau77, Yau78] of E. Calabi's conjecture [Cal57]), it provides obstructions to the existence of Kähler structures on differentiable manifolds, by means of the notion of formality introduced by D.P. Sullivan, [Sul77, Sect. 12].

1.4.1 Kähler Metrics

Let X be a compact complex manifold of complex dimension n, and denote by J its natural integrable almost-complex structure.

Definition 1.4 ([SvD30, Sch29, Käh33, Wei58]). Let X be a compact complex manifold of complex dimension n and J be its natural integrable almost-complex structure. A *Kähler metric* on X is a Hermitian metric g such that the associated $(1, 1)$-form $\omega := g(J\cdot, \cdot\cdot)$ is d-closed (that is, ω is a symplectic form on X).

Remark 1.8. Let X be a complex manifold endowed with a Kähler metric g, and denote the associated $(1, 1)$-form to g by ω. By the Poincaré lemma, see, e.g., [Dem12, I.1.22, Theorem I.2.24], and the Dolbeault and Grothendieck lemma, see, e.g., [Dem12, I.3.29], the property that $d\omega = 0$ is equivalent to ask that, for every $x \in X$, there exist an open neighbourhood U in X with $x \in U$ and a smooth function $u \in C^\infty(U; \mathbb{R})$ such that $\omega \overset{\text{loc}}{=} i\partial\bar{\partial}u$ in U, that is, the metric has a local potential, [Käh33] (see, e.g., [Mor07, Proposition 8.8]).

Remark 1.9. For every $n \in \mathbb{N}$, the complex projective space $\mathbb{CP}^n$ admits a Kähler metric, the so-called *Fubini and Study metric*, [Fub04, Stu05], which is induced by the fibration $\mathbb{S}^1 \hookrightarrow \mathbb{S}^{2n+1} \to \mathbb{CP}^n$; more precisely, by using the homogeneous coordinates $[z_0 : \cdots : z_n]$, one has that the associated $(1, 1)$-form ω_{FS} to the Fubini and Study metric is

$$\omega_{\text{FS}} = \frac{i}{2\pi} \partial\bar{\partial} \log\left(\sum_{\ell=0}^{n} |z^\ell|^2\right).$$

It follows that complex projective manifolds provide examples of Kähler manifolds. Conversely, by the Kodaira embedding theorem [Kod54, Theorem 4], if X is a compact complex manifold endowed with a Kähler metric ω such that $[\omega] \in H^2_{dR}(X; \mathbb{R}) \cap \text{im}\left(H^2(X; \mathbb{Z}) \to H^2_{dR}(X; \mathbb{R})\right)$, then there exists a complex-analytic embedding of X into a complex projective space $\mathbb{CP}^N$ for some $N \in \mathbb{N}$. In a sense, this suggest that projective manifolds are to Kähler manifolds as $\mathbb{Q}$ is to $\mathbb{R}$. Hence, it is natural to ask if every compact Kähler manifold is a deformation of a projective manifold (which is known as the *Kodaira problem*). Since Riemann surfaces are projective, this is trivially true in complex dimension 1. Furthermore, K. Kodaira proved in [Kod63, Theorem 16.1] that every compact Kähler surface is a deformation of an algebraic surface, as conjectured by W. Hodge; another proof, which does not make use of the classification of elliptic surfaces, has been given by N. Buchdahl, [Buc08, Theorem]. In higher dimension, a negative answer to the Kodaira problem has been given by C. Voisin, who constructed examples of compact Kähler manifolds, of any complex dimension greater than or equal to 4, which do not have the homotopy type of a complex projective manifold, [Voi04, Theorem 2] (indeed, recall that, by Ehresmann's theorem, if two compact complex manifolds can be obtained by deformation, then they are homeomorphic, and hence they have the same homotopy type). The examples in [Voi04] being, by construction, bimeromorphic to manifolds that can be deformed to projective manifolds, one could ask (as done by N. Buchdahl, F. Campana, S.-T. Yau) whether, in higher dimension, a birational version of the Kodaira problem may hold true; in [Voi06, Theorem 3], C. Voisin provided a negative answer to the birational version of the Kodaira problem, proving that, in any even complex dimension greater that or equal to 10, there exist compact Kähler manifolds X such that, for any compact

Kähler manifold X' bimeromorphic to X, X' does not have the homotopy type of a projective complex manifold. Positive results under an additional semi-positivity or semi-negativity condition on the canonical bundle have been provided by J. Cao in [Cao12].

In the definition of a Kähler manifold, three different structures are involved: a complex structure, a symplectic structure, and a metric structure. Therefore, changing the point of view allows to give several equivalent definitions of Kähler structure (see, e.g., [Bal06, Theorem 4.17]): we review here two of these characterizations.

Firstly, it is straightforward to prove that a Hermitian metric g on a compact complex manifold X is a Kähler metric if and only if, for every point $x \in X$, there exists a holomorphic coordinate chart $\left(U, \{z^j\}_{j \in \{1,\dots,n\}}\right)$, with $x \in U$, such that

$$g = \sum_{\alpha,\beta=1}^{n} \left(\delta_{\alpha\beta} + \mathrm{o}\left(|z|\right)\right) \, \mathrm{d}z^\alpha \odot \mathrm{d}\bar{z}^\beta \qquad \text{at } x \,,$$

that is, g *osculates to order* 2 the standard Hermitian metric of $\mathbb{C}^n$ (see, e.g., [GH94, pp. 107–108], [Huy05, Proposition 1.3.12], [Mor07, Theorem 11.6]).

As regards the second characterization, we recall that, on a compact complex manifold X endowed with a Hermitian metric g, there is a unique connection ∇^C such that

(i) $\nabla^C g = 0$,
(ii) $\nabla^C J = 0$, and
(iii) $\pi_{\wedge^{0,1} X} \nabla^C \lfloor_{\mathcal{C}^\infty(X;\mathbb{C})} = \bar{\partial} \lfloor_{\mathcal{C}^\infty(X;\mathbb{C})}$;

such a connection is called the *Chern connection* of X (see, e.g., [Huy05, Proposition 4.2.14], [Bal06, Theorem 3.18], [Mor07, Theorem 10.3], [Dem12, Sect. V.12]). Let g be a Hermitian metric on a compact complex manifold X, and set $\omega := g(J\cdot, \cdot\cdot)$ its associated $(1, 1)$-form, where J is the natural integrable almost-complex structure on X; consider the Levi Civita connection ∇^{LC}. One can prove that, for every $x, y, z \in \mathcal{C}^\infty(X; TX)$,

$$\mathrm{d}\omega(x,y,z) = g\left(\left(\nabla^{LC}_x J\right) y, z\right) + g\left(\left(\nabla^{LC}_y J\right) z, x\right) + g\left(\left(\nabla^{LC}_z J\right) x, y\right) \,,$$

and

$$2\,g\left(\left(\nabla^{LC}_x J\right) y, z\right) = \mathrm{d}\omega\left(x, y, z\right) - \mathrm{d}\omega\left(x, Jy, Jz\right) - g\left(\mathrm{Nij}_J\left(y, Jz\right), x\right) \,;$$

(see, e.g., [Bal06, Theorem 4.16], [Tia00, Proposition 1.5]); in particular, it follows that g is a Kähler metric if and only if $\nabla^{LC} J = 0$ if and only if the Chern connection is the Levi Civita connection (see, e.g., [Bal06, Theorem 4.17], [Mor07, Proposition 11.8]).

Remark 1.10. We recall that, given a d-closed 3-form H on a manifold X, an H-*twisted generalized Kähler structure* on X is a pair $(\mathcal{J}_1, \mathcal{J}_2)$ of H-twisted generalized complex structures on X such that *(i)* $\mathcal{J}_1$ and $\mathcal{J}_2$ commute, and *(ii)* the symmetric pairing $\langle \mathcal{J}_1 \cdot, \mathcal{J}_2 \cdot \cdot \rangle$ is positive definite. Generalized Kähler geometry is equivalent to a bi-Hermitian geometry with torsion, [Gua04b, Theorem 2.18].

We recall that a compact manifold X endowed with an H-twisted generalized Kähler structure $(\mathcal{J}_1, \mathcal{J}_2)$ satisfies both the $d_H \, d_H^{\mathcal{J}_1}$-Lemma and the $d_H \, d_H^{\mathcal{J}_2}$-Lemma, [Gua04b, Corollary 4.2].

Any Kähler structure provide an example of a 0-twisted generalized Kähler structure. A left-invariant non-trivial twisted generalized Kähler structure on a (non-completely solvable) solvmanifold (which is the total space of a $\mathbb{T}^2$-bundle over the Inoue surface, [FT09, Proposition 3.2]) has been constructed by A.M. Fino and A. Tomassini, [FT09, Theorem 3.5].

Furthermore, we note that A. Tomasiello proved in [Tom08, Sect. B] satisfying the $d \, d^{\mathcal{J}}$-Lemma is a stable property under small deformations.

1.4.2 Hodge Theory for Kähler Manifolds

The complex, symplectic, and metric structures being related on a Kähler manifold, one gets the following identities concerning the corresponding operators (see, e.g., [Huy05, Proposition 3.1.12]); see also [Hod35, Hod89]. (In [Dem86, Theorems 1.1 and 2.12], commutation relations on arbitrary Hermitian manifolds are provided; see also [Gri66], [Dem12, Sect. VI.6.2].)

Theorem 1.19 (Kähler Identities, [Wei58, Théorème II.1, II.2, Corollaire II.1]).
Let X be a compact Kähler manifold. Consider the differential operators ∂ and $\bar{\partial}$ associated to the complex structure, the symplectic operators L and Λ associated to the symplectic structure, and the Hodge-$*$-operator associated to the Hermitian metric. Then, these operators are related as follows:

(i) $\left[\bar{\partial}, L \right] = [\partial, L] = 0$ and $\left[\Lambda, \bar{\partial}^* \right] = [\Lambda, \partial^*] = 0$;

(ii) $\left[\bar{\partial}^*, L \right] = \mathrm{i} \, \partial$ and $[\partial^*, L] = -\mathrm{i} \, \bar{\partial}$, and $\left[\Lambda, \bar{\partial} \right] = -\mathrm{i} \, \partial^*$ and $[\Lambda, \partial] = \mathrm{i} \, \bar{\partial}^*$.

Therefore, considering the 2nd order self-adjoint elliptic differential operators $\square := [\partial, \partial^*]$, $\overline{\square} := \left[\bar{\partial}, \bar{\partial}^* \right]$, and $\Delta := [d, d^*]$, one gets that

(iii) $\square = \overline{\square} = \frac{1}{2} \Delta$, and Δ commutes with $*$, ∂, $\bar{\partial}$, ∂^*, $\bar{\partial}^*$, L, Λ.

The previous identities can be proven either using the $\mathfrak{sl}(2; \mathbb{C})$ representation $\langle L, \Lambda, H \rangle \to \mathrm{End}^{\bullet}(\wedge^{\bullet} X \otimes \mathbb{C})$, or reducing to prove the corresponding identities on $\mathbb{C}^n$ with the standard Kähler structure (which are known as Y. Akizuki and S. Nakano's identities, [AN54, Sect. 3]) and hence using that every Kähler metric osculates to order 2 the standard Hermitian metric on $\mathbb{C}^n$.

As a consequence, one gets the following theorems, stating a decomposition of the de Rham cohomology of a Kähler manifold related to the complex, respectively symplectic, structure (see, e.g., [Huy05, Corollary 3.2.12], respectively [Huy05, Proposition 3.2.13]).

Theorem 1.20 (Hodge Decomposition Theorem, [Wei58, Théorème IV.3]). *Let X be a compact complex manifold endowed with a Kähler structure. Then there exist a decomposition*

$$H^\bullet_{dR}(X;\mathbb{C}) \simeq \bigoplus_{p+q=\bullet} H^{p,q}_{\bar\partial}(X) \,,$$

and, for every $(p,q) \in \mathbb{N}^2$, *an isomorphism*

$$H^{p,q}_{\bar\partial}(X) \simeq \overline{H^{q,p}_{\bar\partial}(X)} \,.$$

Theorem 1.21 (Lefschetz Decomposition Theorem, [Wei58, Théorème IV.5]). *Let X be a compact complex manifold, of complex dimension n, endowed with a Kähler structure. Then there exist a decomposition*

$$H^\bullet_{dR}(X;\mathbb{C}) = \bigoplus_{r\in\mathbb{N}} L^r \left(\ker \left(\Lambda \colon H^{\bullet-2r}_{dR}(X;\mathbb{C}) \to H^{\bullet-2r-2}_{dR}(X;\mathbb{C}) \right) \right) \,,$$

and, for every $k \in \mathbb{N}$, *an isomorphism*

$$L^k \colon H^{n-k}_{dR}(X;\mathbb{C}) \xrightarrow{\simeq} H^{n+k}_{dR}(X;\mathbb{C}) \,.$$

1.4.3 $\partial\bar\partial$-Lemma and Formality for Compact Kähler Manifolds

The Hodge decomposition theorem and the Lefschetz decomposition theorem provide obstructions to the existence of Kähler structures on a compact complex manifold. In this section, we study another property of compact Kähler manifolds, namely, formality, which provides an obstruction to the existence of a Kähler structure on a compact (differentiable) manifold. Such a property turns out to be a consequence of the validity of the $\partial\bar\partial$-Lemma on compact complex manifolds.

Firstly, we need to recall some general notions regarding homotopy theory of differential algebras; we will then summarize some results concerning the homotopy type of Kähler manifolds: by the classical result by P. Deligne, Ph.A. Griffiths, J. Morgan, and D.P. Sullivan, [DGMS75, Main Theorem], the real homotopy type of a Kähler manifold X is a formal consequence of its cohomology ring $H^\bullet_{dR}(X;\mathbb{R})$.

We recall that a *differential graded algebra* (shortly, *dga*) over a field $\mathbb{K}$ is a graded $\mathbb{K}$-algebra $A^\bullet$ (where the structure of $\mathbb{K}$-algebra is induced by an

inclusion $\mathbb{K} \subseteq A^0$) being graded-commutative[6] and endowed with a differential
$\mathrm{d} \colon A^\bullet \to A^{\bullet+1}$ satisfying the graded-Leibniz rule.[7] A *morphism of differential
graded algebras* $F \colon (A^\bullet, \mathrm{d}_{A^\bullet}) \to (B^\bullet, \mathrm{d}_{B^\bullet})$ is a morphism $A^\bullet \to B^\bullet$ of $\mathbb{K}$-algebras
such that $F \circ \mathrm{d}_{A^\bullet} = \mathrm{d}_{B^\bullet} \circ F$.

Given a dga $(A^\bullet, \mathrm{d})$ over $\mathbb{K}$, the cohomology $H^\bullet(A^\bullet, \mathrm{d}) := \frac{\ker \mathrm{d}}{\operatorname{im} \mathrm{d}}$ endowed
with the zero differential has a natural structure of dga over $\mathbb{K}$; furthermore,
every morphism $F \colon (A^\bullet, \mathrm{d}_{A^\bullet}) \to (B^\bullet, \mathrm{d}_{B^\bullet})$ of dgas induces a morphism
$F^* \colon (H^\bullet(A^\bullet, \mathrm{d}_{A^\bullet}), 0) \to (H^\bullet(B^\bullet, \mathrm{d}_{B^\bullet}), 0)$ of dgas in cohomology; a morphism
$F \colon (A^\bullet, \mathrm{d}_{A^\bullet}) \to (B^\bullet, \mathrm{d}_{B^\bullet})$ of dgas is called a *quasi-isomorphism* (shortly, *qis*) if
the corresponding morphism $F^* \colon (H^\bullet(A^\bullet, \mathrm{d}_{A^\bullet}), 0) \to (H^\bullet(B^\bullet, \mathrm{d}_{B^\bullet}), 0)$ is an
isomorphism.

The de Rham complex $(\wedge^\bullet X, \mathrm{d})$ of a compact (differentiable) manifold X has a
structure of dga over $\mathbb{R}$, whose cohomology is the dga $\left(H^\bullet_{dR}(X; \mathbb{R}), 0\right)$.

Given a dga $(A^\bullet, \mathrm{d}_{A^\bullet})$ over $\mathbb{K}$, the differential $\mathrm{d}_{A^\bullet}$ is called *decomposable* if

$$\mathrm{d}_{A^\bullet}\left(A^\bullet\right) \subseteq \left(\bigoplus_{k \in \mathbb{N} \setminus \{0\}} A^k\right) \cdot \left(\bigoplus_{k \in \mathbb{N} \setminus \{0\}} A^k\right).$$

Given a dga $(A^\bullet, \mathrm{d}_{A^\bullet})$ over $\mathbb{K}$, an *elementary extension* of $(A^\bullet, \mathrm{d}_{A^\bullet})$ is a dga
$(B^\bullet, \mathrm{d}_{B^\bullet})$ over $\mathbb{K}$ such that

(i) $B^\bullet = A^\bullet \otimes_{\mathbb{K}} \wedge^\bullet V_k$ for V_k a finite-dimensional $\mathbb{K}$-vector space and $k > 0$,
     where $\wedge^\bullet V_k$ is the free graded $\mathbb{K}$-algebra generated by V_k, the elements of V_k
     having degree k, and
(ii) $\mathrm{d}_{B^\bullet} \lfloor_{A^\bullet} = \mathrm{d}_{A^\bullet}$ and $\mathrm{d}_{B^\bullet}(V_k) \subseteq A^\bullet$.

A dga $(M^\bullet, \mathrm{d}_{M^\bullet})$ over $\mathbb{K}$ is called *minimal* if it can be written as an increasing
union of sub-dgas,

$$(\mathbb{K}, 0) = \left(M_0^\bullet, \mathrm{d}_{M_0^\bullet}\right) \subset \left(M_1^\bullet, \mathrm{d}_{M_1^\bullet}\right) \subset \left(M_2^\bullet, \mathrm{d}_{M_2^\bullet}\right) \subseteq \cdots,$$

$$\left(M^\bullet, \mathrm{d}_{M^\bullet}\right) = \bigcup_{j \in \mathbb{N}} \left(M_j^\bullet, \mathrm{d}_{M_j^\bullet}\right),$$

such that

(i) for any $j \in \mathbb{N}$, the dga $\left(M_{j+1}^\bullet, \mathrm{d}_{M_{j+1}^\bullet}\right)$ is an elementary extension of the dga
     $\left(M_j^\bullet, \mathrm{d}_{M_j^\bullet}\right)$, and
(ii) $\mathrm{d}_{M^\bullet}$ is decomposable.

---

[6] We recall that a graded $\mathbb{K}$-algebra $A^\bullet$ is *graded-commutative* if, for every $x \in A^{\deg x}$ and $y \in A^{\deg y}$, it holds $x \cdot y = (-1)^{\deg x \cdot \deg y} y \cdot x$.

[7] We recall that a differential $\mathrm{d} \colon A^\bullet \to A^{\bullet+1}$ on a graded $\mathbb{K}$-algebra $A^\bullet$ satisfies the *graded-Leibniz rule* if, for every $x \in A^{\deg x}$ and $y \in A^{\deg y}$, it holds $\mathrm{d}(x \cdot y) = \mathrm{d} x \cdot y + (-1)^{\deg x} x \cdot \mathrm{d} y$.

A *minimal model* for a dga $(A^\bullet, d_{A^\bullet})$ over $\mathbb{K}$ is the datum of a minimal dga $(M^\bullet, d_{M^\bullet})$ over $\mathbb{K}$ and a quasi-isomorphism $\rho\colon (M^\bullet, d_{M^\bullet}) \xrightarrow{\text{qis}} (A^\bullet, d_{A^\bullet})$ of dgas.

Two dgas $(A^\bullet, d_{A^\bullet})$ and $(B^\bullet, d_{B^\bullet})$ over $\mathbb{K}$ are *equivalent* if there exist an integer $n \in \mathbb{N} \setminus \{0\}$, a family $\left\{\left(C_j^\bullet, d_{C_j^\bullet}\right)\right\}_{j \in \{0,\dots,2n\}}$ of dgas over $\mathbb{K}$ with $\left(C_0^\bullet, d_{C_0^\bullet}\right) = (A^\bullet, d_{A^\bullet})$ and $\left(C_{2n}^\bullet, d_{C_{2n}^\bullet}\right) = (B^\bullet, d_{B^\bullet})$, and a family

$$\left(C_{2j+1}^\bullet, d_{C_{2j+1}^\bullet}\right)$$

$$\text{qis} \swarrow \qquad\qquad \searrow \text{qis}$$

$$\left(C_{2j}^\bullet, d_{C_{2j}^\bullet}\right) \qquad\qquad\qquad \left(C_{2j+2}^\bullet, d_{C_{2j+2}^\bullet}\right)$$

of quasi-isomorphisms, varying $j \in \{0,\dots,n-1\}$. A dga $(A^\bullet, d_{A^\bullet})$ over $\mathbb{K}$ is called *formal* if it is equivalent to a dga $(B^\bullet, 0)$ over $\mathbb{K}$ with zero differential, that is, if it is equivalent to $(H^\bullet(A^\bullet, d_{A^\bullet}), 0)$.

A compact manifold X is called *formal* if its de Rham complex $(\wedge^\bullet X, d)$ is a formal dga over $\mathbb{R}$.

Let $(A^\bullet, d_{A^\bullet})$ be a dga over $\mathbb{K}$. Given

$$[\alpha_{12}] \in H^{\deg \alpha_{12}}\left(A^\bullet, d_{A^\bullet}\right), \qquad [\alpha_{23}] \in H^{\deg \alpha_{23}}\left(A^\bullet, d_{A^\bullet}\right), \qquad \text{and}$$

$$[\alpha_{34}] \in H^{\deg \alpha_{34}}\left(A^\bullet, d_{A^\bullet}\right)$$

such that

$$[\alpha_{12}] \cdot [\alpha_{23}] = 0 \qquad \text{and} \qquad [\alpha_{23}] \cdot [\alpha_{34}] = 0,$$

let $\alpha_{13} \in A^{\deg \alpha_{12} + \deg \alpha_{23} - 1}$ and $\alpha_{24} \in A^{\deg \alpha_{23} + \deg \alpha_{34} - 1}$ be such that

$$(-1)^{\deg \alpha_{12}} \alpha_{12} \cdot \alpha_{23} = d_{A^\bullet} \alpha_{13} \qquad \text{and} \qquad (-1)^{\deg \alpha_{23}} \alpha_{23} \cdot \alpha_{34} = d_{A^\bullet} \alpha_{24};$$

one can then define the *triple Massey product* $\langle [\alpha_{12}], [\alpha_{23}], [\alpha_{34}] \rangle$ as

$$\langle [\alpha_{12}], [\alpha_{23}], [\alpha_{34}] \rangle := \left[(-1)^{\deg \alpha_{12}} \alpha_{12} \cdot \alpha_{24} + (-1)^{\deg \alpha_{13}} \alpha_{13} \cdot \alpha_{34}\right]$$

$$\in \frac{H^{\deg \alpha_{12} + \deg \alpha_{23} + \deg \alpha_{34} - 1}\left(A^\bullet, d_{A^\bullet}\right)}{H^{\deg \alpha_{12}}\left(A^\bullet, d_{A^\bullet}\right) \cdot H^{\deg \alpha_{23} + \deg \alpha_{34} - 1}\left(A^\bullet, d_{A^\bullet}\right) + H^{\deg \alpha_{34}}\left(A^\bullet, d_{A^\bullet}\right) \cdot H^{\deg \alpha_{12} + \deg \alpha_{23} - 1}\left(A^\bullet, d_{A^\bullet}\right)}.$$

One can define the higher order Massey product by induction. Fixed $m \in \mathbb{N}$ such that $m \geq 4$, and given

$$[\alpha_{12}] \in H^{\deg \alpha_{12}}\left(A^\bullet, d_{A^\bullet}\right), \qquad \dots, \qquad [\alpha_{m,m+1}] \in H^{\deg \alpha_{m,m+1}}\left(A^\bullet, d_{A^\bullet}\right)$$

such that all the Massey products of order lower than or equal to $m - 1$ vanish, let $\{\alpha_{rs}\}_{1 \leq r < s \leq m+1} \subseteq A^\bullet$ be such that

$$\sum_{h < \ell < k} (-1)^{\deg \alpha_{h\ell}} \, \alpha_{h\ell} \cdot \alpha_{\ell k} \; = \; \mathrm{d}\, \alpha_{hk} \, ,$$

for any $h, k \in \{1, \ldots, m + 1\}$ with $k - h < m$. Then define the mth order *Massey product* as

$$\langle [\alpha_{12}], \ldots, [\alpha_{m,m+1}] \rangle \; := \; \left[\sum_{1 < \ell < m+1} (-1)^{\deg \alpha_{1k}} \, \alpha_{1k} \cdot \alpha_{k,m+1} \right]$$

belonging to a quotient of $H^\bullet (A^\bullet, \mathrm{d}_{A^\bullet})$.

As a direct consequence of the definitions, the Massey products (of any order) on a formal dga are zero.

Now, let X be a compact manifold endowed with a Kähler structure.

The Kähler identities allow to prove the following result, known as $\partial\bar\partial$-*Lemma* (see, e.g., [Huy05, Corollary 3.2.10]), which, in a sense, summarizes many of the cohomological properties of compact Kähler manifolds.

Theorem 1.22 ($\partial\bar\partial$-Lemma for Compact Kähler Manifolds, [DGMS75, Lemma 5.11]). *Let X be a compact Kähler manifold. Then every ∂-closed, $\bar\partial$-closed, d-exact form is also $\partial\bar\partial$-exact.*

Proof. We recall the idea of the proof, as can be found, e.g., in [Huy05, Corollary 3.2.10], see also [DGMS75, pp. 266–267].

Let $\alpha \in \wedge^{p,q} X$ be a ∂-closed $\bar\partial$-closed d-exact form on X. In particular, by the Hodge decomposition theorem for the de Rham cohomology (see, e.g., [War83, 6.8]), the form α is orthogonal to the space of Δ-harmonic forms. Note that, by the Kähler identities, the space of Δ-harmonic forms coincide with the space of $\square$-harmonic forms and with the space of $\overline{\square}$-harmonic forms. Since α is ∂-closed and orthogonal to the space of $\square$-harmonic forms, the conjugate version of the Hodge decomposition theorem for the Dolbeault cohomology, [Hod89], yields $\alpha = \partial\gamma$ for some $\gamma \in \wedge^{p-1,q} X$. By applying the Hodge decomposition theorem for the Dolbeault cohomology, [Hod89], to the form γ, one gets a $\overline{\square}$-harmonic form h_γ, a form $\beta \in \wedge^{p-1,q-1} X$, and a form $\eta \in \wedge^{p-1,q+1} X$ such that $\gamma = h_\gamma + \bar\partial \beta + \bar\partial^* \eta$. By the Kähler identities, one has $\left[\partial, \bar\partial^* \right] = 0$ and that h_γ is also $\square$-harmonic. Hence $\alpha = \partial\gamma = \partial\bar\partial\beta - \bar\partial^* \partial\eta$. It suffices to prove that $\bar\partial^* \partial\eta = 0$. Indeed, since α is $\bar\partial$-closed, one has $\bar\partial\bar\partial^* \partial\eta = 0$, and hence $\left\| \bar\partial^* \partial\eta \right\|^2 = \left\langle \bar\partial^* \partial\eta, \bar\partial^* \partial\eta \right\rangle = \left\langle \bar\partial\bar\partial^* \partial\eta, \partial\eta \right\rangle = 0$. Hence $\alpha = \partial\bar\partial\beta$ is $\partial\bar\partial$-exact. $\qquad\square$

Using the differential operator $\mathrm{d}^c := J^{-1} \, \mathrm{d} \, J = -\mathrm{i} \left(\partial - \bar\partial \right)$ (where J is the integrable almost-complex structure naturally associated to the structure of complex manifold on X), and noting that $\ker \partial \cap \ker \overline{\partial} = \ker \mathrm{d} \cap \ker \mathrm{d}^c$ and $\operatorname{im} \partial\overline{\partial} = \operatorname{im} \mathrm{d}\, \mathrm{d}^c$, the following equivalent formulation can be provided.

Theorem 1.23 ($\mathrm{d}\,\mathrm{d}^c$**-Lemma for Compact Kähler Manifolds, [DGMS75, Lemma 5.11]**). *Let X be a compact Kähler manifold. Then every* d*-closed,* d^c*-closed,* d*-exact form is also* $\mathrm{d}\,\mathrm{d}^c$*-exact.*

Actually, the $\partial\overline{\partial}$-Lemma holds true for a larger class of compact complex manifolds than the compact Kähler manifolds: indeed, it holds, for examples, for any compact complex manifold that can be blown up to a Kähler manifold, [DGMS75, Theorem 5.22], e.g., for compact complex manifolds in class $\mathcal{C}$ of Fujiki, or for Moĭšezon manifolds; we refer to Sect. 2.1.3 for further results concerning the $\partial\overline{\partial}$-Lemma for compact complex manifolds.

If X is a compact Kähler manifold (or, more in general, any compact complex manifold for which the $\partial\overline{\partial}$-Lemma, equivalently the $\mathrm{d}\,\mathrm{d}^c$-Lemma, holds), then one has the following quasi-isomorphisms of dgas:

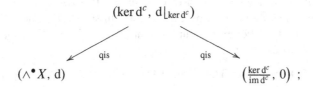

in particular, the dga $(\wedge^\bullet X, \mathrm{d})$ is equivalent to a dga with zero differential, and hence it is formal. This proves the following result by P. Deligne, Ph.A. Griffiths, J. Morgan, and D.P. Sullivan.

Theorem 1.24 ([DGMS75, Main Theorem]). *Let X be a compact complex manifold for which the $\partial\overline{\partial}$-Lemma holds (e.g., a compact Kähler manifold, or a manifold in class $\mathcal{C}$ of Fujiki). Then the differentiable manifold underlying X is formal (that is, the differential graded algebra $(\wedge^\bullet X, \mathrm{d})$ is formal).*

In particular, all Massey products (of any order) on a compact complex manifold satisfying the $\partial\overline{\partial}$-Lemma are zero, [DGMS75, Corollary 1]. This provide an obstruction to the existence of Kähler structures on compact differentiable manifolds.

1.5 Deformations of Complex Structures

A natural way to construct new complex structures on a manifold is by "deforming" a given complex structure. Natural questions arise naturally from this construction, concerning, for example, what properties (e.g., the existence of some special metric) remain still valid after such a small deformation.

We recall in this section the basic notions and the classical results concerning the K. Kodaira, D.C. Spencer, L. Nirenberg, and M. Kuranishi theory of deformations of complex manifolds, [KS58, KS60, KNS58, Kur62], referring to [Huy05], see also, e.g., [Kod05, MK06].

Let B be a complex (respectively, differentiable) manifold. A family $\{X_t\}_{t \in B}$ of compact complex manifolds is said to be a *complex-analytic* (respectively, *differentiable*) *family of compact complex manifolds* if there exist a complex (respectively, differentiable) manifold $\mathcal{X}$ and a surjective holomorphic (respectively, smooth) map $\pi: \mathcal{X} \to B$ such that *(i)* $\pi^{-1}(t) = X_t$ for any $t \in B$, and *(ii)* π is a proper holomorphic (respectively, smooth) submersion. A compact complex manifold X is said to be a *deformation* of a compact complex manifold Y if there exist a complex-analytic family $\{X_t\}_{t \in B}$ of compact complex manifolds, and $b_0, b_1 \in B$ such that $X_{b_0} = X_s$ and $X_{b_1} = X_t$.

A complex-analytic (respectively, differentiable) family $\mathcal{X} \xrightarrow{\pi} B$ of compact complex manifolds is said to be *trivial* if $\mathcal{X}$ is bi-holomorphic (respectively, diffeomorphic) to $B \times X_b \xrightarrow{\pi_B} B$ for some $b \in B$ (where $\pi_B: B \times X_b \to B$ denotes the natural projection onto B); it is said to be *locally trivial* if, for any $b \in B$, there exists an open neighbourhood U of b in B such that $\pi^{-1}(U) \xrightarrow{\pi \lfloor_{\pi^{-1}(U)}} U$ is trivial. The following theorem by Ch. Ehresmann states the local triviality of a differentiable family of compact complex manifolds (see, e.g., [Kod05, Theorems 2.3 and 2.5], [MK06, Theorem 1.4.1]).

Theorem 1.25 (Ehresmann Theorem, [Ehr47]). *Let $\{X_t\}_{t \in B}$ be a differentiable family of compact complex manifolds. For any $s, t \in B$, the manifolds X_s and X_t are diffeomorphic.*

As a consequence of Ehresmann's theorem, a complex-analytic family $\{X_t\}_{t \in B}$ of compact complex manifolds with B contractible can be viewed as a family of complex structures on a compact differentiable manifold.

We recall some other useful definitions, see, e.g., [Huy05, Sect. 6.2]. Let $\pi: \mathcal{X} \to B$ be a complex-analytic family of compact complex manifolds, deformations of $X := \pi^{-1}(0)$. We recall that, given $f: (B', 0') \to (B, 0)$ a morphism of germs with a distinguished point, the pull-back $f^*\mathcal{X} := \mathcal{X} \times_B B'$ gives a complex-analytic family of deformations of X. The complex-analytic family $\pi: \mathcal{X} \to B$ of deformations of X is called *complete* if, for any complex-analytic family $\pi': \mathcal{X}' \to B'$ of deformations of X, there exists a morphism $f: B' \to B$ of germs with a distinguished point such that $\mathcal{X}' = f^*\mathcal{X}$. The complex-analytic family $\pi: \mathcal{X} \to B$ of deformations of X is called *universal* if, for any complex-analytic family $\pi': \mathcal{X}' \to B'$ of deformations of X, there exists a unique morphism $f: B' \to B$ of germs with a distinguished point such that $\mathcal{X}' = f^*\mathcal{X}$. The complex-analytic family $\pi: \mathcal{X} \to B$ of deformations of X is called *versal* if, for any complex-analytic family $\pi': \mathcal{X}' \to B'$ of deformations of X, there exists a morphism $f: B' \to B$ of germs with a distinguished point such that $\mathcal{X}' = f^*\mathcal{X}$ and such that $\mathrm{d} f: T_{0'} B' \to T_0 S$ is uniquely determined.

The theory of complex-analytic deformations of compact complex manifolds has been introduced by K. Kodaira and D.C. Spencer, [KS58, KS60], and developed also by L. Nirenberg, [KNS58], and M. Kuranishi, [Kur62, Kur65], see also

[Kod05, MK06]. In recalling the main results of this theory, we follow the approach in [Huy05], based on the construction of a differential graded Lie algebra structure on $\mathcal{C}^\infty\left(X; T^{1,0}X \otimes \wedge^{0,\bullet}X\right)$, see also [Man04].

Let X be a compact manifold endowed with an integrable almost-complex structure J. Every section $s \in \mathcal{C}^\infty\left(X; T_J^{1,0}X \otimes \wedge_J^{0,1}X\right)$ near to the zero section determines an almost-complex structure J', defined in such a way that $\wedge_{J'}^{1,0}X$ is the graph of $-s\colon \wedge_J^{1,0}X \to \wedge_J^{0,1}X$; it turns out that J' is integrable if and only if the *Maurer and Cartan equation*

$$\bar{\partial}s + \frac{1}{2}\,[s, s] \;=\; 0 \qquad\qquad\qquad \text{(MC)}$$

holds (see, e.g., [Huy05, Lemma 6.1.2]), where

- $[\cdot,\cdot]\colon \mathcal{C}^\infty\left(X; T_J^{1,0}X \otimes \wedge_J^{0,p}X\right) \times \mathcal{C}^\infty\left(X; T_J^{1,0}X \otimes \wedge_J^{0,q}X\right) \to \mathcal{C}^\infty\left(X; T_J^{1,0}X \otimes \wedge_J^{0,p+q}X\right)$ is defined as

$$\left[X \otimes \bar{\alpha},\, Y \otimes \bar{\beta}\right] \;:=\; X \otimes \left(\bar{\beta} \wedge \mathcal{L}_Y\bar{\alpha}\right) + Y \otimes \left(\bar{\alpha} \wedge \mathcal{L}_X\bar{\beta}\right) + [X, Y] \otimes \left(\bar{\alpha} \wedge \bar{\beta}\right) ,$$

where $\mathcal{L}_W\varphi := \iota_W\,\mathrm{d}\varphi + \mathrm{d}\left(\iota_W\varphi\right)$ is the Lie derivative of φ along W; locally, in a chart with holomorphic coordinates $\left\{z^j\right\}_j$, one has

$$\left[w \otimes \mathrm{d}\bar{z}^{\ell_1} \wedge \cdots \wedge \mathrm{d}\bar{z}^{\ell_p},\, w' \otimes \mathrm{d}\bar{z}^{m_1} \wedge \cdots \wedge \mathrm{d}\bar{z}^{m_q} \wedge\right]$$
$$\overset{\text{loc}}{=} \left[w, w'\right] \otimes \mathrm{d}\bar{z}^{\ell_1} \wedge \cdots \wedge \mathrm{d}\bar{z}^{\ell_p} \wedge \mathrm{d}\bar{z}^{m_1} \wedge \cdots \wedge \mathrm{d}\bar{z}^{m_q} ;$$

- $\bar{\partial}\colon \mathcal{C}^\infty\left(X; T_J^{1,0}X \otimes \wedge_J^{0,p}X\right) \to \mathcal{C}^\infty\left(X; T_J^{1,0}X \otimes \wedge_J^{0,p+1}X\right)$ is defined as

$$\bar{\partial}\varphi\left(\bar{Z}, \bar{W}\right) \;:=\; \left[\bar{Z}, \varphi\left(\bar{W}\right)\right]^{1,0} - \left[\bar{W}, \varphi\left(\bar{Z}\right)\right]^{1,0} - \varphi\left(\left[\bar{Z}, \bar{W}\right]\right) ,$$

where $X^{1,0} := X - \mathrm{i}\,J\,X$ is the $(1,0)$-component of X; locally, in a chart with holomorphic coordinates $\left\{z^j\right\}_j$, one has

$$\bar{\partial}\left(\frac{\partial}{\partial z^\ell} \otimes \alpha\right) \overset{\text{loc}}{=} \frac{\partial}{\partial z^\ell} \otimes \bar{\partial}\alpha .$$

Hence, to study complex-analytic families of infinitesimal deformations of a compact complex manifold X, it suffices to study complex-analytic families $\{s(\mathbf{t})\}_{\mathbf{t}\in\Delta(0,\varepsilon)\subset\subset\mathbb{C}^m} \subseteq \mathcal{C}^\infty\left(X; T^{1,0}X \otimes \wedge^{0,1}X\right)$ (where $\varepsilon > 0$ is small enough) with $s(0) = 0$. Consider the power series expansion in $\mathbf{t}$ of $s(\mathbf{t})$,

$$s(\mathbf{t}) \;=:\; \sum_{k\in\mathbb{N}} s_k(\mathbf{t}) ,$$

where $s_k(\mathbf{t}) \in \mathcal{C}^{\infty}\left(X; T^{1,0}X \otimes \wedge^{0,1}X\right)$ is homogeneous of degree k in $\mathbf{t}$, and $s_0(\mathbf{t}) = 0$. Then the Maurer and Cartan equation (MC) can be rewritten, for every $\mathbf{t} \in \Delta(0, \varepsilon)$, as the system

$$\begin{cases} \bar{\partial} s_1(\mathbf{t}) = 0 \\ \bar{\partial} s_k(\mathbf{t}) = -\sum_{1 \leq j \leq k-1} \left[s_j(\mathbf{t}), s_{k-j}(\mathbf{t})\right] \text{ for } k \geq 2 \end{cases} ;$$

in particular, $s_1(\mathbf{t})$ defines a class in $H^{0,1}(X; \Theta_X)$, where Θ_X denotes the sheaf of the germs of holomorphic vector fields on X; up to the action of $\mathrm{Diff}(X)$, one has that $s_1(\mathbf{t})$ is uniquely determined by its class in $H^{0,1}(X; \Theta_X)$ (see, e.g., [Huy05, Lemma 6.14]).

Fix now a Hermitian metric g on X. Consider the decomposition

$$T^{1,0}X \otimes \wedge^{0,1}X = \left(T^{1,0}X \otimes \ker \overline{\square}\big|_{\wedge^{0,1}X}\right) \oplus \left(T^{1,0}X \otimes \bar{\partial} \wedge^{0,0}X\right)$$

$$\oplus \left(T^{1,0}X \otimes \bar{\partial}^* \wedge^{0,2}X\right),$$

and the corresponding projections

$$H_{\bar{\partial}} \colon T^{1,0}X \otimes \wedge^{0,1}X \to T^{1,0}X \otimes \ker \overline{\square}\big|_{\wedge^{0,1}X}, \quad P_{\bar{\partial}} \colon T^{1,0}X \otimes \wedge^{0,1}X \to T^{1,0}X \otimes \bar{\partial} \wedge^{0,0}X.$$

In order that $s(\mathbf{t})$ satisfies (MC), for every $\mathbf{t} \in \Delta(0, \varepsilon)$, one should have

$$\bar{\partial} s_k(\mathbf{t}) = -P_{\bar{\partial}}\left(\sum_{1 \leq j \leq k-1} \left[s_j(\mathbf{t}), s_{k-j}(\mathbf{t})\right]\right).$$

Hence, one gets

$$\bar{\partial} s(\mathbf{t}) + [s(\mathbf{t}), s(\mathbf{t})] = H_{\bar{\partial}}([s(\mathbf{t}), s(\mathbf{t})]).$$

Therefore, define the map

$$\mathrm{obs}\colon H^{0,1}(X; \Theta_X) \to H^{0,2}(X; \Theta_X)$$

as follows. Let $\left\{X_j \otimes \bar{\omega}^k\right\}_{\substack{j \in \{1,\ldots,n\} \\ k \in \{1,\ldots,m\}}}$ be a basis of $H^{0,1}(X; \Theta_X)$. Given $\mu = \sum_{\substack{j \in \{1,\ldots,n\} \\ k \in \{1,\ldots,m\}}} t_k^j X_j \otimes \bar{\omega}^k$, denote $\mathbf{t} = \left(t_k^j\right)_{\substack{j \in \{1,\ldots,n\} \\ k \in \{1,\ldots,m\}}}$, and define $s_1(\mathbf{t}) := \mu$ and $s_k(\mathbf{t})$ such that $\bar{\partial} s_k(\mathbf{t}) := -P_{\bar{\partial}}\left(\sum_{1 \leq j \leq k-1} \left[s_j(\mathbf{t}), s_{k-j}(\mathbf{t})\right]\right)$ for $k \geq 2$; hence, define the formal power series $s(\mathbf{t}) := \sum_{k \in \mathbb{N}} s_k(\mathbf{t})$. Define

$$\mathrm{obs}(\mu) := H_{\bar{\partial}}([s(\mathbf{t}), s(\mathbf{t})]).$$

Hence, one has then that $\{s(\mathbf{t})\}_{\mathbf{t} \in \Delta(0,\varepsilon) \subset \mathbb{C}^m} \subseteq \mathcal{C}^\infty \left(X ; T_J^{1,0} X \otimes \wedge_J^{0,1} X \right)$ (where $\varepsilon > 0$ is small enough) defines an infinitesimal family of compact complex manifolds if obs $(s_1 (\mathbf{t})) = 0$ for every $\mathbf{t} \in \Delta(0, \varepsilon)$ (indeed, for $\varepsilon > 0$ small enough, the formal power series converges, see, e.g., [Kod05, Sect. 5.3], [MK06, Sect. 2.3]).

One gets the following result by M. Kuranishi.

Theorem 1.26 ([Kur62, Theorem 2]). *Let X be a compact complex manifold. Then X admits a versal complex-analytic family of deformations.*

Fixed a Hermitian metric on X, such a family of deformations, which is called the *Kuranishi space* Kur(X) of X, is parametrized by

$$\mathrm{Kur}(X) = \left\{ \mu \in H^{0,1} (X; \Theta_X) : \|\mu\| \ll 1, \ \mathrm{obs}(\mu) = 0 \right\} .$$

Remark 1.11. A compact complex manifold X is called *non-obstructed* if Kur(X) is non-singular. In particular, if $H^{0,2} (X; \Theta_X) = \{0\}$, then X is non-obstructed. There are other interesting cases in which the Kuranishi space turns out to be non-singular: as announced by F.A. Bogomolov, [Bog78], and proven by G. Tian, [Tia87], and, independently, by A.N. Todorov, [Tod89, Theorem 1], this happens for *Calabi-Yau manifolds* (that is, compact complex manifolds X of complex dimension n endowed with a Kähler structure (J, ω, g) and with a nowhere vanishing $\epsilon \in \wedge^{n,0} X$ such that *(i)* $\nabla^{LC} \epsilon = 0$, where ∇^{LC} denotes the Levi Civita connection associated to g, and *(ii)* $\epsilon \wedge \bar\epsilon = (-1)^{\frac{n(n+1)}{2}} \, i^n \, \frac{\omega^n}{n!}$). In [dBT12], P. de Bartolomeis and A. Tomassini introduced the notion of *quantum inner state manifold*, [dBT12, Definition 2.2], as a possible generalization of Calabi-Yau manifolds, proving that, under a suitable hypothesis, the moduli space of quantum inner state deformations of a compact Calabi-Yau manifold is unobstructed, [dBT12, Theorem 3.6]. On the other hand, in [Rol11b], S. Rollenske studied the Kuranishi space of holomorphically parallelizable nilmanifolds, proving that it is cut out by polynomial equations of degree at most equal to the step of nilpotency of the nilmanifold, [Rol11b, Theorem 4.5], and it is smooth if and only if the associated Lie algebra is a free 2-step nilpotent Lie algebra, [Rol11b, Corollary 4.9].

Remark 1.12. We refer to [Rol11b, Rol09b, Rol09a] by S. Rollenske and to [Kas12c] by H. Kasuya for results concerning deformations of nilmanifolds and, respectively, solvmanifolds.

It could be interesting to study what properties of a complex manifold are, in a sense, compatible with the construction of small deformations of the complex structure. In such a context, a property $\mathcal{P}$ concerning compact complex manifolds is called *open under (holomorphic) deformations of the complex structure* (or *stable under small deformations of the complex structure*) if, for every complex-analytic family $\{X_t\}_{t \in B}$ of compact complex manifolds, and for every $b_0 \in B$, if X_{b_0} has the property $\mathcal{P}$, then X_b has the property $\mathcal{P}$ for every b in an open neighbourhood of b_0. A property $\mathcal{P}$ is called is called *closed under (holomorphic) deformations of the complex structure* if, for every complex-analytic family $\{X_t\}_{t \in \Delta_\varepsilon(0)}$ of compact complex manifolds, where $\Delta_\varepsilon(0) := \{z \in \mathbb{C}^m : |z| < \varepsilon\}$ for some $m \in \mathbb{N} \setminus \{0\}$ and for some $\varepsilon > 0$, if X_t has the property $\mathcal{P}$ for every $t \in \Delta_\varepsilon(0) \setminus \{0\}$, then also X_0 has the property $\mathcal{P}$.

Remark 1.13 (Gunnar Þór Magnússon[8]). In a sense, the notion of closedness is considered in the Zariski sense. In fact, as suggested to us by Gunnar Þór Magnússon, by considering, for example, moduli space of complex tori of complex dimension 2, and by using [Kod63, Theorem 16.1] (compare also [Cao12, Main Theorem]), one can construct examples of complex-analytic families $\{X_t\}_{t \in B}$ of compact complex manifolds such that there exists a converging sequence $\{b_k\}_{k \in \mathbb{N}} \subset B$ with $b_\infty := \lim_{k \to +\infty} b_k \in B$ for which X_{b_k} is Moĭšezon for every $k \in \mathbb{N}$, while X_{b_∞} is not.

We recall here the following classical result by K. Kodaira and D.C. Spencer, stating that admitting a Kähler metric is a stable property under deformations of the complex structure.

Theorem 1.27 ([KS60, Theorem 15]). *Let $\{X_t\}_{t \in B}$ be a differentiable family of compact complex manifolds. If X_{t_0} admits a Kähler metric for some $t_0 \in B$, then X_s admits a Kähler metric for every s in an open neighbourhood of t_0 in B. Moreover, given any Kähler metric ω on X_{t_0}, one can choose an open neighbourhood U of t_0 in B and a Kähler metric ω_s on X_s for any $s \in U$ such that ω_s depends differentiably in s and $\omega_{t_0} = \omega$.*

Proof (sketch). We just sketch the idea of the proof, as one can found, e.g., in [Sch07, Lemme 3.3], see also [Voi02, Théorème 9.23] and [KS60, Sect. 6].

Consider a family of compact complex manifolds $\{X_t\}_t$. By the Ehresmann theorem, [Ehr47], for every $k \in \mathbb{Z}$, the function $t \mapsto \dim_\mathbb{C} H_{dR}^k(X_t; \mathbb{C})$ is locally constant. By the theory of families of elliptic differential operators (see, e.g., [Kod05, Sect. 7.1]), for every $(p, q) \in \mathbb{Z}^2$, the functions $t \mapsto \dim_\mathbb{C} H_{BC}^{p,q}(X_t)$ and $t \mapsto \dim_\mathbb{C} H_A^{p,q}(X_t)$ are upper-semi-continuous, see, e.g., [Sch07, Lemme 3.2]. Recall also that, by the inequality *à la* Frölicher for the Bott-Chern cohomology, [AT13b, Theorem A], it holds that $\sum_{p+q=k} \left(\dim_\mathbb{C} H_{BC}^{p,q}(X_t) + \dim_\mathbb{C} H_A^{p,q}(X_t) \right) \geq 2 \dim_\mathbb{C} H_{dR}^k(X_t; \mathbb{C})$ for any $k \in \mathbb{Z}$ and for any t.

If we assume that X_{t_0} is Kähler for some t_0, then in particular X_{t_0} satisfies the $\partial\bar{\partial}$-Lemma. Therefore, for any $k \in \mathbb{Z}$, it holds $\sum_{p+q=k} \left(\dim_\mathbb{C} H_{BC}^{p,q}(X_{t_0}) + \dim_\mathbb{C} H_A^{p,q}(X_{t_0}) \right) = 2 \dim_\mathbb{C} H_{dR}^k(X_{t_0}; \mathbb{C})$. In particular, from the facts remarked above, it follows that, for any $(p, q) \in \mathbb{Z}^2$, the functions $t \mapsto \dim_\mathbb{C} H_{BC}^{p,q}(X_t)$ and $t \mapsto \dim_\mathbb{C} H_A^{p,q}(X_t)$ are locally constant at t_0.

In particular, we will use that the function $t \mapsto \dim_\mathbb{C} H_{BC}^{1,1}(X_t)$ is locally constant at t_0. In fact, by the theory of families of elliptic differential operators (see, e.g., [Kod05, Sect. 7.1]), this implies that the map $h_s \colon \wedge^2 X \otimes \mathbb{C} \to \ker \tilde{\Delta}_{BC_s}\big|_{\wedge^{1,1} X_s}$

---

[8]The author would like to thank Luis Ugarte for having pointed out the subject, and Junyan Cao and Gunnar Þór Magnússon for useful discussions on the matter.

depends smoothly in s in a neighbourhood of t_0, [KS60, Theorem 7], see also [Kod05, Theorem 7.4]. (For simplicity, we identify the smooth structures on X_{t_0} and X_s, for s in a neighbourhood of t_0, by means of the diffeomorphism given by the Ehresmann theorem, [Ehr47].) Hence define, for s in a neighbourhood of t_0,

$$\omega_s := \frac{1}{2} \left(h_s \omega + \overline{h_s \omega} \right).$$

For any s, the form ω_s is a real d-closed $(1,1)$-form on X_s. Furthermore, the family $\{\omega_s\}_s$ depends smoothly in s. Since $\omega_{t_0} = \omega$, it follows that the form ω_s is positive for s in a neighbourhood of X_{t_0}. In particular, ω_s is a Kähler form on X_s depending smoothly in s in a neighbourhood of t_0. □

Remark 1.14. In [Hir62], it is proven that admitting a Kähler structure is not a closed property under deformations of the complex structure: indeed, H. Hironaka provided an explicit example of a complex-analytic family of compact complex manifolds of complex dimension 3 such that *(i)* one of the complex manifold is non-Kähler (indeed, it carries a positive 1-cycle algebraically equivalent to zero), and *(ii)* the others are Kähler and, in fact, bi-regularly embedded in a projective space (and hence projective, [Moi66, Theorem 11]), [Hir62, Theorem]. (Note that, in complex dimension 2, the Kähler property is closed under deformations of the complex structure, since a compact complex surface is Kähler if and only if its 1st Betti number is even, by [Kod64, Miy74, Siu83], or [Lam99, Corollaire 5.7], or [Buc99, Theorem 11].) It is not known whether the limit of compact Kähler manifolds admits some special structure; J.-P. Demailly and M. Păun conjectured that, given a complex-analytic family $\{X_t\}_{t \in S}$ of compact complex manifolds such that one of the fibers, X_{t_0}, is endowed with a Kähler structure, then there exists a countable union $S' \subsetneq S$ of analytic subsets in the base such that X_t admits a Kähler structure for $t \in S \setminus S'$, [DP04, Conjecture 5.1]; they also guessed that a "natural expectation" is that the remaining fibres, X_t for $t \in S'$, are in class $\mathcal{C}$ of Fujiki, [DP04, p. 1272]. In [Pop13, Pop09, Pop10], D. Popovici studied limits of projective and Moïšezon manifolds under holomorphic deformations of complex structures, proving, in particular, that the limit of projective manifolds is Moïšezon under some additional conditions either on Hodge numbers or on some special Hermitian metrics called strongly-Gauduchon metrics, [Pop13]. C. LeBrun and Y.S. Poon [LP92], and F. Campana [Cam91] showed that being in class $\mathcal{C}$ of Fujiki is not a stable property under small deformations of the complex structures, [LP92, Theorem 1], [Cam91, Corollary 3.13], studying twistor spaces. It is conjectured that being in class $\mathcal{C}$ of Fujiki is a closed property under deformations of the complex structure, see, e.g., [Pop11, Standard Conjecture 1.17].

We refer to [Pop11] for a review on the behaviour under holomorphic deformations of properties concerning, e.g., the existence of various types of Hermitian metrics on compact complex manifolds. See also Corollary 2.2, Theorem 4.7 for some results concerning stability or instability of special properties of complex manifolds.

1.6   Currents and de Rham Homology

In this section, we recall the basic notions and results concerning currents on (differentiable) manifolds and de Rham homology: they turn out to be a useful tool to study the geometry of complex manifolds (as an example, we recall F.R. Harvey and H.B. Lawson's intrinsic characterization of Kähler manifolds by means of currents, [HL83, Proposition 12, Theorem 14], or M.L. Michelsohn's intrinsic characterization of balanced manifolds by means of currents [Mic82, Theorem 4.7], see also Theorem 4.18, or J.P. Demailly and M. Păun's characterization of compact complex manifolds in class $\mathcal{C}$ of Fujiki by means of Kähler currents[DP04, Theorem 3.4]). We refer, e.g., to [dR84, Chap. 3], [Dem12, Sect. I.2], and [Fed69] (see also [Ale98, Ale10]) for further details.

Let X be a m-dimensional oriented differentiable manifold.

For every compact set $L \subseteq X$ and for every $s \in \mathbb{N}$, define the semi-norm ρ_L^s on $\wedge^\bullet X$ as follows: chosen $\left(U, \{x^j\}_{j \in \{1,...,m\}} \right)$ a coordinate chart with $U \supset L$, and given

$$\varphi \overset{\text{loc}}{=} \sum_{\substack{\{i_1,...,i_k\} \subseteq \{1,...,m\} \\ i_1 < \cdots < i_k}} \varphi_I \, d x^{i_1} \wedge \cdots \wedge d x^{i_k} \in \wedge^\bullet X \, ,$$

set

$$\rho_L^s(\varphi) := \sup_L \sup_{\substack{\{i_1,...,i_k\} \subseteq \{1,...,m\} \\ i_1 < \cdots < i_k}} \sup_{\substack{(\alpha_1,...,\alpha_m) \in \mathbb{N}^m \\ \alpha_1 + \cdots + \alpha_m \leq s}} \left| \frac{\partial^{\alpha_1 + \cdots + \alpha_m} \varphi_I}{(\partial x^1)^{\alpha_1} \cdots (\partial x^m)^{\alpha_m}} \right| \in \mathbb{R} \, .$$

Consider $\wedge^\bullet X$ endowed with the topology induced by the family of semi-norms ρ_L^s, varying L among the compact sets in X, and $s \in \mathbb{N}$: the manifold X being second-countable, $\wedge^\bullet X$ has a structure of a Fréchet space. Let $\wedge_c^\bullet X$ be the topological subspace of $\wedge^\bullet X$ consisting of differential forms with compact support in X.

For any $k \in \mathbb{N}$, the space of *currents* of dimension k (or degree $m-k$), denoted by

$$\mathcal{D}_k X := \mathcal{D}^{m-k} X \, ,$$

is defined as the topological dual space of $\wedge_c^k X$; the space $\mathcal{D}_\bullet X$ is endowed with the weak-$*$ topology.

Two basic examples of currents are the following.

- If Z is a (possibly non-closed) k-dimensional oriented compact submanifold of X, then

$$[Z] := \int_Z \cdot \in \mathcal{D}_k X$$

is a current of dimension k.

- If $\varphi \in \wedge^k X$, then

$$T_\varphi := \int_X \varphi \wedge \cdot \ \in \mathcal{D}^k X$$

is a current of degree k.

The exterior differential $d: \wedge^\bullet X \to \wedge^{\bullet+1} X$ induces a differential on $\mathcal{D}_\bullet X$ by duality:

$$d: \mathcal{D}_\bullet X \to \mathcal{D}_{\bullet-1} X$$

is defined, for every $T \in \mathcal{D}^k X$, as

$$dT := (-1)^{k+1} T(d\cdot) \ .$$

In particular, if Z is a k-dimensional oriented closed submanifold of X, then $d[Z] = (-1)^{m-k+1}[bZ]$, where b is the boundary operator; if $\varphi \in \wedge^k X$, then $dT_\varphi = T_{d\varphi}$.

By definition, the *de Rham homology* $H_\bullet^{dR}(X;\mathbb{R})$ of X is the homology of the differential complex $(\mathcal{D}_\bullet X, d)$. By means of a regularization process, [dR84, Theorem 12], (see also [Dem12, Sects. 2.D.3 and 2.D.4],) one can prove, [dR84, Theorem 14], that

$$H_{dR}^\bullet(X;\mathbb{R}) \simeq H_{2n-\bullet}^{dR}(X;\mathbb{R}) \ .$$

Since, for every $k \in \mathbb{N}$, the sheaf $\mathcal{A}_X^k$ of germs of k-forms is a sheaf of $\mathcal{C}_X^\infty$-module over a paracompact space (where $\mathcal{C}_X^\infty$ denotes the sheaf of germs of smooth functions over X), and by the Poincaré lemma for forms, see, e.g., [Dem12, I.1.22], one has that

$$0 \to \underline{\mathbb{R}}_X \to \left(\mathcal{A}_X^\bullet, d\right)$$

is a fine (and hence acyclic, see, e.g., [Dem12, Corollary IV.4.19]) resolution of the constant sheaf $\underline{\mathbb{R}}_X$, and hence

$$\check{H}^\bullet(X;\underline{\mathbb{R}}_X) \simeq \frac{\ker\left(d: \wedge^\bullet X \to \wedge^{\bullet+1} X\right)}{\mathrm{im}\left(d: \wedge^{\bullet-1} X \to \wedge^\bullet X\right)} =: H_{dR}^\bullet(X;\mathbb{R}) \ ,$$

see, e.g., [Dem12, IV.6.4].

Analogously, the regularization process [dR84, Theorem 12] allows to prove the analogue of the Poincaré lemma for currents, see, e.g., [Dem12, Theorem I.2.24], and hence, the sheaf $\mathcal{D}_X^k$ of germs of currents of degree k being fine for every $k \in \mathbb{N}$ since it is a sheaf of $\mathcal{C}_X^\infty$-module over a paracompact space, one has that

$$0 \to \underline{\mathbb{R}}_X \to \left(\mathcal{D}_X^\bullet, d\right)$$

is a fine (and hence acyclic, see, e.g., [Dem12, Corollary IV.4.19]) resolution of the constant sheaf $\mathbb{R}_X$ over X, and hence

$$\check{H}^\bullet(X; \mathbb{R}_X) \simeq \frac{\ker\left(\mathrm{d}: \mathcal{D}^\bullet X \to \mathcal{D}^{\bullet+1} X\right)}{\mathrm{im}\left(\mathrm{d}: \mathcal{D}^{\bullet-1} X \to \mathcal{D}^\bullet X\right)} =: H^{dR}_{2n-\bullet}(X; \mathbb{R}) \,,$$

see, e.g., [Dem12, IV.6.4].

If X is compact, then it follows that the map $T_\cdot: \wedge^\bullet X \to \mathcal{D}^\bullet X$ is injective and a quasi-isomorphism of differential complexes: indeed, fixed a Riemannian metric g on X, if α is a Δ-harmonic form (i.e., a d-closed d*-closed form), then $T_\alpha(*\alpha) = \|\alpha\|^2$.

Suppose now that X is a $2n$-dimensional manifold endowed with an almost-complex structure $J \in \mathrm{End}(TX)$. Considering the induced endomorphisms $J \in \mathrm{End}(\wedge^\bullet X)$ and $J \in \mathrm{End}(\wedge^\bullet_c X)$, one can define $J \in \mathrm{End}(\mathcal{D}^\bullet X)$ by duality. In the same way as $J \in \mathrm{End}(\wedge^\bullet X)$ defines a bi-graduation on $\wedge^\bullet X \otimes \mathbb{C}$, one has that $J \in \mathrm{End}(\mathcal{D}^\bullet X)$ defines the splitting

$$\mathcal{D}_\bullet X \otimes \mathbb{C} = \bigoplus_{(p,q)\in\mathbb{N}^2} \mathcal{D}_{p,q} X \,;$$

note that $\mathcal{D}_{p,q} X := \mathcal{D}^{n-p,n-q} X$ is the topological dual of $\wedge^{p,q} X \cap \left(\wedge^\bullet_c X \otimes_\mathbb{R} \mathbb{C}\right)$, for every $(p,q) \in \mathbb{N}^2$.

1.7 Solvmanifolds

Nilmanifolds and solvmanifolds provide an important class of examples in non-Kähler geometry. Indeed, on the one hand, in studying their properties, one often can reduce to study left-invariant objects on them, which is the same to study linear objects on the corresponding Lie algebra (this allows, for example, to reduce the study of the de Rham cohomology of a nilmanifold to the study of the cohomology of a complex of finite-dimensional vector spaces, [Nom54, Theorem 1]); on the other hand, they do not admit too strong structures, e.g., they do not admit any Kähler structure.

In this section, we recall the main definitions and results concerning the theory of nilmanifolds and solvmanifolds, setting also the notation for the following chapters.

1.7.1 Lie Groups and Lie Algebras

We briefly recall the notions of Lie groups and Lie algebras and the relations between them; see, e.g., [War83, Kir08, TO97, FOT08], for further results on the subject.

A *Lie group* (respectively a *Lie group of class* C^k, for $k \in \mathbb{N} \cup \{\infty, \omega\}$, respectively a *complex Lie group*) is[9] a smooth (respectively C^k, respectively complex) manifold endowed with a structure of group given by differentiable (respectively C^k, respectively holomorphic) maps $\cdot : G \times G \to G$ and $(\cdot)^{-1} : G \to G$. A *homomorphism of Lie groups* (respectively, of Lie groups of class C^k, respectively of complex Lie groups) is a smooth (respectively C^k, respectively holomorphic) map being also a homomorphism of groups.

Remark 1.15. According to A.M. Gleason's, and D. Montgomery and L. Zippin's theorems, [Gle52, MZ52], answering to Hilbert's fifth problem, every topological Lie group (that is, a Lie group of class C^0) admits a unique real-analytic structure with respect to which it is a Lie group of class C^ω.

A *Lie algebra* over a field $\mathbb{K}$ (with characteristic char $\mathbb{K} \neq 2$) is a $\mathbb{K}$-vector space $\mathfrak{g}$ endowed with an operation denoted by the *bracket* $[\cdot, \cdot] : \mathfrak{g} \times \mathfrak{g} \to \mathfrak{g}$ being bi-$\mathbb{K}$-linear, skew-symmetric (i.e., for every $x, y \in \mathfrak{g}$, it holds $[x, y] + [y, x] = 0$), and satisfying the *Jacobi identity* (i.e., for every $x, y, z \in \mathfrak{g}$, it holds $[[x, y], z] + [[y, z], x] + [[z, x], y] = 0$). An *homomorphism of Lie algebras* between the Lie algebra $\mathfrak{g}$ endowed with the bracket $[\cdot, \cdot]_\mathfrak{g}$ and the Lie algebra $\mathfrak{h}$ endowed with the bracket $[\cdot, \cdot]_\mathfrak{h}$ is a $\mathbb{K}$-linear map $\varphi : \mathfrak{g} \to \mathfrak{h}$ preserving the brackets (namely, such that $\varphi [\cdot, \cdot]_\mathfrak{g} = [\varphi(\cdot), \varphi(\cdot)]_\mathfrak{h}$).

Given a $\mathbb{K}$-vector space V, the $\mathbb{K}$-vector space $\mathrm{End}(V)$ endowed with the bracket $[x, y] := x \circ y - y \circ x$ has a structure of Lie algebra over $\mathbb{K}$. Given a Lie algebra $\mathfrak{g}$ endowed with the bracket $[\cdot, \cdot]$, define the *adjoint representation*

$$\mathrm{ad} : \mathfrak{g} \to \mathrm{End}(\mathfrak{g}) , \qquad \mathfrak{g} \ni x \overset{\mathrm{ad}}{\to} \mathrm{ad}_x := [x, \cdot] \in \mathrm{End}(\mathfrak{g}) ;$$

in other words, the Jacobi identity states that $\mathrm{ad} : \mathfrak{g} \to \mathrm{End}(\mathfrak{g})$ is a representation of the Lie algebra $\mathfrak{g}$, that is, a homomorphism of Lie algebras.

Let G be a Lie group. For any $g \in G$, consider the *left-translation* homomorphism of Lie groups $L_g : G \to G$ defined by $L_g(h) := g \cdot h$, where $\cdot$ is the group operation on G. An object on G is called G-*left-invariant* if it is invariant under the actions of left-translations L_g varying $g \in G$. The Lie bracket of vector fields yields a structure of Lie algebra on the space of G-left-invariant vector fields on G, see, e.g., [War83, Proposition 3.7]: such a Lie algebra is the *Lie algebra naturally associated to the Lie group G*. By S. Lie's and E. Cartan's theorem, [Lie80, Car30], the functor that associates to every Lie group its naturally associated Lie algebra

---

[9]Let $\mathbf{C}$ be a category with finite products, and consider $\mathbf{Grp}$ the category of groups; a *group-object* G in $\mathbf{C}$ is an object of $\mathbf{C}$ such that $\mathrm{Hom}(\cdot, G) : \mathbf{C} \to \mathbf{Grp}$ is a contravariant functor.

In such a notation, a Lie group (respectively a Lie group of class C^k, for $k \in \mathbb{N} \cup \{\infty, \omega\}$, respectively a complex Lie group) is a group-object in the category of differentiable manifolds with differentiable maps (respectively in the category of manifolds of class C^k with maps of class C^k, respectively in the category of complex manifolds with holomorphic maps). A homomorphism of Lie groups between G and H is a morphism $f : G \to H$ of manifolds inducing, for every manifold X, a homomorphism of groups between $\mathrm{Hom}(X, G) \to \mathrm{Hom}(X, H)$.

is an equivalence between the full sub-category of connected simply-connected Lie groups and the full sub-category of finite-dimensional Lie algebras, see, e.g., [Kir08, Corollary 3.44].

Let $\mathfrak{g}$ be a Lie algebra. Define the *descending central series* $\{\mathfrak{g}^{[n]}\}_{n \in \mathbb{N}}$ of $\mathfrak{g}$ as

$$\begin{cases} \mathfrak{g}^{[0]} := \mathfrak{g} \\ \mathfrak{g}^{[n]} := [\mathfrak{g}, \mathfrak{g}^{[n-1]}] \text{ for } n \in \mathbb{N} \setminus \{0\} \end{cases} ;$$

a Lie algebra is called *nilpotent* if there exists $n \in \mathbb{N}$ such that $\mathfrak{g}^{[n]} = \{0\}$, and the *step of nilpotency* of $\mathfrak{g}$ is defined as $\mathrm{nilstep}(\mathfrak{g}) := \inf\{n \in \mathbb{N} : \mathfrak{g}^{[n]} = \{0\}\}$; a Lie group is called *nilpotent* if its associated Lie algebra is nilpotent.

Let $\mathfrak{g}$ be a Lie algebra. Define the *descending derived series* $\{\mathfrak{g}^{\{n\}}\}_{n \in \mathbb{N}}$ of $\mathfrak{g}$ as

$$\begin{cases} \mathfrak{g}^{\{0\}} := \mathfrak{g} \\ \mathfrak{g}^{\{n\}} := [\mathfrak{g}^{\{n-1\}}, \mathfrak{g}^{\{n-1\}}] \text{ for } n \in \mathbb{N} \setminus \{0\} \end{cases} ;$$

a Lie algebra is called *solvable* if there exists $n \in \mathbb{N}$ such that $\mathfrak{g}^{\{n\}} = \{0\}$; a Lie group is called *solvable* if its associated Lie algebra is solvable. One has that $\mathfrak{g}$ is solvable if and only if $[\mathfrak{g}, \mathfrak{g}]$ is nilpotent, see, e.g., [Kir08, Corollary 5.32]; in particular, every nilpotent Lie algebra is also solvable.

Let $\mathfrak{g}$ be a solvable Lie algebra, and consider the adjoint representation $\mathrm{ad} \colon \mathfrak{g} \ni x \mapsto [x, \cdot] \in \mathrm{End}(\mathfrak{g})$. One says that $\mathfrak{g}$ is *completely-solvable* (or *of type (R)*) if, for any $x \in \mathfrak{g}$, all the eigen-values of the endomorphism $\mathrm{ad}_x \in \mathrm{End}(\mathfrak{g})$ are in $\mathbb{R}$. One says that $\mathfrak{g}$ is *of rigid type* (or *of type (I)*) if, for any $x \in \mathfrak{g}$, all the eigen-values of the endomorphism $\mathrm{ad}_x \in \mathrm{End}(\mathfrak{g})$ are in $i\mathbb{R}$, [Vin94, p. 65].

1.7.2 Nilmanifolds and Solvmanifolds

A *nilmanifold*, [Mal49, p. 278], respectively *solvmanifold*, $X = \Gamma \backslash G$ is a compact quotient of a connected simply-connected nilpotent, respectively solvable, Lie group G by a co-compact discrete subgroup Γ (see, e.g., [TO97, Definitions 2.1.1 and 3.1.1]). A solvmanifold $X = \Gamma \backslash G$ is called *completely-solvable* if its naturally associated Lie algebra is completely-solvable, that is, if, for any $g \in G$, all the eigenvalues of $\mathrm{Ad}\,g \in \mathrm{End}(\mathfrak{g})$ are real, equivalently, if, for any $X \in \mathfrak{g}$, all the eigenvalues of $\mathrm{ad}\,X \in \mathrm{End}(\mathfrak{g})$ are real.

In particular, nilmanifolds and solvmanifolds are *homogeneous spaces*, that is, differentiable manifolds endowed with a transitive action of a Lie group.

Recall that, given an homogeneous manifold X together with a transitive action $\psi \colon G \to \mathrm{Hom}_{\mathrm{Diff}}(X; X)$ of a Lie group G, then X is diffeomorphic to the quotient $H_x \backslash G$ where $x \in X$ and $H_x := \{g \in G : \psi(g)(x) = x\}$ is the *isotropy group* of G at x, see, e.g., [War83, Theorem 3.62].

Remark 1.16 ([Aus61,Has06]). We notice that in the literature, different definitions of solvmanifolds are also considered: for example, L. Auslander defines a solvmanifold as a "homogeneous space of a connected solvable Lie group", [Aus61, p. 398], see also [Aus73a, Aus73b]; K. Hasegawa defines a solvmanifold X as "a compact homogeneous space of solvable Lie group, that is, a compact differentiable manifold on which a connected solvable (nilpotent) Lie group G acts transitively", [Has06, p. 132]; as noticed in [Has06, p. 132], one can assume that $X = D \backslash \tilde{G}$, where $\tilde{G}$ is a connected simply-connected Lie group (namely, the universal covering group of G) and D is a closed subgroup of $\tilde{G}$. Note that a closed subgroup of a solvable non-nilpotent group is not necessarily discrete; however, every compact solvmanifold in the Auslander sense has a solvmanifold $\Gamma \backslash \tilde{G}$ with discrete isotropy subgroup Γ as a finite covering, [Aus61].

Remark 1.17. Nilmanifolds and solvmanifolds $\Gamma \backslash G$ are *aspherical*, that is, $\pi_j (\Gamma \backslash G) = \{0\}$ for $j \geq 2$; (in fact, they are *Eilenberg and MacLane spaces of type $K(\pi; 1)$* with $\pi = \Gamma$). Furthermore, they are uniquely determined, up to diffeomorphism, by their fundamental group, as proven by A.I. Mal'tsev for nilmanifolds, [Mal49, Corollary, p. 293], and by G.D. Mostow for solvmanifolds, [Mos54, Mos57, Theorem A], see also, e.g., [Rag72, Corollary 2, p. 34, Theorem 3.6], [Aus73a, pp. 235, 244], [GOV97, Theorems II.1.3(i) and II.2.6].

Obviously, every nilmanifold is in particular a solvmanifold; on the other hand, there exist solvmanifolds which are not diffeomorphic to any nilmanifold, since their fundamental group is not nilpotent.[10] In fact, given a solvmanifold $\Gamma \backslash G$, if $\pi_1 (\Gamma \backslash G)$ is nilpotent, then $\Gamma \backslash G$ is diffeomorphic to some nilmanifold, see, e.g., [GOV97, Corollary II.2]; if $\pi_1 (\Gamma \backslash G)$ is Abelian, then $\Gamma \backslash G$ is diffeomorphic to a torus, [Mos54, Mos57, Theorem 1], see, e.g., [GOV97, Corollary II.2]. More precisely, one can prove that a compact aspherical homogeneous manifold having nilpotent fundamental group is diffeomorphic to some nilmanifold, which is unique up to diffeomorphism, see, e.g., [GOV97, Theorem II.2.7].

Remark 1.18. Note that nilmanifolds and solvmanifolds are parallelizable,[11] [TO97, Theorem 2.3.11].

Given a $2n$-dimensional solvmanifold $X = \Gamma \backslash G$, consider $(\mathfrak{g}, [\cdot, \cdot])$ the Lie algebra naturally associated to the Lie group G; given a basis $\{e_1, \ldots, e_{2n}\}$ of $\mathfrak{g}$, the Lie algebra structure of $\mathfrak{g}$ is characterized by the *structure constants* $\{c_{\ell m}^k\}_{\ell, m, k \in \{1, \ldots, 2n\}} \subset \mathbb{R}$: namely, for any $k \in \{1, \ldots, 2n\}$,

$$d_{\mathfrak{g}} e^k =: \sum_{\ell, m} c_{\ell m}^k \, e^\ell \wedge e^m \, ,$$

---

[10] We recall that the following theorem by A.I. Mal'tsev: a group Γ is isomorphic to a discrete co-compact subgroup in a simply-connected connected nilpotent Lie group if and only if it is finitely-generated, nilpotent, and torsion-free, [Mal49, Corollary 1], see, e.g., [Rag72, Theorem 2.18].

[11] Recall that a manifold is called *parallelizable* if its tangent bundle is trivial.

where $\{e^1, \ldots, e^{2n}\}$ is the dual basis of $\mathfrak{g}^*$ of $\{e_1, \ldots, e_{2n}\}$ and $d_{\mathfrak{g}} \colon \mathfrak{g}^* \to \wedge^2 \mathfrak{g}^*$ is defined by

$$\mathfrak{g}^* \ni \alpha \mapsto d_{\mathfrak{g}} \alpha(\cdot, \cdot\cdot) := -\alpha\left(\left[\cdot, \cdot\cdot\right]\right) \in \wedge^2 \mathfrak{g}^* .$$

To shorten the notation, as in [Sal01], we will refer to a given solvmanifold $X = \Gamma \backslash G$ writing the structure equations of its Lie algebra: for example, writing

$$X := \left(0^4, \ 12, \ 13\right), \qquad \left(\text{or } \mathfrak{g} := \left(0^4, \ 12, \ 13\right),\right)$$

we mean that $X = \Gamma \backslash G$ and there exists a basis of the Lie algebra $\mathfrak{g}$ naturally associated to G, let us say $\{e_1, \ldots, e_6\}$, whose dual will be denoted by $\{e^1, \ldots, e^6\}$, such that the structure equations with respect to such basis are

$$\begin{cases} d e^1 = d e^2 = d e^3 = d e^4 = 0 \\ d e^5 = e^1 \wedge e^2 =: e^{12} \\ d e^6 = e^1 \wedge e^3 =: e^{13} \end{cases} ,$$

where we also shorten $e^{AB} := e^A \wedge e^B$.

The following theorem by A.I. Mal'tsev characterizes the nilpotent Lie algebras $\mathfrak{g}$ for which the naturally associated connected simply-connected Lie group admits a co-compact discrete subgroup, and hence such that there exists a nilmanifold with $\mathfrak{g}$ as Lie algebra; (see also, e.g., [Rag72, Theorem 2.12]).

Theorem 1.28 ([Mal49, Theorem 7]). *In order that a simply-connected connected nilpotent Lie group contain a discrete co-compact Lie group it is necessary and sufficient that the Lie algebra of this group have rational constant structures with respect to an appropriate basis.*

As regards the existence of discrete co-compact subgroups in solvable Lie groups, one has the following obstruction provided by J. Milnor. (Recall that a Lie group G, with associated Lie algebra $\mathfrak{g}$, is called *unimodular* if, for all $X \in \mathfrak{g}$, it holds $\operatorname{tr} \operatorname{ad} X = 0$; see also [Mil76, Lemmas 6.1, 7.1, and 6.3] for other equivalent formulations.)

Lemma 1.1 ([Mil76, Lemma 6.2]). *Any connected Lie group that admits a discrete subgroup with compact quotient is unimodular and in particular admits a bi-invariant volume form η.*

Dealing with G-*left-invariant* objects on X, we mean objects induced by objects on G being invariant under the left-action of G on itself given by left-translations. By means of left-translations, G-left-invariant objects will be identified with objects on the Lie algebra $\mathfrak{g}$.

For example, a G-left-invariant complex structure $J \in \operatorname{End}(TX)$ on X is uniquely determined by a linear complex structure $J \in \operatorname{End}(\mathfrak{g})$ on $\mathfrak{g}$ satisfying the integrability condition $\operatorname{Nij}_J = 0$, [NN57, Theorem 1.1], where

$$\mathrm{Nij}_J(\cdot,\cdot\cdot) := [\cdot,\cdot\cdot] + J\,[J\cdot,\cdot\cdot] + J\,[\cdot,J\cdot\cdot] - [J\cdot,J\cdot\cdot] \in \wedge^2\mathfrak{g}^* \otimes_\mathbb{R} \mathfrak{g}\,;$$

we will denote the set of G-left-invariant complex structures on X by

$$\mathcal{C}(\mathfrak{g}) := \left\{ J \in \mathrm{End}(\mathfrak{g}) \,:\, J^2 = -\mathrm{id}_\mathfrak{g} \text{ and } \mathrm{Nij}_J = 0 \right\}\,.$$

By the Leibniz rule, the map $d_\mathfrak{g}: \wedge^1\mathfrak{g}^* \to \wedge^2\mathfrak{g}^*$ induces a differential operator $d: \wedge^\bullet\mathfrak{g}^* \to \wedge^{\bullet+1}\mathfrak{g}^*$ giving a graded differential algebra $(\wedge^\bullet\mathfrak{g}^*, d)$, and hence a differential complex $(\wedge^\bullet\mathfrak{g}^*, d)$, which is usually called the *Chevalley and Eilenberg complex* of $\mathfrak{g}$, [CE48]; we will denote by $H_{dR}^\bullet(\mathfrak{g};\mathbb{R}) := H^\bullet(\wedge^\bullet\mathfrak{g}^*, d)$ the cohomology of such a differential complex.

In general, on a solvmanifold, the inclusion $(\wedge^\bullet\mathfrak{g}^*, d) \hookrightarrow (\wedge^\bullet X, d)$ induces an injective map in cohomology, $\iota: H_{dR}^\bullet(\mathfrak{g};\mathbb{R}) \hookrightarrow H_{dR}^\bullet(X;\mathbb{R})$ (see [Rag72, Remark 7.30], [TO97, Theorem 3.2.10], and compare [CF01, Lemma 9] and Lemma 3.2 for the Dolbeault and, respectively, Bott-Chern cohomologies), which is not always an isomorphism, as the example in [dBT06, Corollary 4.2, Remark 4.3] shows. On the other hand, the following theorem by K. Nomizu says that the de Rham cohomology of a nilmanifold can be computed as the cohomology of the subcomplex of left-invariant forms (some results in this direction have been provided also by Y. Matsushima in [Mat51, Theorems 5 and 6]).

Theorem 1.29 ([Nom54, Theorem 1]). *Let $X = \Gamma\backslash G$ be a nilmanifold and denote the Lie algebra naturally associated to G by $\mathfrak{g}$. The complex $(\wedge^\bullet\mathfrak{g}^*, d)$ is a minimal model for $(\wedge^\bullet X, d)$. In particular, the map $(\wedge^\bullet\mathfrak{g}^*, d) \to (\wedge^\bullet X, d)$ of differential complexes is a quasi-isomorphism:*

$$\iota: H_{dR}^\bullet(\mathfrak{g};\mathbb{R}) \overset{\simeq}{\to} H_{dR}^\bullet(X;\mathbb{R})\,.$$

The proof rests on an inductive argument, which can be performed since every nilmanifold can be seen as a principal torus-bundle over a lower dimensional nilmanifold, see [Mal49, Lemma 4], [Mat51, Theorem 3]; see also [Rag72, Corollary 7.28].

A similar result holds also in the case of completely-solvable solvmanifolds, as proven by A. Hattori, as a consequence of the Mostow structure theorem,[12] [Mos54, Mos57].

---

[12]We recall the following, referring, e.g., to [TO97, Theorem 1.2]; see also [Rag72, Theorem 3.3, Corollary 3.5] for a different proof.

Theorem 1.30 (Mostow Structure Theorem, [Mos54, Mos57, Theorem 2]). *Any solvmanifold can be naturally fibred over a torus with a nilmanifold as a fibre. More precisely, let $\Gamma\backslash G$ be a solvmanifold, and consider the maximal connected nilpotent subgroup N of G. Then (i) $N\Gamma$ is a closed subgroup in G, (i) $N \cap \Gamma$ is a lattice in N, and (i) $N\Gamma\backslash G$ is a torus. Hence, one gets the* Mostow bundle

$$N \cap \Gamma\backslash N = \Gamma\backslash N\Gamma \hookrightarrow \Gamma\backslash G \to N\Gamma\backslash G\,.$$

Theorem 1.31 ([Hat60, Corollary 4.2]). *Let $X = \Gamma \backslash G$ be a completely-solvable solvmanifold and denote the Lie algebra naturally associated to G by $\mathfrak{g}$. The map $(\wedge^\bullet \mathfrak{g}^*, \mathrm{d}) \to (\wedge^\bullet X, \mathrm{d})$ of differential complexes is a quasi-isomorphism:*

$$i \colon H^\bullet_{dR}(\mathfrak{g}; \mathbb{R}) \overset{\simeq}{\to} H^\bullet_{dR}(X; \mathbb{R}) \ .$$

The previous result holds true also for solvmanifolds satisfying the so-called Mostow condition, [Mos54, Theorem 8.2, Corollary 8.1], see also [Rag72, Corollary 7.29]. For some results concerning the de Rham cohomology of (non-necessarily completely-solvable) solvmanifolds, see [Gua07, CF11, Kas13a, Kas12a, CFK13]. We refer also to Sect. 3.1.2 for further classical results concerning the computation of the de Rham and Dolbeault cohomologies of nilmanifolds and solvmanifolds.

In some cases, we will see that the study of (properties of) geometric structures on a solvmanifold is reduced to the study of the corresponding (properties of) geometric structures on the associated Lie algebra (see, e.g., Theorem 4.13, Proposition 4.2, Theorem 4.6). To this aim, we need the following trick by F.A. Belgun (see also [FG04, Theorem 2.1]).

Lemma 1.2 (F.A. Belgun's Symmetrization Trick, [Bel00, Theorem 7]). *Let $X = \Gamma \backslash G$ be a solvmanifold, and denote the Lie algebra naturally associated to G by $\mathfrak{g}$. Let η be a G-bi-invariant volume form on G such that $\int_X \eta = 1$, whose existence follows from J. Milnor's lemma [Mil76, Lemma 6.2]. Up to identifying G-left-invariant forms on X and linear forms over $\mathfrak{g}^*$ through left-translations, define the F. A. Belgun's symmetrization map*

$$\mu \colon \wedge^\bullet X \to \wedge^\bullet \mathfrak{g}^* \ , \qquad \mu(\alpha) := \int_X \alpha \lfloor_m \eta(m) \ .$$

One has that

$$\mu \lfloor_{\wedge^\bullet \mathfrak{g}^*} = \mathrm{id} \lfloor_{\wedge^\bullet \mathfrak{g}^*} \ ,$$

and that

$$\mathrm{d} \circ \mu = \mu \circ \mathrm{d} \ .$$

In particular, the symmetrization map μ induces a map $\mu \colon (\wedge^\bullet X, \mathrm{d}) \to (\wedge^\bullet \mathfrak{g}^*, \mathrm{d})$ of differential complexes, and hence a map $\mu \colon H^\bullet_{dR}(X; \mathbb{R}) \to H^\bullet_{dR}(\mathfrak{g}; \mathbb{R})$ in cohomology. Since $\mu \lfloor_{\wedge^\bullet \mathfrak{g}^*} = \mathrm{id} \lfloor_{\wedge^\bullet \mathfrak{g}^*}$, if the inclusion $(\wedge^\bullet \mathfrak{g}^*, \mathrm{d}) \hookrightarrow (\wedge^\bullet X, \mathrm{d})$ is a quasi-isomorphism (for example, if X is a nilmanifold, by [Nom54, Theorem 1],

---

(See also [Boc09, Theorem 3.6], which gives a sufficient condition for the Mostow bundle to be a principal bundle.)

or a completely-solvable solvmanifold, by [Hat60, Corollary 4.2]), then the map $\mu\colon (\wedge^\bullet X, d) \to (\wedge^\bullet \mathfrak{g}^*, d)$ turns out to be a quasi-isomorphism.

K. Nomizu's theorem [Nom54, Theorem 1], A. Hattori's theorem [Hat60, Corollary 4.2], and F.A. Belgun's theorem [Bel00, Theorem 7] suggest that nilmanifolds, and, more in general, solvmanifolds, may provide a very useful and interesting class of examples in non-Kähler geometry. On the other hand, another reason for this statement is given by the following result by Ch. Benson and C.S. Gordon, and by K. Hasegawa: it answers the *Weinstein and Thurston problem* for nilmanifolds, namely, it characterizes the nilmanifolds admitting a Kähler structure.

Theorem 1.32 ([BG88, Theorem A]). *Let X be a nilmanifold endowed with a symplectic structure ω such that the Hard Lefschetz Condition holds. Then X is diffeomorphic to a torus.*

Actually, one can prove that any $2n$-dimensional nilmanifold X endowed with a symplectic structure ω such that the map $[\omega]^{n-1}\colon H^1_{dR}(X;\mathbb{R}) \to H^{2n-1}_{dR}(X;\mathbb{R})$ is an isomorphism is diffeomorphic to a torus, [LO94], see, e.g., [FOT08, Theorem 4.98]. A minimal model proof of Ch. Benson and C.S. Gordon's theorem [BG88, Theorem A] is due to G. Lupton and J. Oprea, [LO94, Theorem 3.5].

Remark 1.19. For some results on Hard Lefschetz Condition for solvmanifolds, obtained by H. Kasuya, see [Kas13a, Kas09].

Theorem 1.33 ([Has89, Theorem 1, Corollary]). *Let X be a nilmanifold. If X is formal, then X is diffeomorphic to a torus.*

Remark 1.20. For some results on formality for solvmanifolds, obtained by H. Kasuya, see [Kas13a, Kas12d, Kas09].

In particular, since compact Kähler manifolds satisfy the Hard Lefschetz Condition, [Wei58, Théorème IV.5], and are formal, [DGMS75, Main Theorem], it follows that a nilmanifold admits a Kähler structure if and only if it is diffeomorphic to a torus (compare also [Han57, Theorem II, Footnote 1]). More in general, compact completely-solvable Kähler solvmanifolds are tori, as proven by A. Tralle and J. Kedra in [TK97, Theorem 1], solving a conjecture by Ch. Benson and C.S. Gordon, [BG90, p. 972]. In fact, the following result by K. Hasegawa gives a complete characterization of Kähler solvmanifolds.

Theorem 1.34 ([Has06, Main Theorem]). *Let X be a compact homogeneous space of solvable Lie group, that is, a compact differentiable manifold on which a connected solvable Lie group acts transitively. Then X admits a Kähler structure if and only if it is a finite quotient of a complex torus which has a structure of a complex torus-bundle over a complex torus. In particular, a completely-solvable solvmanifold has a Kähler structure if and only if it is a complex torus.*

Remark 1.21. For some results on Hodge decomposition and $\partial\bar{\partial}$-Lemma for solvmanifolds, obtained by H. Kasuya, see [Kas11, Kas12b].

Remark 1.22. In this context, another class that could provide several interesting examples is given by complex orbifolds[13] of the type $\tilde{X} = X/G$, where X is a complex manifold (possibly, a nilmanifold or a solvmanifold[14]) and G is a finite group of biholomorphisms of X (compare the Bochner linearization theorem, [Boc45, Theorem 1], see also [Rai06, Theorem 1.7.2]). Orbifolds of such a global-quotient-type have been considered and studied, e.g., by D.D. Joyce in constructing examples of compact seven-dimensional manifolds with holonomy G_2, [Joy96b] and [Joy00, Chaps. 11–12], and examples of compact eight-dimensional manifolds with holonomy Spin(7), [Joy96a, Joy99] and [Joy00, Chaps. 13–14]. We refer to [Sat56, Bai56, Bai54, Ang13a] for results concerning the de Rham, Dolbeault, and Bott-Chern cohomologies of orbifolds.

Appendix: Low Dimensional Solvmanifolds and Special Structures

In this section, we summarize some results concerning low-dimensional solvable Lie algebras, especially focusing on their classification and on the geometric properties of the possibly associated solvmanifolds. As regards nilpotent Lie algebras, we recall that, up to isomorphisms, the nilpotent Lie algebras of dimension less than or equal to 6 are finitely-many; on the other hand, in dimension 7, there are infinitely-many non-isomorphic nilpotent Lie algebras.

We refer mainly to [Boc09] for the classification, and to [TO97, Sal01, Mac13] for results on the geometric properties of nilmanifolds and solvmanifolds. (As regards classification of nilpotent Lie algebras up to dimension 7, see also, e.g., [Gon98] and the references therein.)

A.1 Solvmanifolds up to Dimension 4

In Table 1.1, we list the isomorphism classes of solvable Lie algebras of dimension less or equal to 4 and whose connected simply-connected Lie group admit a lattice (at least for some parameter), [Boc09, Sect. 5, Theorem 5.3, Table 1], [Boc09, Sect. 6, Theorem 6.2, Table 2], [Boc09, Table 8]. The notation follows [Mub63c,

---

[13]The notion of orbifold has been introduced by I. Satake in [Sat56], with the name of *V-manifold*, and has been studied, among others, by W.L. Baily, [Bai56, Bai54]. We recall that a *complex orbifold of complex dimension n* is a singular complex space whose singularities are locally isomorphic to quotient singularities $\mathbb{C}^n/G$, for finite subgroups $G \subset \mathrm{GL}(n;\mathbb{C})$, [Sat56, Definition 2]. We refer, e.g., to [Joy07, Joy00, Sat56, Bai56, Bai54] for more results concerning complex orbifolds and their cohomology.

[14]Recall that nilmanifolds and solvmanifolds are Eilenberg and MacLane spaces of type $K(\pi; 1)$; in particular, they are not simply-connected.

Table 1.1 Isomorphism classes of solvable Lie algebras of dimension less or equal to 4 and whose connected simply-connected Lie group admit a lattice (at least for some parameter), [Boc09, Sect. 5, Theorem 5.3, Table 1], [Boc09, Sect. 6, Theorem 6.2, Table 2], [Boc09, Table 8]

Name	Structure equations	Conditions	Type	b_1	Lattice	Complex?	Symplectic?	Formal?	HLC[a]?
$\mathfrak{g}_1$	(0)		Abelian	1	Torus	×	×	✓	×
$2\mathfrak{g}_1$	$(0,0)$		Abelian	2	Torus	Torus	✓	Kähler	✓
$3\mathfrak{g}_1$	$(0,0,0)$		Abelian	3	Torus	×	×	✓	×
$\mathfrak{g}_{3.1}$	$(-23,0,0)$		Nilpotent	2	[Boc09, Theorem 5.4]	×	×	×	×
$\mathfrak{g}_{3.4}^{-1}$	$(-13,23,0)$		Completely-solvable	1	[Boc09, p. 25], [TO97, Theorem 1.9]	×	×	✓	×
$\mathfrak{g}_{3.5}^{0}$	$(-23,13,0)$		Solvable	1	[Boc09, p. 25], [TO97, Theorem 1.9]	×	×	✓	×
$4\mathfrak{g}_1$	$(0,0,0,0)$		Abelian	4	Torus	Torus	✓	Kähler	✓
$\mathfrak{g}_{3.1}\oplus\mathfrak{g}_1$	$(-23,0,0,0)$		Nilpotent	3	[Boc09, p. 34]	Primary Kodaira surface	✓	×	×
$\mathfrak{g}_{3.4}^{-1}\oplus\mathfrak{g}_1$	$(-13,23,0,0)$		Completely-solvable	2	[Boc09, p. 34]	×	✓	✓	✓
$\mathfrak{g}_{3.5}^{0}\oplus\mathfrak{g}_1$	$(-23,13,0,0)$		Solvable	2	[Boc09, p. 35]	Hyper-elliptic surface	✓	Kähler	✓
$\mathfrak{g}_{4.1}$	$(-24,-34,0,0)$		Nilpotent	2	[Boc09, p. 35]	×	✓	×	×
$\mathfrak{g}_{4.5}^{p,-p-1}$	$(-14,-p^{-1}24,(p+1)^{-1}34,0)$	$-\frac{1}{2}\leq p<0$	Completely-solvable	1	[Boc09, p. 35]	×	×	✓	×

(continued)

Table 1.1 (continued)

Name	Structure equations	Conditions	Type	b_1	Lattice	Complex?	Symplectic?	Formal?	HLC[a]?
$\mathfrak{g}_{4.6}^{-2p,-p}$	$((2p)^{-1}14,$ $-p^{-1}24-$ $34, 24-p^{-1}34, 0)$	$p > 0$	Solvable	1	[Boc09, p. 35]	Inoue surface of type S^0	×	✓	×
$\mathfrak{g}_{4.8}^{-1}$	$(-23, -24, 34, 0)$		Completely-solvable	1	[Boc09, p. 35]	Inoue surface of type S^+	×	✓	×
$\mathfrak{g}_{4.9}^{0}$	$(-23, -34, 24, 0)$		Solvable	1	[Boc09, p. 36]	Secondary Kodaira surface	×	✓	×

[a] By [Boc09, Theorem 9.2], a four-dimensional symplectic solvmanifold does not satisfy the Hard Lefschetz Condition if and only if it is a non-torus nilmanifold; in particular, the Hard Lefschetz Condition does not depend on the symplectic structure for four-dimensional solvmanifolds

Boc09]. (For the classification of three-dimensional solvable Lie algebras and, respectively, of three-dimensional solvmanifolds, see also, e.g., [TO97, Theorems 1.6 and 1.9]. For results concerning four-dimensional solvmanifolds, see also, e.g., [TO97, Theorem 1.10].)

A.2 Five-Dimensional Solvmanifolds

In Table 1.2, we list the 9 isomorphism classes of five-dimensional nilpotent Lie algebras, [Boc09, Table 9], [Mub63a], and the 19 isomorphism classes of five-dimensional solvable non-nilpotent unimodular Lie algebras, [Boc09, Tables 10–14] (a list of the 24, respectively 33 classes of five-dimensional solvable non-nilpotent decomposable, respectively indecomposable Lie algebras has been obtained by G.M. Mubarakzjanov in [Mub63a]).

A.3 Six-Dimensional Nilmanifolds

V.V. Morozov classified in [Mor58] the six-dimensional nilpotent Lie algebras, up to isomorphism, in 34 different classes, see also [Mag86], see also [Boc09, Table 15], see also [Gon98, Sect. 3]. In [Sal01], S.M. Salamon identified the 18 classes of six-dimensional nilpotent Lie algebras admitting a linear integrable complex structure, [Sal01, Theorems 3.1–3.3, Proposition 3.4], and provided an estimate of the moduli spaces of complex structures and symplectic structures, [Sal01, Eqs. (28) and (31)], see [Sal01, Sect. 5]: up to equivalence, the linear integrable non-nilpotent complex structures have been classified by L. Ugarte and R. Villacampa in [UV09], the linear integrable Abelian complex structures have been classified by A. Andrada, M.L. Barberis, and I.G. Dotti in [ABDM11], and lastly the linear integrable nilpotent non-Abelian complex structures have been classified by M. Ceballos, A. Otal, L. Ugarte, and R. Villacampa in [COUV11]. The 26 classes of six-dimensional nilmanifolds admitting left-invariant symplectic structures have been classified by M. Goze and Y. Khakimdjanov, [GK96], see also [Sal01, Sect. 5]. There are 16 classes of nilmanifolds admitting both left-invariant complex structures and left-invariant symplectic structures (only one admitting a Kähler structure). Hence, there are five classes of six-dimensional nilmanifolds admitting neither left-invariant complex structures nor left-invariant symplectic structures, [Sal01, Theorem 5.1]: in fact, G.R. Cavalcanti and M. Gualtieri proved that such classes admit left-invariant generalized complex structures, [CG04, Theorems 4.1 and 4.2], see also [CG04, Table 1]. In Table 1.3, we list the classes of six-dimensional nilpotent Lie algebras following the notation [Boc09, Tables 4 and 15], and [Mag86, CFGU97, CG04, Sal01]. Moreover, in Table 1.4, we recall the classification of linear integrable complex structures on six-dimensional nilpotent Lie algebras up to equivalence by M. Ceballos, A. Otal, L. Ugarte, and R. Villacampa, [COUV11].

Table 1.2 Isomorphism classes of five-dimensional nilpotent Lie algebras [Boc09, Table 9], see also [Mub63a], and isomorphism classes of five-dimensional solvable non-nilpotent unimodular Lie algebras [Boc09, Tables 10–14, 3], see [Boc09, Sect. 7]

Name	Structure equations	Conditions	Type[a]	b_1	b_2	Lattice[b]?	Formal?	Note
$5\mathfrak{g}_1$	$(0,0,0,0,0)$		Ab	5		✓	✓	
$\mathfrak{g}_{3.1} \oplus 2\mathfrak{g}_1$	$(-23,0,0,0,0)$		nil	4		✓	✗	
$\mathfrak{g}_{4.1} \oplus \mathfrak{g}_1$	$(-24,-34,0,0,0)$		nil	3		✓	✗	
$\mathfrak{g}_{5.1}$	$(-35,-45,0,0,0)$		nil	3		✓	✗	
$\mathfrak{g}_{5.2}$	$(-25,-35,-45,0,0)$		nil	2		✓	✗	
$\mathfrak{g}_{5.3}$	$(-25,-45,-24,0,0)$		nil	2		✓	✗	
$\mathfrak{g}_{5.4}$	$(-24,-35,0,0,0)$		nil	4		✓	✗	
$\mathfrak{g}_{5.5}$	$(-34-25,-35,0,0,0)$		nil	3		✓	✗	
$\mathfrak{g}_{5.6}$	$(-34-25,-35,-45,0,0)$		nil	2		✓	✗	
$\mathfrak{g}_{3.4}^{-1} \oplus 2\mathfrak{g}_1$	$(-13,23,0,0,0)$		cplt-sol			✓	✓	[Boc09, p. 36]
$\mathfrak{g}_{3.5}^{0} \oplus 2\mathfrak{g}_1$	$(-23,13,0,0,0)$		sol			✓	✓	[Boc09, p. 36]
$\mathfrak{g}_{4.2}^{-2} \oplus \mathfrak{g}_1$	$(\frac{1}{2}14,-24-34,-34,0,0)$		cplt-sol	–	–	✗	–	[Boc09, Th 7.11]
$\mathfrak{g}_{4.5}^{p,-p-1} \oplus \mathfrak{g}_1$	$(-14,-p^{-1}24,(p+1)^{-1}34,0,0)$	$-\frac{1}{2} \leq p < 0$	cplt-sol			✓	✓	[Boc09, p. 36]
$\mathfrak{g}_{4.6}^{-2p,-p} \oplus \mathfrak{g}_1$	$((2p)^{-1}14,-p^{-1}24-34,24-34,0,0)$ $p^{-1}34,0,0)$	$p > 0$	sol			✓	✓	[Boc09, p. 36]
$\mathfrak{g}_{4.8}^{-1} \oplus \mathfrak{g}_1$	$(-23,-24,34,0,0)$		cplt-sol			✓	✓	[Boc09, p. 36]
$\mathfrak{g}_{4.9}^{0} \oplus \mathfrak{g}_1$	$(-23,-34,24,0,0)$		sol			✓	✓	[Boc09, p. 36]
$\mathfrak{g}_{5.7}^{p,q,r}$	$(-15,-p^{-1}25,-q^{-1}35,-r^{-1}45,0)$	$-1 \leq r \leq q \leq p \leq 1,$ $pqr \neq 0,\ p+q+r = -1$	cplt-sol	1	0,2,4	✓	✓	[Boc09, Pr 7.2.1]
$\mathfrak{g}_{5.8}^{-1}$	$(-25,0,-35,45,0)$		cplt-sol	2	3	✓	✗	[Boc09, Pr 7.2.2, Table 3]
$\mathfrak{g}_{5.9}^{p,-2-p}$	$(-15-25,-25,-p^{-1}35,(2+p)^{-1}45,0)$	$p \geq -1$	cplt-sol	–	–	✗	–	[Boc09, Pr 7.2.3]

$\mathfrak{g}_{5.11}^{-3}$	$(-15-25,\ -25-35,\ -35,\ \frac{1}{3}45,\ 0)$		cplt-sol	–	–	×	–	[Boc09, Pr 7.2.4] ([Har96])
$\mathfrak{g}_{5.13}^{-1-2q,q,r}$	$(-15,\ (1+2q)^{-1}25,\ -q^{-1}35-r^{-1}45,\ r^{-1}35-q^{-1}45,\ 0)$	$-1 \leq q \leq 0,\ q \neq -\frac{1}{2},\ r \neq 0$	sol	≥ 1	≥ 0	✓	?	[Boc09, Pr 7.2.5-6, Table 3] ([Har96])
$\mathfrak{g}_{5.14}^{0}$	$(-25,\ 0,\ -45,\ 35,\ 0)$		sol	≥ 2	≥ 3	✓	?	[Boc09, Pr 7.2.7-8, Table 3]
$\mathfrak{g}_{5.15}^{-1}$	$(-15-25,\ -25,\ 35-45,\ 45,\ 0)$		cplt-sol	1	2	✓	×	[Boc09, Pr 7.2.9, Table 3]
$\mathfrak{g}_{5.16}^{-1,q}$	$(-15-25,\ -25,\ 35-q^{-1}45,\ q^{-1}35+45,\ 0)$	$q \neq 0$	sol	–	–	×	–	[Boc09, Pr 7.2.10] ([Har96])
$\mathfrak{g}_{5.17}^{p,-p,r}$	$(-p^{-1}15-25,\ 15-p^{-1}25,\ p^{-1}35-r^{-1}45,\ r^{-1}35+p^{-1}45,\ 0)$	$r \neq 0$	sol	≥ 1	≥ 0	✓	?	[Boc09, Pr 7.2.11-12-13-14, Table 3] ([Har96])
$\mathfrak{g}_{5.18}^{0}$	$(-25-35,\ 15-45,\ -45,\ 35,\ 0)$		sol	≥ 1	≥ 2	✓	?	[Boc09, Pr 7.2.15, Table 3]
$\mathfrak{g}_{5.19}^{p,-2p-2}$	$(-23-(1+p)^{-1}15,\ -25,\ -p^{-1}35,\ (2p+2)^{-1}45,\ 0)$	$p \neq -1$	cplt-sol	–	–	×	–	[Boc09, Pr 7.2.16]
$\mathfrak{g}_{5.20}^{-1}$	$(-23-45,\ -25,\ 35,\ 0,\ 0)$		cplt-sol	2	1	✓	✓	[Boc09, Pr 7.2.16-17, Table 3]
$\mathfrak{g}_{5.23}^{-4}$	$(-23-\frac{1}{2}15,\ -25,\ -25-35,\ \frac{1}{4}45,\ 0)$		cplt-sol	–	–	×	–	[Boc09, Pr 7.2.16]

(continued)

Table 1.2 (continued)

Name	Structure equations	Conditions	Type[a]	b_1	b_2	Lattice[b]?	Formal?	Note
$\mathfrak{g}_{5.25}^{p,4p}$	$(-23-(2p)^{-1}15, -p^{-1}25+35, -25-p^{-1}35, (4p)^{-1}45, 0)$	$p \neq 0$	sol	–	–	×	–	[Boc09, Pr 7.2.16]
$\mathfrak{g}_{5.26}^{0,\varepsilon}$	$(-23-\varepsilon^{-1}45, 35, -25, 0, 0)$	$\varepsilon \in \{-1,1\}$	sol	≥ 2	≥ 1	✓	?	[Boc09, Pr 7.2.16-18, Table 3]
$\mathfrak{g}_{5.28}^{-\frac{3}{2}}$	$(-23+(\frac{1}{2})^{-1}15, \frac{2}{3}25, -35, -35-45, 0)$		cplt-sol	–	–	×	–	[Boc09, Pr 7.2.16]
$\mathfrak{g}_{5.30}^{-\frac{4}{3}}$	$(-24-\frac{3}{2}15, -34+(\frac{1}{3})^{-1}25, \frac{3}{4}35, -45, 0)$		cplt-sol	–	–	×	–	[Boc09, Pr 7.2.19]
$\mathfrak{g}_{5.33}^{-1,-1}$	$(-14, -25, 34+35, 0, 0)$		cplt-sol	2	1	✓	✓	[DF09, Theorem 3.1, Claim 3, Table 3]
$\mathfrak{g}_{5.35}^{-2,0}$	$(\frac{1}{2}14, -24-35, -34+25, 0, 0)$		sol	≥ 2	≥ 1	✓	?	[Boc09, Pr 7.2.20, Table 3]

a "Ab" means Abelian; "nil" means nilpotent; "cplt-sol" means completely-solvable; "sol" means solvable

b "✓" means that there exists a lattice at least for some values of the parameters; "×" means that there is no lattice for any value of the parameters

Table 1.3 Isomorphism classes of six-dimensional nilpotent Lie algebras [Boc09, Tables 4, 15, and 16], see also [Sal01, Sect. 5], [CG04, Table 1], see [Boc09, Sect. 8]

♯ [Boc09]	Name	Structure equations [Boc09]	b_1	b_2	b_3	nilstep	Type 3 (complex)	Type 2	Type 1	Type 0 (symplectic)	[Mag86]	[CFGU97]	[CG04]	[Sal01]	Structure equations [Sal01]
B01	$6\mathfrak{g}_1$	(0,0,0,0,0,0)	6	15	20	1	✓	✓	✓	✓	–	CFGU01	CG34	S34	(0,0,0,0,0,0)
B02	$\mathfrak{g}_{3.1} \oplus 3\mathfrak{g}_1$	(−23,0,0,0,0,0)	5	11	14	2	✓	✓	✓	✓	–	CFGU08	CG33	S33	(0,0,0,0,0,12)
B03	$\mathfrak{g}_{5.4} \oplus \mathfrak{g}_1$	(−24−35,0,0,0,0,0)	5	9	10	2	✓	✓	✓	×	–	CFGU03	CG32	S32	(0,0,0,0,0,12+34)
B04	$\mathfrak{g}_{5.1} \oplus \mathfrak{g}_1$	(−35,−45,0,0,0,0)	4	9	12	2	✓	✓	✓	✓	–	CFGU06	CG30	S31	(0,0,0,0,12,13)
B05	$2\mathfrak{g}_{3.1}$	(−23,0,0,−56,0,0)	4	8	10	2	✓	✓	✓	✓	–	CFGU02	CG29	S30	(0,0,0,0,12,34)
B06	$\mathfrak{g}_{6.N4}$	(0,0,0,−12,−13−24)	4	8	10	2	✓	✓	✓	✓	M04	CFGU04	CG28	S29	(0,0,0,0,12,14+23)
B07	$\mathfrak{g}_{6.N5}$	(0,0,0,−13−24,−14+23)	4	8	10	2	✓	✓	✓	✓	M05	CFGU05	CG31	S28	(0,0,0,0,13+42,14+23)
B08	$\mathfrak{g}_{5.5} \oplus \mathfrak{g}_1$	(−34−25,−35,0,0,0,0)	4	7	8	3	✓	✓	✓	✓	–	CFGU09	CG27	S27	(0,0,0,0,12,14+25)
B09	$\mathfrak{g}_{4.1} \oplus 2\mathfrak{g}_1$	(−24,−34,0,0,0,0)	4	7	8	3	×	✓	✓	✓	–	CFGU17	CG26	S26	(0,0,0,0,12,15)
B10	$\mathfrak{g}_{6.N12}$	(0,0,0,−13,0,−14−25)	4	6	6	3	×	✓	✓	×	M12	CFGU20	CG25	S25	(0,0,0,0,12,15+34)
B11	$\mathfrak{g}_{6.N3}$	(0,0,0,−13,−23,−12)	3	8	12	2	✓	✓	✓	✓	M03	CFGU07	CG13	S24	(0,0,0,12,13,23)
B12	$\mathfrak{g}_{6.N1}$	(0,0,−12,−13,0,−15)	3	6	8	3	✓	✓	✓	✓	M01	CFGU10	CG12	S23	(0,0,0,12,13,14)
B13	$\mathfrak{g}_{6.N6}$	(0,0,0,−13,−14−23,−12)	3	6	8	3	✓	✓	✓	✓	M06	CFGU11	CG10	S21	(0,0,0,12,13,14+23)
B14	$\mathfrak{g}_{6.N7}$	(0,0,0,−13,−14,−23)	3	6	8	3	✓	✓	✓	✓	M07	CFGU12	CG11	S22	(0,0,0,12,13,24)
B15	$\mathfrak{g}_{5.2} \oplus \mathfrak{g}_1$	(−25,−35,−45,0,0,0)	3	5	6	4	×	×	✓	✓	–	CFGU21	CG17	S13	(0,0,0,12,14,15)
B16	$\mathfrak{g}_{5.3} \oplus \mathfrak{g}_1$	(−25,−45,−24,0,0,0)	3	5	6	3	✓	✓	✓	×	–	CFGU16	CG18	S17	(0,0,0,12,14,24)
B17	$\mathfrak{g}_{5.6} \oplus \mathfrak{g}_1$	(−34−25,−35,−45,0,0,0)	3	5	6	4	×	×	✓	✓	–	CFGU22	CG16	S12	(0,0,0,12,14,15+24)
B18	$\mathfrak{g}_{6.N8}$	(0,0,−12,−13,−12,−25)	3	5	6	3	✓	×	✓	✓	M08	CFGU13	CG20[a]	S20	(0,0,0,12,13+14,24)
B19	$\mathfrak{g}_{6.N9}$	(0,0,−12,−13,−23,−15)	3	5	6	3	✓	×	✓	✓	M09	CFGU14	CG19	S19	(0,0,0,12,14,13+42)
B20	$\mathfrak{g}_{6.N10}$	(0,0,−12,0,−13−24,−14+23)	3	5	6	3	✓	✓	✓	✓	M10	CFGU15	CG24[b]	S18	(0,0,0,12,13+42,14+23)
B21	$\mathfrak{g}_{6.N13}$	(0,0,0,−13,−12,−14−25)	3	5	6	3	×	✓	✓	×	M13	CFGU18	CG09	S14	(0,0,0,12,13,14+35)
B22	$\mathfrak{g}_{6.N14}^{1}$	(0,0,0,−13,−23,−14−25)	3	5	6	3	✓	✓	✓	×	M14	CFGU19	CG22	S16	(0,0,0,12,23,14−35)

(continued)

Table 1.3 (continued)

# [Boc09]	Name	Structure equations [Boc09]	b_1	b_2	b_3	nilstep	Type 3 (complex)	Type 2	Type 1	Type 0 (symplectic)	[Mag86]	[CFGU97]	[CG04]	[Sal01]	Structure equations [Sal01]
B23	$g_{6,N14}^{-1}$	$(0,0,0,-13,-23,-14+25)$	3	5	6	3	×	✓	✓	×	M14	CFGU19	CG21	S15	$(0,0,0,12,23,14+35)$
B24	$g_{6,N15}$	$(0,0,-12,-13,-12,-14-25)$	3	5	6	4	×	×	✓	✓	M15	CFGU24	CG15	S11	$(0,0,0,12,14,15+23+24)$
B25	$g_{6,N17}$	$(0,0,-12,-13,0,-14-25)$	3	5	6	4	×	×	✓	✓	M17	CFGU25	CG14	S10	$(0,0,0,12,14,15+23)$
B26	$g_{6,N16}$	$(0,0,0,-13,-14-23,-15-24)$	3	4	4	4	×	✓	✓	✓	M16	CFGU27	CG23	S09	$(0,0,0,12,14-23,15+34)$
B27	$g_{6,N11}$	$(0,0,-12,-13,-14,-23)$	2	4	6	4	×	×	×	✓	M11	CFGU23	CG04	S06	$(0,0,12,13,23,14)$
B28	$g_{6,N18}^{1}$	$(0,0,-12,-13,-23,-14-25)$	2	4	6	4	✓	✓	✓	✓	M18	CFGU26	CG06	S08	$(0,0,12,13,23,14+25)$
B29	$g_{6,N18}^{-1}$	$(0,0,-12,-13,-23,-14+25)$	2	4	6	4	×	×	×	✓	M18	CFGU26	CG05	S07	$(0,0,12,13,23,14-25)$
B30	$g_{6,N2}$	$(0,0,-12,-13,-14,-15)$	2	3	4	5	×	×	✓	✓	M02	CFGU28	CG01	S03	$(0,0,12,13,14,15)$
B31	$g_{6,N19}$	$(0,0,-12,-13,-14,-15-23)$	2	3	4	5	×	×	✓	✓	M19	CFGU29	CG03	S05	$(0,0,12,13,14,23+15)$
B32	$g_{6,N20}$	$(0,0,-12,-13,-14-23,-15-24)$	2	3	4	5	×	×	✓	✓	M20	CFGU30	CG08	S04	$(0,0,12,13,14+23,24+15)$
B33	$g_{6,N21}$	$(0,0,-12,-23,-24,-15-34)$	2	2	2	5	×	×	✓	×	M21	CFGU31	CG02	S02	$(0,0,12,13,14,34+52)$
B34	$g_{6,N22}$	$(0,0,-12,-23,-24+13,-15-34)$	2	2	2	5	×	×	✓	×	M22	CFGU32	CG07	S01	$(0,0,12,13,14+23,34+52)$

a The structure equations in [CG04] are written as $(0, 0, 0, 12, 14, 23 + 24)$

b The structure equations in [CG04] are written as $(0, 0, 0, 12, 14 + 23, 13 + 42)$

Table 1.4 M. Ceballos, A. Otal, L. Ugarte, and R. Villacampa's classification of linear integrable complex structures on six-dimensional nilpotent Lie algebras up to equivalence [COUV11]

#	Algebra	b_1	b_2	b_3	Complex structure(s)	Conditions ($\lambda \geq 0$, $c \geq 0$, $B \in \mathbb{C}$, $D \in \mathbb{C}$) $\left(S(B,c) = c^4 - 2\left(|B|^2 + 1\right)c^2 + \left(|B|^2 - 1\right)^2\right)$
00	$\mathfrak{h}_1 := (0, 0, 0, 0, 0, 0)$	6	15	20	$J := (0, 0, 0)$	
01	$\mathfrak{h}_2 := (0, 0, 0, 0, 12, 34)$	4	8	10	$J_1^D := \left(0, 0, \omega^{1\bar{1}} + D\,\omega^{2\bar{2}}\right)$,	$\mathfrak{Im}\,D = 1$
02					$J_2^D := \left(0, 0, \omega^{12} + \omega^{1\bar{1}} + \omega^{1\bar{2}} + D\,\omega^{2\bar{2}}\right)$,	$\mathfrak{Im}\,D > 0$
03	$\mathfrak{h}_3 := (0, 0, 0, 0, 0, 12 + 34)$	5	9	10	$J_1 := \left(0, 0, \omega^{1\bar{1}} + \omega^{2\bar{2}}\right)$	
04					$J_2 := \left(0, 0, \omega^{1\bar{1}} - \omega^{2\bar{2}}\right)$	
05	$\mathfrak{h}_4 := (0, 0, 0, 0, 12, 14 + 23)$	4	8	10	$J_1 := \left(0, 0, \omega^{1\bar{1}} + \omega^{1\bar{2}} + \frac{1}{4}\omega^{2\bar{2}}\right)$	
06					$J_2^D := \left(0, 0, \omega^{12} + \omega^{1\bar{1}} + \omega^{1\bar{2}} + D\,\omega^{2\bar{2}}\right)$,	$D \in \mathbb{R} \setminus \{0\}$
07	$\mathfrak{h}_5 := (0, 0, 0, 13 + 42, 14 + 23)$	4	8	10	$J_1^D := \left(0, 0, \omega^{1\bar{1}} + \omega^{1\bar{2}} + D\,\omega^{2\bar{2}}\right)$,	$D \in [0, \tfrac{1}{4}]$
08					$J_2 := (0, 0, \omega^{12})$	
09					$J_3^{(\lambda, D)} := \left(0, 0, \omega^{12} + \omega^{1\bar{1}} + \lambda\,\omega^{1\bar{2}} + D\,\omega^{2\bar{2}}\right)$,	$(\lambda, D) \in \{(0, x + \mathrm{i}\,y) \in \mathbb{R} \times \mathbb{C} : y \geq 0, 4y^2 < 1 + 4x\}$ $\cup \left\{(\lambda, \mathrm{i}\,y) \in \mathbb{R} \times \mathbb{C} : 0 < \lambda^2 < \tfrac{1}{2}, 0 \leq y < \tfrac{\lambda^2}{2}\right\}$ $\cup \left\{(\lambda, \mathrm{i}\,y) \in \mathbb{R} \times \mathbb{C} : \tfrac{1}{2} \leq \lambda^2 < 1, 0 \leq y < \tfrac{1-\lambda^2}{2}\right\}$ $\cup \left\{(\lambda, \mathrm{i}\,y) \in \mathbb{R} \times \mathbb{C} : \lambda^2 > 1, 0 \leq y < \tfrac{\lambda^2-1}{2}\right\}$
10	$\mathfrak{h}_6 := (0, 0, 0, 0, 12, 13)$	4	9	12	$J := \left(0, 0, \omega^{12} + \omega^{1\bar{1}} + \omega^{1\bar{2}}\right)$	
11	$\mathfrak{h}_7 := (0, 0, 0, 12, 13, 23)$	3	8	12	$J := \left(0, \omega^{1\bar{1}}, \omega^{12} + \omega^{2\bar{1}}\right)$	
12	$\mathfrak{h}_8 := (0, 0, 0, 0, 0, 12)$	5	11	14	$J := \left(0, 0, \omega^{1\bar{1}}\right)$	
13	$\mathfrak{h}_9 := (0, 0, 0, 0, 12, 14 + 25)$	4	7	8	$J := \left(0, \omega^{1\bar{1}}, \omega^{12} + \omega^{2\bar{1}}\right)$	
14	$\mathfrak{h}_{10} := (0, 0, 0, 12, 13, 14)$	3	6	8	$J := \left(0, \omega^{1\bar{1}}, \omega^{12} + \omega^{2\bar{1}}\right)$	

(continued)

Table 1.4 (continued)

#	Algebra	b_1	b_2	b_3	Complex structure(s)	Conditions ($\lambda \geq 0$, $c \geq 0$, $B \in \mathbb{C}$, $D \in \mathbb{C}$) $\left(S(B,c) := c^4 - 2\left(|B|^2 + 1\right)c^2 + \left(|B|^2 - 1\right)^2\right)$
15	$\mathfrak{h}_{11} := (0, 0, 0, 12, 13, 14+23)$	3	6	8	$J^B := \left(0,\ \omega^{1\bar1},\ \omega^{12} + B\omega^{1\bar2} + |B-1|\,\omega^{2\bar1}\right)$,	$B \in \mathbb{R} \setminus \{0, 1\}$
16	$\mathfrak{h}_{12} := (0, 0, 0, 12, 13, 24)$	3	6	8	$J^B := \left(0,\ \omega^{1\bar1},\ \omega^{12} + B\omega^{1\bar2} + |B-1|\,\omega^{2\bar1}\right)$,	$\Im m\, B \neq 0$
17	$\mathfrak{h}_{13} := (0, 0, 0, 12, 13+14, 24)$	3	5	6	$J^{(B,c)} := \left(0,\ \omega^{1\bar1},\ \omega^{12} + B\omega^{1\bar2} + c\omega^{2\bar1}\right)$,	$c \neq |B-1|$, $(c, |B|) \neq (0, 1)$, $S(B,c) < 0$
18	$\mathfrak{h}_{14} := (0, 0, 0, 12, 14, 13+42)$	3	5	6	$J^{(B,c)} := \left(0,\ \omega^{1\bar1},\ \omega^{12} + B\omega^{1\bar2} + c\omega^{2\bar1}\right)$,	$c \neq |B-1|$, $(c, |B|) \neq (0, 1)$, $S(B,c) = 0$
19	$\mathfrak{h}_{15} := (0, 0, 0, 12, 13+42, 14+23)$	3	5	6	$J_1 := \left(0,\ \omega^{1\bar1},\ \omega^{2\bar1}\right)$	
20					$J_2^c := \left(0,\ \omega^{1\bar1},\ \omega^{12} + c\omega^{2\bar1}\right)$,	$c \neq 1$
21					$J_3^{(B,c)} := \left(0,\ \omega^{1\bar1},\ \omega^{12} + B\omega^{1\bar2} + c\omega^{2\bar1}\right)$,	$c \neq |B-1|$, $(c, |B|) \neq (0, 1)$, $S(B,c) > 0$
22	$\mathfrak{h}_{16} := (0, 0, 0, 12, 14, 24)$	3	5	6	$J^B := \left(0,\ \omega^{1\bar1},\ \omega^{12} + B\omega^{1\bar3}\right)$,	$|B| = 1$, $B \neq 1$
23	$\mathfrak{h}_{19}^- := (0, 0, 0, 12, 23, 14-35)$	3	5	6	$J_1 := \left(0,\ \omega^{13} + \omega^{1\bar3},\ i\left(\omega^{12} - \omega^{2\bar1}\right)\right)$	
24					$J_2 := \left(0,\ \omega^{13} + \omega^{1\bar3},\ -i\left(\omega^{12} - \omega^{2\bar1}\right)\right)$	
25	$\mathfrak{h}_{26}^+ := (0, 0, 12, 13, 23, 14+25)$	2	4	6	$J_1 := \left(0,\ \omega^{13} + \omega^{1\bar3},\ i\omega^{1\bar1} + i\left(\omega^{1\bar2} - \omega^{2\bar1}\right)\right)$	
26					$J_2 := \left(0,\ \omega^{13} + \omega^{1\bar3},\ i\omega^{1\bar1} - i\left(\omega^{1\bar2} - \omega^{2\bar1}\right)\right)$	

A.4 Six-Dimensional Solvmanifolds

The six-dimensional solvable non-decomposable Lie algebras were classified by G.M. Mubarakzjanov, [Mub63b, CS05], and by P. Turkowski, [Tur90]; see also [Boc09, Sect. 8]. In addition to [Boc09, Table 15], which lists the six-dimensional nilpotent non-decomposable Lie algebras, [Mor58], we refer to [Boc09, Tables 17/33], and also [Mac13, Table A.1], for a list of the six-dimensional unimodular solvable non-nilpotent non-decomposable Lie algebras; see [Boc09, Sect. 8]. The Betti numbers of six-dimensional unimodular solvable non-nilpotent Lie algebras can be found in [Mac13, Table 1] (see also [Boc09, Tables 34/36] for the first Betti numbers). See [Boc09, Tables 5–6] for a list of decomposable non-nilpotent solvmanifolds, together with their Betti numbers and results concerning formality and symplectic structures. In [CM12], six-dimensional non-completely-solvable almost-Abelian solvmanifolds for which Mostow condition does not hold are investigated; in particular, see [CM12, Table 1]. The six-dimensional unimodular solvable non-nilpotent Lie algebras admitting a symplectic structure are listed in [Mac13, Theorem 1], and their symplectic structures are listed in [Mac13, Table B.1]. In [Boc09, Sect. 9] and [Mac13], the Hard Lefschetz Condition on solvmanifolds has been investigated; in particular, [Boc09, Table 7] summarizes the relations between formality, Hard Lefschetz Condition, and even odd-degree Betti numbers, providing some known examples, see also [IRTU03, Theorem 3.1, Table 1] (see also [FMS03, dBT06, Kas13a, Kas09, Kas11, Kas12d] for examples and results on cohomologically-Kähler manifolds); furthermore, in [Mac13, Theorem 4], M. Macrì provided three solvmanifolds endowed with symplectic structures that satisfy the Hard Lefschetz Condition, more precisely, $\Gamma \backslash \left(G_{5.7}^{p,-1,-1} \times \mathbb{R} \right)$ (see also [Boc09, Theorem 9.1]), and $\Gamma \backslash \left(G_{3.4} \times \mathbb{R}^3 \right)$, and $\Gamma \backslash (G_{3.4} \times G_{3.4})$.

Chapter 2
Cohomology of Complex Manifolds

Abstract In this chapter, we study cohomological properties of compact complex manifolds. In particular, we are concerned with studying the *Bott-Chern cohomology*, which, in a sense, constitutes a bridge between the de Rham cohomology and the Dolbeault cohomology of a complex manifold.

In Sect. 2.1, we recall some definitions and results on the *Bott-Chern* and *Aeppli* cohomologies, see, e.g., Schweitzer (*Autour de la cohomologie de Bott-Chern*, arXiv:0709.3528 [math.AG], 2007), and on the $\partial\bar{\partial}$-*Lemma*, referring to Deligne et al. (Invent. Math. 29(3):245–274, 1975). In Sect. 2.2, we provide an inequality *à la* Frölicher for the Bott-Chern cohomology, Theorem 2.13, which also allows to characterize the validity of the $\partial\bar{\partial}$-*Lemma* in terms of the dimensions of the Bott-Chern cohomology groups, Theorem 2.14; the proof of such inequality is based on two exact sequences, firstly considered by J. Varouchas in (Propriétés cohomologiques d'une classe de variétés analytiques complexes compactes, Séminaire d'analyse P. Lelong-P. Dolbeault-H. Skoda, années 1983/1984, Lecture Notes in Math., vol. 1198, Springer, Berlin, 1986, pp. 233–243). Finally, in Appendix: Cohomological Properties of Generalized Complex Manifolds, we consider how to extend such results to the symplectic and generalized complex contexts.

2.1 Cohomologies of Complex Manifolds

The *Bott-Chern cohomology* and the *Aeppli cohomology* provide important invariants for the study of the geometry of compact (especially, non-Kähler) complex manifolds. These cohomology groups have been introduced by R. Bott and S.S. Chern in [BC65], and by A. Aeppli in [Aep65], and hence studied by many authors, e.g., B. Bigolin [Big69, Big70] (both from the sheaf-theoretic and from the analytic viewpoints), A. Andreotti and F. Norguet [AN71] (to study cycles of algebraic manifolds), J. Varouchas [Var86] (to study the cohomological properties of a certain class of compact complex manifolds), M. Abate [Aba88] (to study annular bundles), L. Alessandrini and G. Bassanelli [AB96] (to investigate the

D. Angella, *Cohomological Aspects in Complex Non-Kähler Geometry*,
Lecture Notes in Mathematics 2095, DOI 10.1007/978-3-319-02441-7_2,
© Springer International Publishing Switzerland 2014

properties of balanced metrics), S. Ofman [Ofm85a, Ofm85b, Ofm88] (in view of applications to integration on analytic cycles), S. Boucksom [Bou04] (in order to extend divisorial Zariski decompositions to compact complex manifolds), J.-P. Demailly [Dem12] (as a tool in complex geometry), M. Schweitzer [Sch07] (in the context of cohomology theories), L. Lussardi [Lus10] (in the non-compact Kähler case), R. Kooistra [Koo11] (in the framework of cohomology theories), J.-M. Bismut [Bis11b, Bis11a] (in the context of Chern characters), L.-S. Tseng and S.-T. Yau [TY11] (in the framework of generalized geometry and type II String Theory).

In this preliminary section, we recall the basic notions and classical results concerning cohomologies of complex manifolds. More precisely, we recall the definitions of the Bott-Chern and Aeppli cohomologies, and some results on Hodge theory, referring to [Sch07], see also [Dem12, Sects. VI.8.1 and VI.12.1]; then, we recall the notion of $\partial\bar{\partial}$-Lemma, referring to [DGMS75].

2.1.1 The Bott-Chern Cohomology

Let X be a complex manifold.

Definition 2.1 ([BC65]). The *Bott-Chern cohomology* of a complex manifold X is the bi-graded algebra

$$H_{BC}^{\bullet,\bullet}(X) := \frac{\ker \partial \cap \ker \bar{\partial}}{\operatorname{im} \partial\bar{\partial}} .$$

Unlike in the case of the Dolbeault cohomology groups, for every $(p,q) \in \mathbb{N}^2$, the conjugation induces an isomorphism

$$H_{BC}^{p,q}(X) \xrightarrow{\simeq} H_{BC}^{q,p}(X) .$$

Furthermore, since $\ker \partial \cap \ker \bar{\partial} \subseteq \ker d$ and $\operatorname{im} \partial\bar{\partial} \subseteq \operatorname{im} d$, one has the natural map of graded $\mathbb{C}$-vector spaces

$$\bigoplus_{p+q=\bullet} H_{BC}^{p,q}(X) \to H_{dR}^{\bullet}(X;\mathbb{C}) ,$$

and, since $\ker \partial \cap \ker \bar{\partial} \subseteq \ker \bar{\partial}$ and $\operatorname{im} \partial\bar{\partial} \subseteq \operatorname{im} \bar{\partial}$, one has the natural map of bi-graded $\mathbb{C}$-vector spaces

$$H_{BC}^{\bullet,\bullet}(X) \to H_{\bar{\partial}}^{\bullet,\bullet}(X) .$$

In general, even for compact complex manifolds, these maps are neither injective nor surjective: see, e.g., the examples in [Sch07, Sect. 1.c] or in Sect. 3.2.4. A case of special interest is when X is a compact complex manifold satisfying the $\partial\bar{\partial}$-Lemma, namely, the property that every ∂-closed $\bar{\partial}$-closed d-exact form is also $\partial\bar{\partial}$-exact, [DGMS75], that is, the natural map $H_{BC}^{\bullet,\bullet}(X) \to H_{dR}^{\bullet}(X;\mathbb{C})$ is injective (we recall that compact Kähler manifolds and, more in general, manifolds in class $\mathcal{C}$ of Fujiki, [Fuj78], that is, compact complex manifolds admitting a proper modification from a Kähler manifold, satisfy the $\partial\bar{\partial}$-Lemma, [DGMS75], Lemma 5.11, Corollary 5.23]; we refer to Sect. 2.1.3 for further details). In fact, we recall the following result.

Theorem 2.1 ([DGMS75, Lemma 5.15, Remarks 5.16, 5.21]). *Let X be a compact complex manifold. If X satisfies the $\partial\bar{\partial}$-Lemma, then the natural maps*

$$\bigoplus_{p+q=\bullet} H_{BC}^{p,q}(X) \to H_{dR}^{\bullet}(X;\mathbb{C}) \qquad and \qquad H_{BC}^{\bullet,\bullet}(X) \to H_{\bar{\partial}}^{\bullet,\bullet}(X)$$

induced by the identity are isomorphisms.

Proof. We provide here a different proof, which uses [AT13b, Theorem A], see Theorem 2.13.

More precisely, we recall that, by [AT13b, Theorem A] and [Sch07, Sect. 2.c], we have that, for any compact complex manifold X of complex dimension n, for any $k \in \mathbb{Z}$, the inequality $\sum_{p+q=k} \left(\dim_{\mathbb{C}} H_{BC}^{p,q}(X) + \dim_{\mathbb{C}} H_{BC}^{n-p,n-q}(X)\right) \geq 2 \sum_{p+q=k} \dim_{\mathbb{C}} H_{\bar{\partial}}^{p,q}(X) \geq 2 \dim_{\mathbb{C}} H_{dR}^{k}(X;\mathbb{C})$ holds.

Suppose now that X satisfies the $\partial\bar{\partial}$-Lemma, namely, the natural map $\bigoplus_{p+q=\bullet} H_{BC}^{p,q}(X) \to H_{dR}^{\bullet}(X;\mathbb{C})$ induced by the identity is injective. In particular, for any $k \in \mathbb{Z}$, it follows that $\sum_{p+q=k} \dim_{\mathbb{C}} H_{BC}^{p,q}(X) \leq \dim_{\mathbb{C}} H_{dR}^{k}(X;\mathbb{C})$.

Hence, one gets that $\sum_{p+q=k} \dim_{\mathbb{C}} H_{BC}^{p,q}(X) = \dim_{\mathbb{C}} H_{dR}^{k}(X;\mathbb{C}) < +\infty$. In particular, the natural map $\bigoplus_{p+q=\bullet} H_{BC}^{p,q}(X) \to H_{dR}^{\bullet}(X;\mathbb{C})$ induced by the identity is actually an isomorphism.

Furthermore, one gets also that $\dim_{\mathbb{C}} H_{dR}^{k}(X;\mathbb{C}) = \sum_{p+q=k} \dim_{\mathbb{C}} H_{\bar{\partial}}^{p,q}(X)$ and hence $\sum_{p+q=k} \dim_{\mathbb{C}} H_{BC}^{p,q}(X) = \sum_{p+q=k} \dim_{\mathbb{C}} H_{\bar{\partial}}^{p,q}(X) < +\infty$. Therefore, to finish the proof, it suffices to prove that, if the natural map $H_{BC}^{\bullet,\bullet}(X) \to H_{dR}^{\bullet}(X;\mathbb{C})$ induced by the identity is injective, then also the natural map $H_{BC}^{\bullet,\bullet}(X) \to H_{\bar{\partial}}^{\bullet,\bullet}(X)$ induced by the identity is injective.

Suppose that the natural map $H_{BC}^{\bullet,\bullet}(X) \to H_{dR}^{\bullet}(X;\mathbb{C})$ induced by the identity is injective. Let $\alpha^{p,q} = \bar{\partial}\beta^{p,q-1} \in \wedge^{p,q}X$ be a ∂-closed $\bar{\partial}$-closed $\bar{\partial}$-exact form. In particular, $\beta^{p,q-1} \in \wedge^{p,q-1}X$ is $\partial\bar{\partial}$-closed. Consider d $\beta^{p,q-1} \in \wedge^{p+1,q-1}X \oplus \wedge^{p,q}X$: it is a ∂-closed $\bar{\partial}$-closed d-exact form. Hence, by the hypothesis, the form d $\beta^{p,q-1}$ is also $\partial\bar{\partial}$-exact. Therefore there exist $\gamma^{p,q-2} \in \wedge^{p,q-2}X$ and $\gamma^{p-1,q-1} \in \wedge^{p-1,q-1}X$ such that d $\beta^{p,q-1} = \partial\bar{\partial}\gamma^{p,q-2} + \partial\bar{\partial}\gamma^{p-1,q-1}$. In particular, it follows that $\alpha^{p,q} = \bar{\partial}\beta^{p,q-1} = \partial\bar{\partial}\gamma^{p-1,q-1}$. Then the natural map $H_{BC}^{\bullet,\bullet}(X) \to H_{\bar{\partial}}^{\bullet,\bullet}(X)$ induced by the identity is injective. $\qquad\square$

As for the de Rham and the Dolbeault cohomologies, a Hodge theory can be developed also for the Bott-Chern cohomology for compact complex manifolds: we recall here some results, referring to [Sch07, Sect. 2] (see also [Big69, Sect. 5], and [Lus10]).

Suppose that X is a compact complex manifold. Fix a Hermitian metric on X, and define the differential operator

$$\tilde{\Delta}_{BC} := \left(\partial\bar{\partial}\right)\left(\partial\bar{\partial}\right)^* + \left(\partial\bar{\partial}\right)^*\left(\partial\bar{\partial}\right) + \left(\bar{\partial}^*\partial\right)\left(\bar{\partial}^*\partial\right)^* + \left(\bar{\partial}^*\partial\right)^*\left(\bar{\partial}^*\partial\right) + \bar{\partial}^*\bar{\partial} + \partial^*\partial\,,$$

see [KS60, Proposition 5] (where it is used to prove the stability of the Kähler property under small deformations of the complex structure), and also [Sch07, Sect. 2.b], [Big69, Sect. 5.1]. One has the following result.

Theorem 2.2 ([KS60, Proposition 5], see also [Sch07, Sect. 2.b]). *Let X be a compact complex manifold endowed with a Hermitian metric. The operator $\tilde{\Delta}_{BC}$ is a 4th order self-adjoint elliptic differential operator, and*

$$\ker \tilde{\Delta}_{BC} = \ker\partial \cap \ker\bar{\partial} \cap \ker\bar{\partial}^*\partial^*\,.$$

Therefore, as a consequence of the general theory of self-adjoint elliptic differential operators, see, e.g., [Kod05, p. 450], the following result holds.

Theorem 2.3 ([Sch07, Théorème 2.2], [Sch07, Corollaire 2.3]). *Let X be a compact complex manifold, endowed with a Hermitian metric. Then there exist an orthogonal decomposition*

$$\wedge^{\bullet,\bullet}X = \ker \tilde{\Delta}_{BC} \oplus \operatorname{im}\partial\bar{\partial} \oplus \left(\operatorname{im}\partial^* + \operatorname{im}\bar{\partial}^*\right)$$

and an isomorphism

$$H_{BC}^{\bullet,\bullet}(X) \simeq \ker \tilde{\Delta}_{BC}\,.$$

In particular, the Bott-Chern cohomology groups of X are finite-dimensional $\mathbb{C}$-vector spaces.

Another consequence of general results in spectral theory, see, e.g., [KS60, Theorem 4], [Kod05, Theorem 7.3], is the semi-continuity property for the dimensions of the Bott-Chern cohomology.

Theorem 2.4 ([Sch07, Lemme 3.2]). *Let $\{X_t\}_{t\in B}$ a complex-analytic family of compact complex manifolds. Then, for every $(p,q) \in \mathbb{N}^2$, the function*

$$B \ni t \mapsto \dim_{\mathbb{C}} H_{BC}^{p,q}(X_t) \in \mathbb{N}$$

is upper-semi-continuous.

By using the Kähler identities (in particular, the fact that $\overline{\square} = \square$ and that $\partial^*\overline\partial + \overline\partial\partial^* = 0 = \overline\partial^*\partial + \partial\overline\partial^*$), one can prove that, on a compact Kähler manifold,

$$\tilde\Delta_{BC} = \overline\square^2 + \partial^*\partial + \overline\partial^*\overline\partial\,,$$

[KS60, Proposition 6], [Sch07, Proposition 2.4].

Indeed, recall that $\overline\square = \square$ and that $\partial^*\overline\partial = \mathrm{i}\left[\Lambda,\overline\partial\right]\overline\partial = -\mathrm{i}\,\overline\partial\Lambda\overline\partial = -\mathrm{i}\,\overline\partial$ $\left[\Lambda,\overline\partial\right] = -\overline\partial\partial^*$, and hence $\overline\partial^*\partial = -\partial\overline\partial^*$ and $\partial^*\overline\partial = -\overline\partial\partial^*$; therefore

$$\overline\square^2 = \overline\square\,\overline\square = \overline\partial\overline\partial^*\partial\partial^* + \overline\partial\overline\partial^*\partial^*\partial + \overline\partial^*\overline\partial\partial\partial^* + \overline\partial^*\overline\partial\partial^*\partial$$

$$= -\overline\partial\partial\overline\partial^*\partial^* - \overline\partial\partial^*\overline\partial^*\partial - \overline\partial^*\partial\overline\partial\partial^* - \overline\partial^*\partial^*\overline\partial\partial$$

$$= \partial\overline\partial\overline\partial^*\partial^* + \overline\partial\partial^*\partial\overline\partial^* + \partial\partial^*\overline\partial\partial^* + \overline\partial^*\partial^*\partial\overline\partial$$

$$= \partial\overline\partial\overline\partial^*\partial^* + \partial^*\overline\partial\overline\partial^*\partial + \overline\partial^*\partial\partial^*\overline\partial + \overline\partial^*\partial^*\partial\overline\partial$$

$$= \tilde\Delta_{BC} - \partial^*\partial - \overline\partial^*\overline\partial\,.$$

Hence $\ker\tilde\Delta_{BC} = \ker\overline\square = \ker\Delta$; in particular, it follows that, on a compact Kähler manifold, the de Rham cohomology, the Dolbeault cohomology, and the Bott-Chern cohomology are isomorphic (actually, since the $\partial\overline\partial$-Lemma holds on every compact Kähler manifold, one gets an isomorphism that does not depend on the choice of the Hermitian metric).

2.1.2 The Aeppli Cohomology

Let X be a complex manifold. Dualizing the definition of the Bott-Chern cohomology, one can define another cohomology on X: the Aeppli cohomology.

Definition 2.2 ([Aep65]). The *Aeppli cohomology* of a complex manifold X is the bi-graded $H_{BC}^{\bullet,\bullet}(X)$-module

$$H_A^{\bullet,\bullet}(X) := \frac{\ker\partial\overline\partial}{\operatorname{im}\partial + \operatorname{im}\overline\partial}\,.$$

As for the Bott-Chern cohomology, the conjugation induces, for every $(p,q) \in \mathbb{N}^2$, the isomorphism

$$H_A^{p,q}(X) \xrightarrow{\;\simeq\;} H_A^{q,p}(X)\,.$$

Furthermore, since $\ker d \subseteq \ker \partial\bar{\partial}$ and $\operatorname{im} d \subseteq \operatorname{im} \partial + \operatorname{im} \bar{\partial}$, one has the natural map of graded $\mathbb{C}$-vector spaces

$$H^{\bullet}_{dR}(X;\mathbb{C}) \to \bigoplus_{p+q=\bullet} H^{p,q}_{A}(X),$$

and, since $\ker \bar{\partial} \subseteq \ker \partial\bar{\partial}$ and $\operatorname{im} \bar{\partial} \subseteq \operatorname{im} \partial + \operatorname{im} \bar{\partial}$, one has the natural map of bi-graded $\mathbb{C}$-vector spaces

$$H^{\bullet,\bullet}_{\bar{\partial}}(X) \to H^{\bullet,\bullet}_{A}(X);$$

as we have noted for the Bott-Chern cohomology, such maps are, in general, neither injective nor surjective, but they are isomorphisms whenever X is compact and satisfies the $\partial\bar{\partial}$-Lemma, [DGMS75, Lemma 5.15, Remarks 5.16, 5.21], and hence, in particular, if X is a compact complex manifold admitting a Kähler structure, [DGMS75, Lemma 5.11], or if X is a compact complex manifold in class $\mathcal{C}$ of Fujiki, [DGMS75, Corollary 5.23].

Remark 2.1. On a compact Kähler manifold X, the associated $(1,1)$-form ω of the Kähler metric defines a non-zero class in $H^2_{dR}(X;\mathbb{R})$. For general Hermitian manifolds, special classes of metrics are often defined in terms of closedness of powers of ω, so they define classes in the Bott-Chern or Aeppli cohomology groups (e.g., a Hermitian metric on a complex manifold of complex dimension n is said *balanced* if $d\omega^{n-1} = 0$ [Mic82], *pluriclosed* if $\partial\bar{\partial}\omega = 0$ [Bis89], *astheno-Kähler* if $\partial\bar{\partial}\omega^{n-2} = 0$ [JY93, JY94], *Gauduchon* if $\partial\bar{\partial}\omega^{n-1} = 0$ [Gau77]). (Note that, they define possibly the zero class in the Bott-Chern or Aeppli cohomologies: for the balanced case, see [FLY12, Corollary 1.3], where it is shown that, for $k \geq 2$, the complex structures on $\#_{j=1}^{k} (\mathbb{S}^3 \times \mathbb{S}^3)$ constructed from the conifold transitions admit balanced metrics.)

We refer to [Sch07, Sect. 2.c] for the following results, concerning Hodge theory for the Aeppli cohomology on compact complex manifolds.

Suppose that X is a compact complex manifold. Once fixed a Hermitian metric on X, one defines the differential operator

$$\tilde{\Delta}_A := \partial\partial^* + \bar{\partial}\bar{\partial}^* + \left(\partial\bar{\partial}\right)^* \left(\partial\bar{\partial}\right) + \left(\partial\bar{\partial}\right)\left(\partial\bar{\partial}\right)^* + \left(\bar{\partial}\partial^*\right)^* \left(\bar{\partial}\partial^*\right) + \left(\bar{\partial}\partial^*\right)\left(\bar{\partial}\partial^*\right)^*,$$

which turns out to be a 4th order self-adjoint elliptic differential operator such that

$$\ker \tilde{\Delta}_A = \ker \partial\bar{\partial} \cap \ker \partial^* \cap \ker \bar{\partial}^*.$$

Hence one has an orthogonal decomposition

$$\wedge^{\bullet,\bullet} X = \ker \tilde{\Delta}_A \oplus \left(\operatorname{im} \partial + \operatorname{im} \bar{\partial}\right) \oplus \operatorname{im} \left(\partial\bar{\partial}\right)^*$$

from which one gets an isomorphism

$$H_A^{\bullet,\bullet}(X) \simeq \ker \tilde{\Delta}_A \; ;$$

in particular, this proves that the Aeppli cohomology groups of a compact complex manifold are finite-dimensional $\mathbb{C}$-vector spaces.

Furthermore, as for the Bott-Chern cohomology, if $\{X_t\}_{t \in B}$ is a complex-analytic family of compact complex manifolds, with B a complex manifold, then, for every $(p,q) \in \mathbb{N}^2$, the function $B \ni t \mapsto \dim_{\mathbb{C}} H_A^{p,q}(X_t) \in \mathbb{N}$ is upper-semi-continuous.

Once again, whenever X is a compact Kähler manifold, by using the Kähler identities, one has

$$\tilde{\Delta}_A = \overline{\Box}^2 + \partial\partial^* + \partial\partial^{*} \; ;$$

indeed, recall that $\overline{\Box} = \Box$ and that $\partial^*\overline{\partial} = \mathrm{i}\left[\Lambda,\overline{\partial}\right]\overline{\partial} = -\mathrm{i}\,\overline{\partial}\,\Lambda\,\overline{\partial} = -\mathrm{i}\,\overline{\partial}$ $\left[\Lambda,\overline{\partial}\right] = -\overline{\partial}\partial^*$, and hence $\overline{\partial}^*\partial = -\partial\overline{\partial}^*$; therefore

$$\overline{\Box}^2 = \overline{\Box}\,\Box = \overline{\partial}\overline{\partial}^*\partial\partial^* + \overline{\partial}\overline{\partial}^*\partial^*\partial + \overline{\partial}^*\overline{\partial}\partial\partial^* + \overline{\partial}^*\overline{\partial}\partial^*\partial$$

$$= -\overline{\partial}\partial\overline{\partial}^*\partial^* - \overline{\partial}\partial^*\overline{\partial}^*\partial - \overline{\partial}^*\partial\overline{\partial}\partial^* - \overline{\partial}^*\partial^*\overline{\partial}\partial$$

$$= \partial\overline{\partial}\overline{\partial}^*\partial^* + \overline{\partial}\partial^*\partial\overline{\partial}^* + \partial\overline{\partial}\overline{\partial}^*\partial^* + \overline{\partial}^*\partial^*\partial\overline{\partial}$$

$$= \tilde{\Delta}_A - \partial\partial^* - \overline{\partial}\overline{\partial}^* \; .$$

In particular, it follows that, on a compact Kähler manifold, $\ker \tilde{\Delta}_A = \ker \overline{\Box} = \ker \Delta$, and hence the de Rham cohomology, the Dolbeault cohomology, and the Aeppli cohomology are isomorphic (actually, since the $\partial\overline{\partial}$-Lemma holds on every compact Kähler manifold, one gets an isomorphism that does not depend on the choice of the Hermitian metric).

In fact, since $\ker \tilde{\Delta}_{BC} = \ker \partial \cap \ker \overline{\partial} \cap \ker \overline{\partial}^*\partial^*$ and $\ker \tilde{\Delta}_A = \ker \partial\overline{\partial} \cap \ker \partial^* \cap \ker \overline{\partial}^*$, one has the following isomorphism between the Bott-Chern cohomology and the Aeppli cohomology.

Theorem 2.5 ([Sch07, Sect. 2.c]). *Let X be a compact complex manifold of complex dimension n. For any $(p,q) \in \mathbb{N}^2$, the Hodge-$*$-operator associated to a Hermitian metric on X induces an isomorphism,*

$$*: H_{BC}^{p,q}(X) \xrightarrow{\simeq} H_A^{n-q,n-p}(X) \; ,$$

between the Bott-Chern and the Aeppli cohomologies.

Remark 2.2. We refer to [Dem12, Sect. VI.12], [Sch07, Sect. 4], [Koo11, Sects. 3.2 and 3.5] for more details on the sheaf-theoretic interpretation of the Bott-Chern and Aeppli cohomologies.

We just recall that, for any $(p, q) \in \mathbb{N}^2$, by defining the complex $\left(\mathcal{L}_{X\,p,q}^{\bullet}, \, d_{\mathcal{L}_{X\,p,q}^{\bullet}} \right)$ of sheaves as

$$\left(\mathcal{L}_{X\,p,q}^{\bullet}, \, d_{\mathcal{L}_{X\,p,q}^{\bullet}} \right) : \mathcal{A}_X^{0,0} \overset{\mathrm{pr} \circ d}{\longrightarrow} \bigoplus_{\substack{r+s=1 \\ r<p,\,s<q}} \mathcal{A}_X^{r,s} \to \cdots \overset{\mathrm{pr} \circ d}{\longrightarrow} \bigoplus_{\substack{r+s=p+q-2 \\ r<p,\,s<q}} \mathcal{A}_X^{r,s}$$

$$\overset{\partial\bar\partial}{\longrightarrow} \bigoplus_{\substack{r+s=p+q \\ r\geq p,\,s\geq q}} \mathcal{A}_X^{r,s} \overset{d}{\to} \bigoplus_{\substack{r+s=p+q \\ r\geq p,\,s\geq q}} \mathcal{A}_X^{r,s} \to \cdots,$$

where $\mathcal{A}_X^{r,s}$ denotes the sheaf of germs of (r, s)-forms over X for $(r, s) \in \mathbb{N}^2$ and pr denotes the projection onto the appropriate space, and since, for every $k \in \mathbb{N}$, the sheaves $\mathcal{L}_{X\,p,q}^{k}$ are fine (indeed, they are sheaves of $\left(\mathcal{C}_X^{\infty} \otimes_{\mathbb{R}} \mathbb{C} \right)$-modules over a para-compact space), one has, see, e.g., [Dem12, Corollary IV.4.19, (IV.12.9)],

$$\mathbb{H}^{p+q-1} \left(X; \left(\mathcal{L}_{X\,p,q}^{\bullet}, \, d_{\mathcal{L}_{X\,p,q}^{\bullet}} \right) \right) \simeq H_{BC}^{p,q}(X)$$

and

$$\mathbb{H}^{p+q-2} \left(X; \left(\mathcal{L}_{X\,p,q}^{\bullet}, \, d_{\mathcal{L}_{X\,p,q}^{\bullet}} \right) \right) \simeq H_{A}^{p-1,q-1}(X) .$$

We note also that, by the Poincaré lemma (see, e.g., [Dem12, I.1.22, Theorem I.2.24]) and by the Dolbeault and Grothendieck lemma (see, e.g., [Dem12, I.3.29]), one gets M. Schweitzer's lemma [Sch07, Lemme 4.1], and therefore, by [Dem12, Corollary IV.12.6], one gets that, for any $(r, s) \in \mathbb{N}^2$, the natural map $\mathcal{A}_X^{r,s} \to \mathcal{D}_X^{r,s}$, where $\mathcal{D}_X^{r,s}$ denotes the sheaf of germs of (r, s)-currents over X, induces the isomorphisms, [Sch07, Sect. 4.d], [Dem12, Lemma VI.12.1], [Koo11, Proposition 3.4.1],

$$H_{BC}^{p,q}(X) \simeq \frac{\ker \left(\partial : \mathcal{D}^{p,q} X \to \mathcal{D}^{p+1,q} X \right) \cap \ker \left(\bar\partial : \mathcal{D}^{p,q} X \to \mathcal{D}^{p,q+1} X \right)}{\mathrm{im} \left(\partial\bar\partial : \mathcal{D}^{p-1,q-1} X \to \mathcal{D}^{p,q} X \right)},$$

and

$$H_A^{p-1,q-1}(X)$$

$$\simeq \frac{\ker \left(\partial\bar\partial : \mathcal{D}^{p-1,q-1} X \to \mathcal{D}^{p,q} X \right)}{\mathrm{im} \left(\partial : \mathcal{D}^{p-2,q-1} X \to \mathcal{D}^{p-1,q-1} X \right) + \mathrm{im} \left(\bar\partial : \mathcal{D}^{p-1,q-2} X \to \mathcal{D}^{p-1,q-1} X \right)},$$

that is, the Bott-Chern and Aeppli cohomologies of a compact complex manifolds can be computed both by using currents and by using just differential forms.

Finally, we recall that, for any $(p, q) \in \mathbb{N}^2$, by defining the complex $\left(\mathcal{B}_{X\,p,q}^{\bullet}, \, d_{\mathcal{B}_{X\,p,q}^{\bullet}} \right)$ of sheaves as (for simplicity of notation, let us assume $p < q$)

$$\left(\mathcal{B}^\bullet_{X\,p,q},\, d_{\mathcal{B}^\bullet_{X\,p,q}}\right):$$

$$
\begin{array}{ccccccccc}
& & \mathcal{O}_X & \xrightarrow{\;\partial\;} & \Omega^1_X & \xrightarrow{\;\partial\;} & \cdots & \xrightarrow{\;\partial\;} & \Omega^{p-1}_X & \xrightarrow{\;0\;} & 0 & \longrightarrow & 0 \\
& {\scriptstyle +}\nearrow & & & & & & & & & & & \\
\mathbb{C} & & \oplus & & \oplus & & \cdots & & \cdots & & \oplus & & \\
& {\scriptstyle -}\searrow & & & & & & & & & & & \\
& & \overline{\mathcal{O}}_X & \xrightarrow{\;\overline{\partial}\;} & \overline{\Omega}^1_X & \xrightarrow{\;\overline{\partial}\;} & \cdots & & \cdots & \xrightarrow{\;\overline{\partial}\;} & \overline{\Omega}^{q-1}_X & \xrightarrow{\;0\;} & 0
\end{array}
$$

where $\mathcal{O}_X := \Omega^0_X$, respectively $\overline{\mathcal{O}}_X := \overline{\Omega}^0_X$, denotes the sheaf of germs of holomorphic, respectively anti-holomorphic, functions over X, and Ω^k_X, respectively $\overline{\Omega}^k_X$, for $k \in \mathbb{N}$, denotes the sheaf of germs of holomorphic, respectively anti-holomorphic, k-forms over X, one has that the natural map

$$\left(\mathcal{B}^\bullet_{X\,p,q},\, d_{\mathcal{B}^\bullet_{X\,p,q}}\right) \to \left(\mathcal{L}^\bullet_{X\,p,q},\, d_{\mathcal{L}^\bullet_{X\,p,q}}\right)$$

of complexes of sheaves is a quasi-isomorphism, [Sch07, Sect. 4.b], and then

$$H^{p,q}_{BC}(X) \simeq \mathbb{H}^{p+q}\left(X;\left(\mathcal{B}^\bullet_{X\,p,q},\, d_{\mathcal{B}^\bullet_{X\,p,q}}\right)\right).$$

2.1.3 The $\partial\overline{\partial}$-Lemma

Let X be a compact complex manifold, and consider its complex de Rham $H^\bullet_{dR}(X;\mathbb{C})$, Dolbeault $H^{\bullet,\bullet}_{\overline{\partial}}(X)$, conjugate Dolbeault $H^{\bullet,\bullet}_{\partial}(X)$, Bott-Chern $H^{\bullet,\bullet}_{BC}(X)$, and Aeppli $H^{\bullet,\bullet}_A(X)$ cohomologies.

The identity map induces the following natural maps of (bi-)graded $\mathbb{C}$-vector spaces:

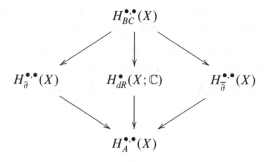

In general, these maps are neither injective nor surjective: see, e.g., the examples in [Sch07, Sect. 1.c] or in Sect. 3.2.4.

By [DGMS75, Lemma 5.15, Proposition 5.17], it turns out that, if one of the above map is an isomorphism, then all the maps are isomorphisms, [DGMS75, Remark 5.16]; this is encoded in the notion of $\partial\bar{\partial}$-*Lemma*, which can be introduced in the more general setting of bounded double complexes of vector spaces.

We start by recalling the following general result by P. Deligne, Ph.A. Griffiths, J. Morgan, and D.P. Sullivan, [DGMS75] (compare also [AT13a, Lemma 1.4]).

Proposition 2.1 ([DGMS75, Lemma 5.15]). *Let* $\left(K^{\bullet,\bullet}, \mathrm{d}', \mathrm{d}''\right)$ *be a bounded double complex of vector spaces (or, more in general, of objects of any Abelian category), and let* $(K^{\bullet}, \mathrm{d})$ *be the associated single complex, where* $\mathrm{d} := \mathrm{d}' + \mathrm{d}''$. *For each* $h \in \mathbb{N}$, *the following conditions are equivalent:*

$(a)_h$ $\ker \mathrm{d}' \cap \ker \mathrm{d}'' \cap \mathrm{im}\, \mathrm{d} = \mathrm{im}\, \mathrm{d}'\, \mathrm{d}''$ *in* K^h;

$(b)_h$ $\ker \mathrm{d}'' \cap \mathrm{im}\, \mathrm{d}' = \mathrm{im}\, \mathrm{d}'\, \mathrm{d}''$ *and* $\ker \mathrm{d}' \cap \mathrm{im}\, \mathrm{d}'' = \mathrm{im}\, \mathrm{d}'\, \mathrm{d}''$ *in* K^h;

$(c)_h$ $\ker \mathrm{d}' \cap \ker \mathrm{d}'' \cap \left(\mathrm{im}\, \mathrm{d}' + \mathrm{im}\, \mathrm{d}''\right) = \mathrm{im}\, \mathrm{d}'\, \mathrm{d}''$ *in* K^h;

$(a^*)_{h-1}$ $\mathrm{im}\, \mathrm{d}' + \mathrm{im}\, \mathrm{d}'' + \ker \mathrm{d} = \ker \mathrm{d}'\, \mathrm{d}''$ *in* K^{h-1};

$(b^*)_{h-1}$ $\mathrm{im}\, \mathrm{d}'' + \ker \mathrm{d}' = \ker \mathrm{d}'\, \mathrm{d}''$ *and* $\mathrm{im}\, \mathrm{d}' + \ker \mathrm{d}'' = \ker \mathrm{d}'\, \mathrm{d}''$ *in* K^{h-1};

$(c^*)_{h-1}$ $\mathrm{im}\, \mathrm{d}' + \mathrm{im}\, \mathrm{d}'' + \left(\ker \mathrm{d}' \cap \ker \mathrm{d}''\right) = \ker \mathrm{d}'\, \mathrm{d}''$ *in* K^{h-1}.

The above equivalent conditions define the validity of the $\mathrm{d}'\, \mathrm{d}''$-Lemma for a double complex.

Definition 2.3 ([DGMS75]). Let $\left(K^{\bullet,\bullet}, \mathrm{d}', \mathrm{d}''\right)$ be a bounded double complex of vector spaces (or, more in general, of objects of any Abelian category), and let $(K^{\bullet}, \mathrm{d})$ be the associated simple complex, where $\mathrm{d} := \mathrm{d}' + \mathrm{d}''$. One says that $\left(K^{\bullet,\bullet}, \mathrm{d}', \mathrm{d}''\right)$ *satisfies the* $\mathrm{d}'\, \mathrm{d}''$-*Lemma* if, for every $h \in \mathbb{N}$,

$$\ker \mathrm{d}' \cap \ker \mathrm{d}'' \cap \mathrm{im}\, \mathrm{d} = \mathrm{im}\, \mathrm{d}'\, \mathrm{d}'' \qquad \text{in } K^h ,$$

equivalently, if, for every $h \in \mathbb{N}$, the equivalent conditions in [DGMS75, Lemma 5.15] hold.

The following result by P. Deligne, Ph.A. Griffiths, J. Morgan, and D.P. Sullivan, [DGMS75], gives a characterization for the validity of the $\mathrm{d}'\, \mathrm{d}''$-Lemma.

Theorem 2.6 ([DGMS75, Proposition 5.17]). *Let* $\left(K^{\bullet,\bullet}, \mathrm{d}', \mathrm{d}''\right)$ *be a bounded double complex of vector spaces, and let* $(K^{\bullet}, \mathrm{d})$ *be the associated simple complex, where* $\mathrm{d} := \mathrm{d}' + \mathrm{d}$. *The following conditions are equivalent:*

(i) $\left(K^{\bullet,\bullet}, \mathrm{d}', \mathrm{d}''\right)$ *satisfies the* $\mathrm{d}'\, \mathrm{d}''$-*Lemma;*

(ii) $K^{\bullet,\bullet}$ *is a sum of double complexes of the following two types (see Figure 2.1):*

    (dots) complexes which have only a single component, with $\mathrm{d}' = 0$ *and* $\mathrm{d}'' = 0$;

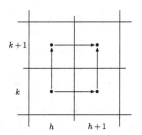

Diagram of a dot in a double complex.   Diagram of a square in a double complex.

Fig. 2.1 Diagrams of (**a**) a dot and of (**b**) a square in a double complex

(squares) complexes which are a square of isomorphisms,

$$
\begin{array}{ccc}
K^{p-1,q} & \xrightarrow{\ d'\ } & K^{p,q} \\[4pt]
d'' \Big\uparrow \simeq & & d'' \Big\uparrow \simeq \\[4pt]
K^{p-1,q-1} & \xrightarrow{\ d'\ } & K^{p,q-1}
\end{array}
$$

(iii) the spectral sequence defined by the filtration associated to either degree (denoted by $'F$ or $''F$) degenerates at E_1 (namely, $E_1 = E_\infty$) and, for every $h \in \mathbb{N}$, the two induced filtrations are h-opposite on $H^h_{dR}(X;\mathbb{C})$, i.e., $'F^p \oplus ''F^q \xrightarrow{\sim} H^h_{dR}(X;\mathbb{C})$ for $p + q - 1 = h$.

In particular, we are interested in dealing with compact complex manifolds X, for which one considers the double complex $\left(\wedge^{\bullet,\bullet}X, \partial, \bar{\partial}\right)$.

Definition 2.4 ([DGMS75]). A compact complex manifold X is said to *satisfy the $\partial\bar{\partial}$-Lemma* if $\left(\wedge^{\bullet,\bullet}X, \partial, \bar{\partial}\right)$ satisfies the $\partial\bar{\partial}$-Lemma, namely, if

$$
\ker \partial \cap \ker \bar{\partial} \cap \operatorname{im} d = \operatorname{im} \partial\bar{\partial} ,
$$

that is, in other words, if the natural map $H^{\bullet,\bullet}_{BC}(X) \to H^\bullet_{dR}(X;\mathbb{C})$ of graded $\mathbb{C}$-vector spaces induced by the identity is injective.

Remark 2.3. Let X be a compact complex manifold. By considering the differential operator

$$
d^c := -i\left(\partial - \bar{\partial}\right) ,
$$

one can say that X *satisfies the $d\,d^c$-Lemma*, by definition, if

$$
\operatorname{im} d \cap \ker d^c = \operatorname{im} d\,d^c .
$$

Since $d\,d^c = 2\,i\,\partial\bar{\partial}$, and $\partial = \frac{1}{2}\,(d + i\,d^c)$ and $\bar{\partial} = \frac{1}{2}\,(d - i\,d^c)$, one has

$$\ker d \cap \ker d^c \;=\; \ker \partial \cap \ker \bar{\partial} \qquad \text{and} \qquad \operatorname{im} d\,d^c = \operatorname{im} \partial\bar{\partial}\,;$$

and hence X satisfies the $d\,d^c$-Lemma if and only if X satisfies the $\partial\bar{\partial}$-Lemma.

For compact complex manifolds, P. Deligne, Ph.A. Griffiths, J. Morgan, and D.P. Sullivan's characterization [DGMS75, Proposition 5.17] is rewritten as follows.

Theorem 2.7 ([DGMS75, 5.21]). *A compact complex manifold X satisfies the $\partial\bar{\partial}$-Lemma if and only if (i) the Hodge and Frölicher spectral sequence degenerates at the first step (that is, $E_1 \simeq E_\infty$), and (ii) the natural filtration on $\left(\wedge^{\bullet,\bullet} X,\, \partial,\, \bar{\partial}\right)$ induces, for every $k \in \mathbb{N}$, a Hodge structure of weight k on $H^k_{dR}(X;\mathbb{C})$ (that is, $H^k_{dR}(X;\mathbb{C}) = \bigoplus_{p+q=k} F^p H^k_{dR}(X;\mathbb{C}) \cap \bar{F}^q H^k_{dR}(X;\mathbb{C})$, where $F^\bullet H^\bullet_{dR}(X;\mathbb{C})$ is the filtration induced by $F^\bullet \wedge^{\bullet_1,\bullet_2} X := \bigoplus_{p \geq \bullet,\, q} \wedge^{p,q} X$ on $H^\bullet_{dR}(X;\mathbb{C})$ and $\bar{F}^\bullet H^\bullet_{dR}(X;\mathbb{C})$ is the conjugated filtration to $F^\bullet H^\bullet_{dR}(X;\mathbb{C})$).*

Another characterization for the validity of the $\partial\bar{\partial}$-Lemma, in terms of the dimensions of the Bott-Chern cohomology, will be given in Theorem 2.14.

Actually, as already mentioned, if a compact complex manifold satisfies the $\partial\bar{\partial}$-Lemma, then all the natural maps between cohomologies induced by the identity turn out to be isomorphisms.

Theorem 2.8 ([DGMS75, Lemma 5.15, Remarks 5.16, 5.21]). *A compact complex manifold X satisfies the $\partial\bar{\partial}$-Lemma if and only if all the natural maps*

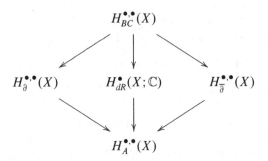

induced by the identity are isomorphisms.

We recall that if X is a compact complex manifold endowed with a Kähler structure, then X satisfies the $\partial\bar{\partial}$-Lemma, [DGMS75, Lemma 5.11]. Moreover, one has the following result.

Theorem 2.9 ([DGMS75, Theorem 5.22]). *Let X and Y be compact complex manifolds of the same dimension, and let $f\colon X \to Y$ be a holomorphic birational map. If X satisfies the $\partial\bar{\partial}$-Lemma, then also Y satisfies the $\partial\bar{\partial}$-Lemma.*

Indeed, one has that, if X and Y are complex manifolds of the same dimension, and $\pi\colon X \to Y$ is a proper surjective holomorphic map, then the maps

$$\pi^*\colon H_{dR}^\bullet(Y;\mathbb{C}) \to H_{dR}^\bullet(X;\mathbb{C}) \qquad \text{and} \qquad \pi^*\colon H_{\bar{\partial}}^{\bullet,\bullet}(Y) \to H_{\bar{\partial}}^{\bullet,\bullet}(X)$$

induced by $\pi\colon X \to Y$ are injective, see, e.g., [Wel74, Theorem 3.1]; then one can use the characterization in [DGMS75, 5.21].

In particular, it follows that *Moĭšezon manifolds* (that is, compact complex manifolds X such that the degree of transcendence over $\mathbb{C}$ of the field of meromorphic functions over X is equal to the complex dimension of X, [Moĭ66], equivalently, compact complex manifolds admitting a proper modification[1] from a projective manifold, [Moĭ66, Theorem 1]), and, more in general, manifolds *in class $\mathcal{C}$ of Fujiki* (that is, compact complex manifolds admitting a proper modification from a Kähler manifold, [Fuj78]) satisfy the $\partial\bar{\partial}$-Lemma.

Corollary 2.1 ([DGMS75, Lemma 5.11, Corollary 5.23]). *The $\partial\bar{\partial}$-Lemma holds for compact Kähler manifolds, for Moĭšezon manifolds, and for manifolds in class $\mathcal{C}$ of Fujiki.*

Remark 2.4. In [Hir62], H. Hironaka provided an example of a non-Kähler Moĭšezon manifold of complex dimension 3 with arbitrary small deformations being projective (in fact, as proven by D. Popovici, the limit of a punctured-disk family of projective manifolds is Moĭšezon under additional conditions, [Pop13]; see also [Pop09, Theorem 1.1], [Pop10, Theorem 1.1]); in particular, H. Hironaka's manifold provides an example of a non-Kähler manifold satisfying the $\partial\bar{\partial}$-Lemma. Studying twistor spaces, C. LeBrun and Y.S. Poon, and F. Campana, showed that being in class $\mathcal{C}$ of Fujiki is not a stable property under small deformations of the complex structures, [LP92, Theorem 1], [Cam91, Corollary 3.13]; since the property of satisfying the $\partial\bar{\partial}$-Lemma is stable under small deformations of the complex structure, Corollary 2.2, or [Voi02, Proposition 9.21], or [Wu06, Theorem 5.12], or [Tom08, Sect. B], C. LeBrun and Y.S. Poon's, and F. Campana's, result yields examples of compact complex manifolds satisfying the $\partial\bar{\partial}$-Lemma and not belonging to class $\mathcal{C}$ of Fujiki.

Remark 2.5. For some results on Hodge decomposition and $\partial\bar{\partial}$-Lemma for solvmanifolds, obtained by H. Kasuya, see [Kas11, Kas12b].

Finally, we recall the following obstructions to the existence of complex structures satisfying the $\partial\bar{\partial}$-Lemma on a compact (differentiable) manifold.

Theorem 2.10 ([DGMS75, Main Theorem, Corollary 1]). *Let X be a compact manifold. If X admits a complex structure such that the $\partial\bar{\partial}$-Lemma holds, then the differential graded algebra $(\wedge^\bullet X, \mathrm{d})$ is formal. In particular, all the Massey products of any order are zero.*

---

[1] We recall that a proper holomorphic map $f\colon X \to Y$ from the complex manifold X to the complex manifold Y is called a *modification* if there exists a nowhere dense closed analytic subset $B \subset Y$ such that $f\lfloor_{X \setminus f^{-1}(B)}\colon X \setminus f^{-1}(B) \to Y \setminus B$ is a biholomorphism.

Indeed, if X satisfies the $\partial\bar{\partial}$-Lemma, equivalently, the $d\,d^c$-Lemma, then the inclusion $\ker d^c \to \wedge^\bullet X$ and the projection $\ker d^c \to \frac{\ker d^c}{\operatorname{im} d^c}$ induce the quasi-isomorphisms

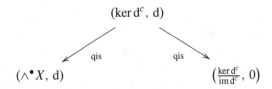

$$(\ker d^c, d)$$
$$\swarrow \text{qis} \qquad\qquad \text{qis} \searrow$$
$$(\wedge^\bullet X, d) \qquad\qquad\qquad\qquad \left(\tfrac{\ker d^c}{\operatorname{im} d^c}, 0\right)$$

of differential graded algebras, proving that $(\wedge^\bullet X, d)$ is equivalent to $\left(\frac{\ker d^c}{\operatorname{im} d^c}, 0\right)$, and hence formal.

2.2   Cohomological Properties of Compact Complex Manifolds and the $\partial\bar{\partial}$-Lemma

In this section, we study some cohomological properties of compact complex manifolds, especially in relation with the $\partial\bar{\partial}$-Lemma, see [AT13b]. More precisely, we prove an inequality *à la* Frölicher for the Bott-Chern cohomology, Theorem 2.13, and we characterize the validity of the $\partial\bar{\partial}$-Lemma in terms of the dimensions of the Bott-Chern cohomology groups, Theorem 2.14.

Let X be a compact complex manifold of complex dimension n.

As a matter of notation, for every $(p,q) \in \mathbb{N}^2$, for every $k \in \mathbb{N}$, and for $\sharp \in \left\{\bar{\partial}, \partial, BC, A\right\}$, we will denote

$$h_\sharp^{p,q} := \dim_\mathbb{C} H_\sharp^{p,q}(X) \in \mathbb{N} \qquad \text{and} \qquad h_\sharp^k := \sum_{p+q=k} h_\sharp^{p,q} \in \mathbb{N},$$

while recall that the Betti numbers are denoted by

$$b_k := \dim_\mathbb{C} H_{dR}^k(X; \mathbb{C}) \in \mathbb{N}.$$

Recall that, for every $(p,q) \in \mathbb{N}^2$, the conjugation induces the isomorphisms $H_{BC}^{p,q}(X) \xrightarrow{\simeq} H_{BC}^{q,p}(X)$, and $H_A^{p,q}(X) \xrightarrow{\simeq} H_A^{q,p}(X)$, and $H_{\bar{\partial}}^{p,q}(X) \xrightarrow{\simeq} H_\partial^{q,p}(X)$, and the Hodge-$*$-operator associated to any given Hermitian metric induces the isomorphisms $H_{BC}^{p,q}(X) \xrightarrow{\simeq} H_A^{n-q,n-p}(X)$ and $H_{\bar{\partial}}^{p,q}(X) \xrightarrow{\simeq} H_\partial^{n-q,n-p}(X)$; hence, for every $(p,q) \in \mathbb{N}^2$, one has the equalities

$$h_{BC}^{p,q} = h_{BC}^{q,p} = h_A^{n-p,n-q} = h_A^{n-q,n-p} \qquad \text{and}$$
$$h_{\bar{\partial}}^{p,q} = h_\partial^{q,p} = h_{\bar{\partial}}^{n-p,n-q} = h_\partial^{n-q,n-p},$$

and therefore, for every $k \in \mathbb{N}$, one has the equalities

$$h_{BC}^k = h_A^{2n-k} \quad \text{and} \quad h_{\bar{\partial}}^k = h_{\partial}^k = h_{\bar{\partial}}^{2n-k} = h_{\partial}^{2n-k} \; ;$$

Recall also that the Hodge-$*$-operator (of any given Riemannian metric and volume form on X) yields, for every $k \in \mathbb{N}$, the isomorphism $H_{dR}^k(X;\mathbb{R}) \xrightarrow{\simeq} H_{dR}^{2n-k}(X;\mathbb{R})$, and hence the equality

$$b_k = b_{2n-k} \, .$$

2.2.1 *J. Varouchas' Exact Sequences*

In order to prove a inequality *à la* Frölicher for the Bott-Chern and Aeppli cohomologies and to give therefore a characterization of compact complex manifolds satisfying the $\partial\bar{\partial}$-Lemma in terms of the dimensions of their Bott-Chern cohomology groups, we need to recall two exact sequences from [Var86].

Following J. Varouchas, one defines the (finite-dimensional) bi-graded $\mathbb{C}$-vector spaces

$$A^{\bullet,\bullet} := \frac{\operatorname{im}\bar{\partial} \cap \operatorname{im}\partial}{\operatorname{im}\partial\bar{\partial}} \, , \qquad B^{\bullet,\bullet} := \frac{\ker\bar{\partial} \cap \operatorname{im}\partial}{\operatorname{im}\partial\bar{\partial}} \, , \qquad C^{\bullet,\bullet} := \frac{\ker\partial\bar{\partial}}{\ker\bar{\partial} + \operatorname{im}\partial}$$

and

$$D^{\bullet,\bullet} := \frac{\operatorname{im}\bar{\partial} \cap \ker\partial}{\operatorname{im}\partial\bar{\partial}} \, , \qquad E^{\bullet,\bullet} := \frac{\ker\partial\bar{\partial}}{\ker\partial + \operatorname{im}\bar{\partial}} \, , \qquad F^{\bullet,\bullet} := \frac{\ker\partial\bar{\partial}}{\ker\bar{\partial} + \ker\partial} \, .$$

For every $(p,q) \in \mathbb{N}^2$ and $k \in \mathbb{N}$, we will denote their dimensions by

$$a^{p,q} := \dim_{\mathbb{C}} A^{p,q} \, , \qquad b^{p,q} := \dim_{\mathbb{C}} B^{p,q} \, , \qquad c^{p,q} := \dim_{\mathbb{C}} C^{p,q} \, ,$$

$$d^{p,q} := \dim_{\mathbb{C}} D^{p,q} \, , \qquad e^{p,q} := \dim_{\mathbb{C}} E^{p,q} \, , \qquad f^{p,q} := \dim_{\mathbb{C}} F^{p,q} \, ,$$

and

$$a^k := \sum_{p+q=k} a^{p,q} \, , \qquad b^k := \sum_{p+q=k} b^{p,q} \, , \qquad c^k := \sum_{p+q=k} c^{p,q} \, ,$$

$$d^k := \sum_{p+q=k} d^{p,q} \, , \qquad e^k := \sum_{p+q=k} e^{p,q} \, , \qquad f^k := \sum_{p+q=k} f^{p,q} \, .$$

The previous vector spaces give the following exact sequences, by J. Varouchas.

Theorem 2.11 ([Var86, Sect. 3.1]). *The sequences*

$$0 \to A^{\bullet,\bullet} \to B^{\bullet,\bullet} \to H_{\bar\partial}^{\bullet,\bullet}(X) \to H_A^{\bullet,\bullet}(X) \to C^{\bullet,\bullet} \to 0 \qquad (2.1)$$

and

$$0 \to D^{\bullet,\bullet} \to H_{BC}^{\bullet,\bullet}(X) \to H_{\bar\partial}^{\bullet,\bullet}(X) \to E^{\bullet,\bullet} \to F^{\bullet,\bullet} \to 0 \qquad (2.2)$$

are exact sequences of finite-dimensional bi-graded $\mathbb{C}$-vector spaces.

Proof. We first prove the exactness of (2.1). Since $\operatorname{im}\bar\partial \subseteq \ker\bar\partial$, the map $A^{\bullet,\bullet} \to B^{\bullet,\bullet}$ is injective. The kernel of the map $B^{\bullet,\bullet} \to H_{\bar\partial}^{\bullet,\bullet}(X)$ is $\frac{\ker\bar\partial \cap \operatorname{im}\partial \cap \operatorname{im}\bar\partial}{\operatorname{im}\partial\bar\partial} = \frac{\operatorname{im}\partial \cap \operatorname{im}\bar\partial}{\operatorname{im}\partial\bar\partial}$, that is, the image of the map $A^{\bullet,\bullet} \to B^{\bullet,\bullet}$. The kernel of the map $H_{\bar\partial}^{\bullet,\bullet}(X) \to H_A^{\bullet,\bullet}(X)$ is $\frac{\ker\bar\partial \cap \operatorname{im}\partial}{\operatorname{im}\bar\partial}$, that is, the image of the map $B^{\bullet,\bullet} \to H_{\bar\partial}^{\bullet,\bullet}(X)$. The kernel of the map $H_A^{\bullet,\bullet}(X) \to C^{\bullet,\bullet}$ is $\frac{\ker\bar\partial \cap \ker\partial\bar\partial}{\operatorname{im}\partial + \operatorname{im}\bar\partial} = \frac{\ker\bar\partial}{\operatorname{im}\partial + \operatorname{im}\bar\partial}$, that is, the image of the map $H_{\bar\partial}^{\bullet,\bullet}(X) \to H_A^{\bullet,\bullet}(X)$. Finally, since $\operatorname{im}\partial + \operatorname{im}\bar\partial \subseteq \ker\bar\partial + \operatorname{im}\partial$, the map $H_A^{\bullet,\bullet}(X) \to C^{\bullet,\bullet}$ is surjective. In particular, since $H_A^{\bullet,\bullet}(X) \to C^{\bullet,\bullet}$ is surjective, then $C^{\bullet,\bullet}$ has finite dimension; since the identity induces an injective map $B^{\bullet,\bullet} \to H_{BC}^{\bullet,\bullet}(X)$, then $B^{\bullet,\bullet}$ has finite dimension; hence, since $A^{\bullet,\bullet} \to B^{\bullet,\bullet}$ is injective, then also $A^{\bullet,\bullet}$ has finite dimension.

We prove now the exactness of (2.2). Since $\operatorname{im}\bar\partial \subseteq \ker\bar\partial$, the map $D^{\bullet,\bullet} \to H_{BC}^{\bullet,\bullet}(X)$ is injective. The kernel of the map $H_{BC}^{\bullet,\bullet}(X) \to H_{\bar\partial}^{\bullet,\bullet}(X)$ is $\frac{\ker\partial \cap \ker\bar\partial \cap \operatorname{im}\bar\partial}{\operatorname{im}\partial\bar\partial} = \frac{\operatorname{im}\bar\partial \cap \ker\partial}{\operatorname{im}\partial\bar\partial}$, that is, the image of the map $D^{\bullet,\bullet} \to H_{BC}^{\bullet,\bullet}(X)$. The kernel of the map $H_{\bar\partial}^{\bullet,\bullet}(X) \to E^{\bullet,\bullet}$ is $\frac{\ker\bar\partial \cap (\ker\partial + \operatorname{im}\bar\partial)}{\operatorname{im}\bar\partial} = \frac{\ker\bar\partial \cap \ker\partial}{\operatorname{im}\bar\partial}$, that is, the image of the map $H_{BC}^{\bullet,\bullet}(X) \to H_{\bar\partial}^{\bullet,\bullet}(X)$. The kernel of the map $E^{\bullet,\bullet} \to F^{\bullet,\bullet}$ is $\frac{\ker\partial\bar\partial \cap (\ker\bar\partial + \ker\partial)}{\ker\partial + \operatorname{im}\bar\partial} = \frac{\ker\partial\bar\partial \cap \ker\bar\partial}{\ker\partial + \operatorname{im}\bar\partial}$, that is, the image of the map $H_{\bar\partial}^{\bullet,\bullet}(X) \to E^{\bullet,\bullet}$. Finally, since $\ker\partial + \operatorname{im}\bar\partial \subseteq \ker\bar\partial + \ker\partial$, the map $E^{\bullet,\bullet} \to F^{\bullet,\bullet}$ is surjective. In particular, since $D^{\bullet,\bullet} \to H_{BC}^{\bullet,\bullet}(X)$ is injective, then $D^{\bullet,\bullet}$ has finite dimension; since the identity induces a surjective map $H_A^{\bullet,\bullet}(X) \to E^{\bullet,\bullet}$, then $E^{\bullet,\bullet}$ has finite dimension; hence, since $E^{\bullet,\bullet} \to F^{\bullet,\bullet}$ is surjective, then also $F^{\bullet,\bullet}$ has finite dimension. □

Note, [Var86, Sect. 3.1], that the conjugation yields, for every $(p,q) \in \mathbb{N}^2$, the equalities

$$a^{p,q} = a^{q,p}, \qquad f^{p,q} = f^{q,p}, \qquad d^{p,q} = b^{q,p}, \qquad e^{p,q} = c^{q,p}, \qquad (2.3)$$

and the isomorphisms $\bar\partial \colon C^{\bullet,\bullet} \xrightarrow{\simeq} D^{\bullet,\bullet+1}$ and $\partial \colon E^{\bullet,\bullet} \xrightarrow{\simeq} B^{\bullet+1,\bullet}$ yield the equalities

$$c^{p,q} = d^{p,q+1}, \qquad e^{p,q} = b^{p+1,q};$$

hence, for every $k \in \mathbb{N}$, one gets the equalities

$$d^k = b^k, \qquad e^k = c^k, \qquad \text{and} \qquad c^k = d^{k+1}, \qquad e^k = b^{k+1}.$$

Remark 2.6. Following the same argument used in [Sch07] to prove the duality between Bott-Chern and Aeppli cohomology groups, we can prove the duality between $A^{\bullet,\bullet}$ and $F^{\bullet,\bullet}$, and, similarly, between $C^{\bullet,\bullet}$ and $\overline{D}^{\bullet,\bullet}$.

Indeed, note that the pairing

$$A^{\bullet,\bullet} \times F^{\bullet,\bullet} \to \mathbb{C}, \qquad ([\alpha], [\beta]) \mapsto \int_X \alpha \wedge \overline{\beta},$$

is non degenerate: choose a Hermitian metric g on X; if $[\alpha] \in A^{\bullet,\bullet} \subseteq H_{BC}^{\bullet,\bullet}(X)$, then there exists a $\tilde\Delta_{BC}$-harmonic representative $\tilde\alpha$ in $[\alpha] \in A^{\bullet,\bullet}$, by [Sch07, Corollaire 2.3], that is, $\partial\tilde\alpha = \bar\partial\tilde\alpha = \partial\bar\partial * \tilde\alpha = 0$; hence, $[*\tilde\alpha] \in F^{\bullet,\bullet}$, and $([\tilde\alpha], [*\tilde\alpha]) = \int_X \tilde\alpha \wedge *\overline{\tilde\alpha}$ is zero if and only if $\tilde\alpha$ is zero if and only if $[\alpha] \in A^{\bullet,\bullet}$ is zero.

Analogously, the pairing

$$C^{\bullet,\bullet} \times \overline{D}^{\bullet,\bullet} \to \mathbb{C}, \qquad ([\alpha], [\beta]) \mapsto \int_X \alpha \wedge \overline{\beta},$$

is non-degenerate: indeed, choose a Hermitian metric g on X; if $[\alpha] \in \overline{D}^{\bullet,\bullet} \subseteq \overline{H_{BC}^{\bullet,\bullet}(X)}$, then there exists a $\tilde\Delta_{BC}$-harmonic representative $\tilde\alpha$ in $[\alpha] \in \overline{D}^{\bullet,\bullet}$, by [Sch07, Corollaire 2.3], that is, $\partial\tilde\alpha = \bar\partial\tilde\alpha = \partial\bar\partial * \tilde\alpha = 0$; hence, $[*\tilde\alpha] \in C^{\bullet,\bullet}$, and $([\tilde\alpha], [*\tilde\alpha]) = \int_X \tilde\alpha \wedge *\overline{\tilde\alpha}$ is zero if and only if $\tilde\alpha$ is zero if and only if $[\alpha] \in \overline{D}^{\bullet,\bullet}$ is zero.

2.2.2 An Inequality à la *Frölicher for the Bott-Chern Cohomology*

We can now state and prove an inequality *à la* Frölicher for the Bott-Chern and Aeppli cohomologies, Theorem 2.13.

Firstly, we recall that, on a compact complex manifold X, the *Frölicher inequality* [Frö55, Theorem 2] relates the Hodge numbers and the Betti numbers.

Theorem 2.12 ([Frö55, Theorem 2]). *Let X be a compact complex manifold. Then, for every $k \in \mathbb{N}$, the following inequality holds:*

$$\sum_{p+q=k} \dim_{\mathbb{C}} H_{\bar\partial}^{p,q}(X) \geq \dim_{\mathbb{C}} H_{dR}^k(X;\mathbb{C}).$$

Table 2.1 Dimensions of cohomologies for the small deformations of the Iwasawa manifold

Classes	$h^1_{\bar{\partial}}$	h^1_{BC}	h^1_A	$h^2_{\bar{\partial}}$	h^2_{BC}	h^2_A	$h^3_{\bar{\partial}}$	h^3_{BC}	h^3_A	$h^4_{\bar{\partial}}$	h^4_{BC}	h^4_A	$h^5_{\bar{\partial}}$	h^5_{BC}	h^5_A
(i)	5	4	6	11	10	12	14	14	14	11	12	10	5	6	4
(ii.a)	4	4	6	9	8	11	12	14	14	9	11	8	4	6	4
(ii.b)	4	4	6	9	8	10	12	14	14	9	10	8	4	6	4
(iii.a)	4	4	6	8	6	11	10	14	14	8	11	6	4	6	4
(iii.b)	4	4	6	8	6	10	10	14	14	8	10	6	4	6	4
	$b_1 = 4$			$b_2 = 8$			$b_3 = 10$			$b_4 = 8$			$b_5 = 4$		

The equality $\sum_{p+q=k} \dim_{\mathbb{C}} H^{p,q}_{\bar{\partial}}(X) = \dim_{\mathbb{C}} H^k_{dR}(X;\mathbb{C})$ holds for every $k \in \mathbb{N}$ if and only if the Hodge and Frölicher spectral sequence $\{(E_r, d_r)\}_{r \in \mathbb{N}}$ degenerates at the first step.

It is in general not true that h^k_{BC} (respectively, h^k_A) is higher than the kth Betti number of X for every $k \in \mathbb{N}$: an example is provided by the small deformations of the Iwasawa manifold $\mathbb{I}_3 := \mathbb{H}(3;\mathbb{Z}[i]) \backslash \mathbb{H}(3;\mathbb{C})$ (see Sect. 3.2.1). In Table 2.1, we summarize the dimensions of the Bott-Chern and Aeppli cohomology groups for $\mathbb{I}_3$ (which have been computed in [Sch07, Proposition 1.2]) and for the small deformations of $\mathbb{I}_3$ (see Sect. 3.2.4). We recall that the small deformations of the Iwasawa manifold, according to I. Nakamura's classification, [Nak75, Sect. 3], are divided into three classes, *(i)*, *(ii)*, and *(iii)*, in terms of their Hodge numbers; it turns out that the Bott-Chern cohomology yields a finer classification of the Kuranishi space of $\mathbb{I}_3$, allowing a further subdivision of class *(ii)*, respectively class *(iii)*, into subclasses *(ii.a)*, *(ii.b)*, respectively *(iii.a)*, *(iii.b)*, see Sect. 3.2.1.2.

The following result gives an inequality *à la* Frölicher for the Bott-Chern cohomology. (We recall that, on a compact complex manifold X of complex dimension n, for any $(p,q) \in \mathbb{N}^2$, one has the equality $\dim_{\mathbb{C}} H^{p,q}_{BC}(X) = \dim_{\mathbb{C}} H^{n-q,n-p}_A(X)$, and, for any $k \in \mathbb{N}$, the equality $\sum_{p+q=k} \dim_{\mathbb{C}} H^{p,q}_{BC}(X) = \sum_{r+s=2n-k} \dim_{\mathbb{C}} H^{r,s}_A(X)$, see [Sch07, Sect. 2.c].)

Theorem 2.13 ([AT13b, Theorem A]). *Let X be a compact complex manifold. Then, for every $(p,q) \in \mathbb{N}^2$, the following inequality holds:*

$$\dim_{\mathbb{C}} H^{p,q}_{BC}(X) + \dim_{\mathbb{C}} H^{p,q}_A(X) \geq \dim_{\mathbb{C}} H^{p,q}_{\bar{\partial}}(X) + \dim_{\mathbb{C}} H^{p,q}_{\partial}(X) . \quad (2.4)$$

In particular, for every $k \in \mathbb{N}$, the following inequality holds:

$$\sum_{p+q=k} \left(\dim_{\mathbb{C}} H^{p,q}_{BC}(X) + \dim_{\mathbb{C}} H^{p,q}_A(X) \right) \geq 2 \dim_{\mathbb{C}} H^k_{dR}(X;\mathbb{C}) . \quad (2.5)$$

Proof. Fix $(p,q) \in \mathbb{N}^2$. The exact sequences (2.1), respectively (2.2), yield the equality

$$h^{p,q}_A = h^{p,q}_{\bar{\partial}} + c^{p,q} + a^{p,q} - b^{p,q} ,$$

respectively

$$h_{BC}^{p,q} = h_{\bar{\partial}}^{p,q} + d^{p,q} + f^{p,q} - e^{p,q} \; ;$$

using also the symmetries $h_A^{p,q} = h_A^{q,p}$ and $h_{\bar{\partial}}^{p,q} = h_{\partial}^{q,p}$, and the equalities (2.3), we get[2]

$$h_{BC}^{p,q} + h_A^{p,q} = h_{BC}^{p,q} + h_A^{q,p}$$

$$= h_{\bar{\partial}}^{p,q} + h_{\bar{\partial}}^{q,p} + f^{p,q} + a^{q,p} + d^{p,q} - b^{q,p} - e^{p,q} + c^{q,p}$$

$$= h_{\bar{\partial}}^{p,q} + h_{\partial}^{p,q} + f^{p,q} + a^{p,q}$$

$$\geq h_{\bar{\partial}}^{p,q} + h_{\partial}^{p,q} \; ,$$

which proves (2.4).

Now, fix $k \in \mathbb{N}$; summing over $(p, q) \in \mathbb{N} \times \mathbb{N}$ such that $p + q = k$, we get

$$h_{BC}^k + h_A^k = \sum_{p+q=k} \left(h_{BC}^{p,q} + h_A^{p,q} \right)$$

$$\geq \sum_{p+q=k} \left(h_{\bar{\partial}}^{p,q} + h_{\partial}^{p,q} \right) = h_{\bar{\partial}}^k + h_{\partial}^k$$

$$\geq 2 b_k \; ,$$

from which we get (2.5). □

Remark 2.7. Note that small deformations of the Iwasawa manifold show that both the inequalities (2.4) and (2.5) can be strict.

For example, for small deformations of $\mathbb{I}_3$ in class *(i)*, one has,

$$h_{BC}^1 + h_A^1 = 10 > 8 = 2 \cdot b_1 \; , \qquad h_{BC}^2 + h_A^2 = 22 > 16 = 2 \cdot b_2 \; ,$$

$$h_{BC}^3 + h_A^3 = 28 > 20 = 2 \cdot b_3 \; ,$$

showing that (2.5) is strict for every $k \in \{1, 2, 3, 4, 5\}$.

On the other hand, for small deformations of $\mathbb{I}_3$ in class *(ii)* or in class *(iii)*, one has

$$h_{BC}^{1,0} + h_A^{1,0} = \frac{1}{2} \left(h_{BC}^1 + h_A^1 \right) = 5 > 4 = h_{\bar{\partial}}^1 = h_{\bar{\partial}}^{1,0} + h_{\bar{\partial}}^{0,1} = h_{\bar{\partial}}^{1,0} + h_{\partial}^{1,0} \; ,$$

showing that (2.4) is strict, for example, for $(p, q) = (1, 0)$.

---

[2] We get actually that, for every $k \in \mathbb{N}$, it holds $h_{BC}^k + h_A^k = 2 h_{\bar{\partial}}^k + a^k + f^k$.

(For further examples among the small deformations of the Iwasawa manifold, compare the computations in Sect. 3.2.4, which are summarized in Table 3.5.)

2.2.3 A Characterization of the $\partial\bar{\partial}$-Lemma in Terms of the Bott-Chern Cohomology

This section is devoted to give a characterization of the validity of the $\partial\bar{\partial}$-Lemma in terms of the Bott-Chern cohomology.

Note that, if a compact complex manifold X satisfies the $\partial\bar{\partial}$-Lemma, then, for every $k \in \mathbb{N}$, it holds $h^k_{BC} = h^k_A = h^k_{\bar{\partial}} = h^k_{\partial} = b_k$, and hence (2.5) is actually an equality. In fact, we prove now that also the converse holds true: more precisely, the equality in (2.5) holds for every $k \in \mathbb{N}$ if and only if the $\partial\bar{\partial}$-Lemma holds; in particular, this gives a characterization of the validity of the $\partial\bar{\partial}$-Lemma just in terms of $\{h^k_{BC}\}_{k\in\mathbb{N}}$.

Theorem 2.14 ([AT13b, Theorem B]). *Let X be a compact complex manifold. The equality*

$$\sum_{p+q=k} \left(\dim_{\mathbb{C}} H^{p,q}_{BC}(X) + \dim_{\mathbb{C}} H^{p,q}_A(X)\right) = 2 \dim_{\mathbb{C}} H^k_{dR}(X;\mathbb{C})$$

holds for every $k \in \mathbb{N}$ if and only if X satisfies the $\partial\bar{\partial}$-Lemma.

Proof. If X satisfies the $\partial\bar{\partial}$-Lemma, then the natural maps $H^{\bullet,\bullet}_{BC}(X) \to H^\bullet_{dR}(X;\mathbb{C})$, $H^{\bullet,\bullet}_{BC}(X) \to H^{\bullet,\bullet}_{\bar{\partial}}(X)$, and $H^{\bullet,\bullet}_{\bar{\partial}}(X) \to H^{\bullet,\bullet}_A(X)$, $H^\bullet_{dR}(X;\mathbb{C}) \to H^{\bullet,\bullet}_A(X)$ induced by the identity are isomorphisms, [DGMS75, Remark 5.16], and hence, for every $k \in \mathbb{N}$, one has

$$h^k_{BC} = h^k_A = h^k_{\bar{\partial}} = b_k$$

and hence, in particular,

$$h^k_{BC} + h^k_A = 2 b_k .$$

We split the proof of the converse into the following claims.

Claim 1—*If $h^k_{BC}+h^k_A = 2 b_k$ holds for every $k \in \mathbb{N}$, then the Hodge and Frölicher spectral sequences degenerate at the first step (namely, $E_1 \simeq E_\infty$, that is, $h^k_{\bar{\partial}} = b_k$ for every $k \in \mathbb{N}$) and $a^k = 0 = f^k$ for every $k \in \mathbb{N}$.*
Since, for every $k \in \mathbb{N}$, we have

$$2 b_k = h^k_{BC} + h^k_A = 2 h^k_{\bar{\partial}} + a^k + f^k \geq 2 b_k ,$$

then $h^k_{\bar{\partial}} = b_k$ and $a^k = 0 = f^k$ for every $k \in \mathbb{N}$.

`Claim 2`—*Fix $k \in \mathbb{N}$. If $a^{k+1} := \sum_{p+q=k+1} \dim_\mathbb{C} A^{p,q} = 0$, then the natural map*

$$\bigoplus_{p+q=k} H^{p,q}_{BC}(X) \to H^k_{dR}(X;\mathbb{C})$$

is surjective.

Let $a = [\alpha] \in H^k_{dR}(X;\mathbb{C})$. We have to prove that a admits a representative whose pure-type components are d-closed. Consider the pure-type decomposition of α:

$$\alpha =: \sum_{j=0}^{k} (-1)^j \, \alpha^{k-j,j} \,,$$

where $\alpha^{k-j,j} \in \wedge^{k-j,j} X$. Since $d\,\alpha = 0$, we get that

$$\partial\alpha^{k,0} = 0\,, \quad \bar\partial\alpha^{k-j,j} - \partial\alpha^{k-j-1,j+1} = 0 \text{ for } j \in \{0,\dots,k-1\}\,, \quad \bar\partial\alpha^{0,k} = 0\,;$$

by the hypothesis $a^{k+1} = 0$, for every $j \in \{0,\dots,k-1\}$, we get that,

$$\bar\partial\alpha^{k-j,j} = \partial\alpha^{k-j-1,j+1} \in \left(\operatorname{im}\bar\partial \cap \operatorname{im}\partial\right) \cap \wedge^{k-j,j+1} X = \operatorname{im}\partial\bar\partial \cap \wedge^{k-j,j+1} X$$

and hence there exists $\eta^{k-j-1,j} \in \wedge^{k-j-1,j} X$ such that

$$\bar\partial\alpha^{k-j,j} = \partial\bar\partial\eta^{k-j-1,j} = \partial\alpha^{k-j-1,j+1} \,.$$

Define

$$\eta := \sum_{j=0}^{k-1} (-1)^j \, \eta^{k-j-1,j} \in \wedge^{k-1} X \otimes_\mathbb{R} \mathbb{C} \,.$$

The claim follows noting that

$$a = [\alpha] = [\alpha + d\,\eta]$$

$$= \Bigg[\left(\alpha^{k,0} + \partial\eta^{k-1,0}\right) + \sum_{j=1}^{k-1} (-1)^j \left(\alpha^{k-j,j} + \partial\eta^{k-j-1,j} - \bar\partial\eta^{k-j,j-1}\right)$$

$$+ (-1)^k \left(\alpha^{0,k} - \bar\partial\eta^{0,k-1}\right) \Bigg]$$

$$= \left[\alpha^{k,0} + \partial\eta^{k-1,0}\right] + \sum_{j=1}^{k-1} (-1)^j \left[\alpha^{k-j,j} + \partial\eta^{k-j-1,j} - \bar{\partial}\eta^{k-j,j-1}\right]$$

$$+ (-1)^k \left[\alpha^{0,k} - \bar{\partial}\eta^{0,k-1}\right],$$

that is, each of the pure-type components of $\alpha + \mathrm{d}\,\eta$ is both ∂-closed and $\bar{\partial}$-closed.

Claim 3—If $h_{BC}^k \geq b_k$ and $h_{BC}^k + h_A^k = 2\,b_k$ for every $k \in \varkappa$, then $h_{BC}^k = b_k$ for every $k \in \mathbb{N}$.

If n is the complex dimension of X, then, for every $k \in \mathbb{N}$, we have

$$b_k \leq h_{BC}^k = h_A^{2n-k} = 2\,b_{2n-k} - h_{BC}^{2n-k} \leq b_{2n-k} = b_k$$

and hence $h_{BC}^k = b_k$ for every $k \in \mathbb{N}$.

Now, by Claim 1, we get that $a^k = 0$ for each $k \in \mathbb{N}$; hence, by Claim 2, for every $k \in \mathbb{N}$ the natural map

$$\bigoplus_{p+q=k} H_{BC}^{p,q}(X) \to H_{dR}^k(X;\mathbb{C})$$

induced by the identity is surjective, and hence, in particular, $h_{BC}^k \geq b_k$. By Claim 3 we get therefore that $h_{BC}^k = b_k$ for every $k \in \mathbb{N}$. Hence, the natural map $H_{BC}^{\bullet,\bullet}(X) \to H_{dR}^\bullet(X;\mathbb{C})$ is actually an isomorphism, which is equivalent to say that X satisfies the $\partial\bar{\partial}$-Lemma. $\square$

Remark 2.8. We note that, using the exact sequences (2.2) and (2.1), one can prove that, on a compact complex manifold X and for every $k \in \mathbb{N}$,

$$e^k = \left(h_{\bar{\partial}}^k - h_{BC}^k\right) + f^k + c^{k-1}$$

$$= \left(h_{\bar{\partial}}^k - h_{BC}^k\right) - \left(h_{\bar{\partial}}^{k-1} - h_A^{k-1}\right) + f^k - a^{k-1} + e^{k-2}.$$

Remark 2.9. Note that $E_1 \simeq E_\infty$ is not sufficient to have the equality $h_{BC}^k + h_A^k = 2\,b_k$ for every $k \in \mathbb{N}$ (and hence the $\partial\bar{\partial}$-Lemma): a counterexample is provided by small deformations of the Iwasawa manifold.

Indeed, for small deformations of $\mathbb{I}_3$ in class *(iii)*, since

$$h_{\bar{\partial}}^1 = 4 = b_1, \qquad h_{\bar{\partial}}^2 = 8 = b_2, h_{\bar{\partial}}^3 = 10 = b_3,$$

the Hodge and Frölicher spectral sequences degenerate at the first step, but

$$h_{BC}^1 + h_A^1 = 10 > 8 = 2\,b_1, \qquad h_{BC}^2 + h_A^2 = 16 = 2\,b_2,$$

$$h_{BC}^3 + h_A^3 = 28 > 20 = 2\,b_3.$$

Using Theorem 2.14, we get another proof of the stability of the $\partial\bar{\partial}$-Lemma under small deformations of the complex structure, [AT13b, Corollary 2.7]; for different proofs of the same result by means of other techniques see, e.g., [Voi02, Proposition 9.21] (by proving that the conditions in the characterization [DGMS75, 5.21] are stable under small deformations), [Wu06, Theorem 5.12] (with an inductive argument), [Tom08, Sect. B] (by an expansion in series argument). (See [AK12, AK13a] for examples on the behaviour of the $\partial\bar{\partial}$-Lemma as for closedness.)

Corollary 2.2 ([AT13b, Corollary 2.7]; [Voi02, Proposition 9.21]; [Wu06, Theorem 5.12]; [Tom08, Sect. B]). *Satisfying the $\partial\bar{\partial}$-Lemma is a stable property under small deformations of the complex structure.*

Proof. Let $\{X_t\}_{t \in B}$ be a complex-analytic family of compact complex manifolds. Since, for every $k \in \mathbb{N}$, the dimensions $h_{BC}^k(X_t)$ and $h_A^k(X_t)$ are upper-semi-continuous functions in t, [Sch07, Lemme 3.2], while the dimensions $b_k(X_t)$ are constant in t by Ehresmann's theorem, one gets that, if X_{t_0} satisfies the equality $h_{BC}^k(X_{t_0}) + h_A^k(X_{t_0}) = 2\,b_k(X_{t_0})$ for every $k \in \mathbb{N}$, the same holds true for X_t with t near t_0. □

Remark 2.10. We recall that [DGMS75, 5.21] by P. Deligne, Ph.A. Griffiths, J. Morgan, and D.P. Sullivan characterizes the validity of the $\partial\bar{\partial}$-Lemma on a compact complex manifold in terms of the degeneracy of the Hodge and Frölicher spectral sequence and of the existence of Hodge structures in cohomology. In particular, if follows that, on a compact complex manifold satisfying the $\partial\bar{\partial}$-Lemma, one has the equality $b_k = \sum_{p+q=k} h_{\bar{\partial}}^{p,q}$ for every $k \in \mathbb{N}$ (which is equivalent to the degeneracy of the Hodge and Frölicher spectral sequence) and the symmetry $h_{\bar{\partial}}^{p,q} = h_{\bar{\partial}}^{q,p}$ for every $(p,q) \in \mathbb{N}^2$.

Note that, on a compact complex surface X, since the Hodge and Frölicher spectral sequence degenerates at the first step (see, e.g., [BHPVdV04, Theorem IV.2.8]) if $h_{\bar{\partial}}^{1,0} = h_{\bar{\partial}}^{0,1}$ then $b_1 = 2\,h_{\bar{\partial}}^{1,0}$ is even, and hence X is Kähler, by [Kod64, Miy74, Siu83], or [Lam99, Corollaire 5.7], or [Buc99, Theorem 11]. As already remarked, the small deformations of $\mathbb{I}_3$ in class *(iii)* satisfy the degeneracy condition of the Hodge and Frölicher spectral sequence, but they do not satisfy either the $\partial\bar{\partial}$-Lemma, or the symmetry of the Hodge numbers.

In studying the six-dimensional nilmanifolds endowed with left-invariant complex structures, M. Ceballos, A. Otal, L. Ugarte, and R. Villacampa found an example of a compact complex manifold of complex dimension 3 such that $E_1 \simeq E_\infty$ and $h_{\bar{\partial}}^{p,q} = h_{\partial}^{p,q}$ for every $(p,q) \in \mathbb{N}^2$ but for which the $\partial\bar{\partial}$-Lemma does not hold, [COUV11, Corollary 6.10]. We recall, for a compact complex manifold X satisfying such a property, the double complex $\left(\wedge^{\bullet,\bullet}X, \partial, \bar{\partial}\right)$ has the form in Fig. 2.2 (where dots denote generators of the $\mathcal{C}^\infty(X; \mathbb{R})$-module $\wedge^{\bullet,\bullet}X$, horizontal arrows are meant as ∂, vertical ones as $\bar{\partial}$ and zero arrows are not depicted).

Fig. 2.2 An abstract example

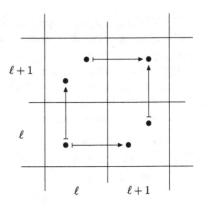

Appendix: Cohomological Properties of Generalized Complex Manifolds

The above results, concerning the characterization of $\partial\bar{\partial}$-Lemma by means of an inequality *à la* Frölicher on compact complex manifolds, can be slightly generalized to a more algebraic context, allowing both to recover the results in the complex setting and to get corollaries in the symplectic and in the generalized complex case, [AT13a]. We briefly recall the results obtained in a joint work with A. Tomassini, referring to [AT13a] for more details and for the proofs.

Cohomological Aspects of Bi-Differential Graded Vector Spaces. Consider a $\mathbb{Z}$-graded $\mathbb{K}$-vector space $A^\bullet$ endowed with two endomorphisms $\delta_1 \in \mathrm{End}^{\delta_1}(A^\bullet)$ and $\delta_2 \in \mathrm{End}^{\delta_2}(A^\bullet)$ such that

$$\delta_1^2 = \delta_2^2 = \delta_1\delta_2 + \delta_2\delta_1 = 0 .$$

Following notation in [DGMS75, Remark 5.16], consider the following cohomologies and the following natural maps between them induced by the identity:

$$H^\bullet_{(\delta_1,\delta_2;\delta_1\delta_2)}(A^\bullet) := \frac{\ker\delta_1 \cap \ker\delta_2}{\mathrm{im}\,\delta_1\delta_2}$$

$$H^\bullet_{(\delta_1;\delta_1)}(A^\bullet) := \frac{\ker\delta_1}{\mathrm{im}\,\delta_1} \qquad H_{(\delta_1+\delta_2;\delta_1+\delta_2)}(\mathrm{Tot}\,A^\bullet) := \frac{\ker(\delta_1+\delta_2)}{\mathrm{im}(\delta_1+\delta_2)} \qquad H^\bullet_{(\delta_2;\delta_2)}(A^\bullet) := \frac{\ker\delta_2}{\mathrm{im}\,\delta_2}$$

$$H^\bullet_{(\delta_1\delta_2;\delta_1,\delta_2)}(A^\bullet) := \frac{\ker\delta_1\delta_2}{\mathrm{im}\,\delta_1 + \mathrm{im}\,\delta_2}$$

By definition, $A^\bullet$ is said to satisfy the $\delta_1\delta_2$-*Lemma* if and only if

$$\ker\delta_1 \cap \ker\delta_2 \cap (\operatorname{im}\delta_1 + \operatorname{im}\delta_2) = \operatorname{im}\delta_1\delta_2 \,,$$

namely, if and only if the natural map $H^\bullet_{(\delta_1,\delta_2;\delta_1\delta_2)}\left(A^\bullet\right) \to H^\bullet_{(\delta_1\delta_2;\delta_1,\delta_2)}\left(A^\bullet\right)$ induced by the identity is injective, see [DGMS75]. (Several equivalent definitions can be given, [DGMS75, Lemma 5.15], see [AT13a, Lemma 1.4].)

In order to generalize Theorems 2.13 and 2.14, the following inequality *à la* Frölicher can be proven, characterizing the validity of the $\delta_1\delta_2$-Lemma. (See also [AT13a, Theorem 3.3, Corollary 3.5] for other statements.)

Theorem 2.15 ([AT13a, Theorem 2.4, Corollary 3.4]). *Let $A^\bullet$ be a bounded $\mathbb{Z}$-graded $\mathbb{K}$-vector space endowed with two endomorphisms $\delta_1 \in \operatorname{End}^{\hat{\delta}_1}\left(A^\bullet\right)$ and $\delta_2 \in \operatorname{End}^{\hat{\delta}_2}\left(A^\bullet\right)$ such that $\delta_1^2 = \delta_2^2 = \delta_1\delta_2 + \delta_2\delta_1 = 0$. Denote the greatest common divisor of $\hat{\delta}_1$ and $\hat{\delta}_2$ by $\operatorname{GCD}(\hat{\delta}_1,\hat{\delta}_2)$. Suppose that*

$$\dim_{\mathbb{K}} H^\bullet_{(\delta_1;\delta_1)}\left(A^\bullet\right) < +\infty \qquad and \qquad \dim_{\mathbb{K}} H^\bullet_{(\delta_2;\delta_2)}\left(A^\bullet\right) < +\infty \,.$$

Then

$$\dim_{\mathbb{K}} H^\bullet_{(\delta_1,\delta_2;\delta_1\delta_2)}\left(A^\bullet\right) + \dim_{\mathbb{K}} H^\bullet_{(\delta_1\delta_2;\delta_1,\delta_2)}\left(A^\bullet\right) \geq \dim_{\mathbb{K}} H^\bullet_{(\delta_1;\delta_1)}\left(A^\bullet\right) + \dim_{\mathbb{K}} H^\bullet_{(\delta_2;\delta_2)}\left(A^\bullet\right)\,.$$

Furthermore, the following conditions are equivalent:

1. *$A^{\operatorname{GCD}(\hat{\delta}_1,\hat{\delta}_2)\bullet}$ satisfies the $\delta_1\delta_2$-Lemma;*
2. *the equality*

$$\sum_{p+q=\bullet}\left(\dim_{\mathbb{K}} H^{\hat{\delta}_1 p + \hat{\delta}_2 q}_{(\delta_1,\delta_2;\delta_1\delta_2)}\left(A^\bullet\right) + \dim_{\mathbb{K}} H^{\hat{\delta}_1 p + \hat{\delta}_2 q}_{(\delta_1\delta_2;\delta_1,\delta_2)}\left(A^\bullet\right)\right)$$

$$= 2 \dim_{\mathbb{K}} H^\bullet_{(\delta_1\otimes_{\mathbb{K}}\operatorname{id} + \delta_2\otimes_{\mathbb{K}}\beta;\, \delta_1\otimes_{\mathbb{K}}\operatorname{id} + \delta_2\otimes_{\mathbb{K}}\beta)}\left(\operatorname{Tot}^\bullet \operatorname{Doub}^{\bullet,\bullet} A^\bullet\right)\,.$$

holds.

Remark 2.11. We recall the following constructions.

Given a $\mathbb{Z}^2$-graded $\mathbb{K}$-vector space $A^{\bullet,\bullet}$ endowed with two endomorphisms $\delta_1 \in \operatorname{End}^{\hat{\delta}_{1,1},\hat{\delta}_{1,2}}\left(A^{\bullet,\bullet}\right)$ and $\delta_2 \in \operatorname{End}^{\hat{\delta}_{2,1},\hat{\delta}_{2,2}}\left(A^{\bullet,\bullet}\right)$ such that $\delta_1^2 = \delta_2^2 = \delta_1\delta_2 + \delta_2\delta_1 = 0$, define the $\mathbb{Z}$-graded $\mathbb{K}$-vector space

$$\operatorname{Tot}^\bullet\left(A^{\bullet,\bullet}\right) := \bigoplus_{p+q=\bullet} A^{p,q}\,,$$

endowed with the endomorphisms

$$\delta_1 \in \operatorname{End}^{\hat{\delta}_{1,1}+\hat{\delta}_{1,2}}\left(\operatorname{Tot}^\bullet\left(A^{\bullet,\bullet}\right)\right) \qquad and \qquad \delta_2 \in \operatorname{End}^{\hat{\delta}_{2,1}+\hat{\delta}_{2,2}}\left(\operatorname{Tot}^\bullet\left(A^{\bullet,\bullet}\right)\right)$$

such that $\delta_1^2 = \delta_2^2 = \delta_1\delta_2 + \delta_2\delta_1 = 0$.

Conversely, given a $\mathbb{Z}$-graded $\mathbb{K}$-vector space $A^\bullet$ endowed with two endomorphisms $\delta_1 \in \mathrm{End}^{\hat{\delta}_1}(A^\bullet)$ and $\delta_2 \in \mathrm{End}^{\hat{\delta}_2}(A^\bullet)$ such that $\delta_1^2 = \delta_2^2 = \delta_1\delta_2 + \delta_2\delta_1 = 0$, following [Bry88, Sect. 1.3], [Cav05, Sect. 4.2], see [Goo85, Sect. II.2], [Con85, Sect. II], take an infinite cyclic multiplicative group $\{\beta^m : m \in \mathbb{Z}\}$ generated by some β, and consider the $\mathbb{Z}$-graded $\mathbb{K}$-vector space $\bigoplus_{\bullet \in \mathbb{Z}} \mathbb{K} \beta^\bullet$, and define the $\mathbb{Z}^2$-graded $\mathbb{K}$-vector space

$$\mathrm{Doub}^{\bullet_1, \bullet_2}(A^\bullet) := A^{\hat{\delta}_1 \bullet_1 + \hat{\delta}_2 \bullet_2} \otimes_{\mathbb{K}} \mathbb{K} \beta^{\bullet_2} \,,$$

endowed with the endomorphisms

$$\delta_1 \otimes_{\mathbb{K}} \mathrm{id} \in \mathrm{End}^{1,0}\left(\mathrm{Doub}^{\bullet, \bullet}(A^\bullet)\right) \qquad \text{and} \qquad \delta_2 \otimes_{\mathbb{K}} \beta \in \mathrm{End}^{1,0}\left(\mathrm{Doub}^{\bullet, \bullet}(A^\bullet)\right) \,,$$

which satisfy $(\delta_1 \otimes_{\mathbb{K}} \mathrm{id})^2 = (\delta_2 \otimes_{\mathbb{K}} \beta)^2 = (\delta_1 \otimes_{\mathbb{K}} \mathrm{id})(\delta_2 \otimes_{\mathbb{K}} \beta) + (\delta_2 \otimes_{\mathbb{K}} \beta)(\delta_1 \otimes_{\mathbb{K}} \mathrm{id}) = 0$; following [Bry88, Sect. 1.3], [Cav05, Sect. 4.2], the double complex $(\mathrm{Doub}^{\bullet, \bullet}(A^\bullet), \delta_1 \otimes_{\mathbb{K}} \mathrm{id}, \delta_2 \otimes_{\mathbb{K}} \beta)$ is called the *canonical double complex* associated to $A^\bullet$.

Cohomological Aspects of Complex Manifolds. Given a compact complex manifold X, by applying Theorem 2.15 to the double complex $\left(\wedge^{\bullet, \bullet} X, \partial, \overline{\partial}\right)$, one straightforwardly recovers the inequality *à la* Frölicher for the Bott-Chern cohomology of compact complex manifolds and the corresponding characterization of the $\partial\overline{\partial}$-Lemma by means of the Bott-Chern cohomology in [AT13b, Theorems A and B], see Theorems 2.13 and 2.14.

Cohomological Aspects of Symplectic Manifolds. Given a compact manifold X endowed with a symplectic structure ω, one can apply Theorem 2.15 to the $\mathbb{Z}$-graded $\mathbb{R}$-vector space $\wedge^\bullet X$ endowed with the endomorphisms $\mathrm{d} \in \mathrm{End}^1$ $(\wedge^\bullet X)$ and $\mathrm{d}^\Lambda \in \mathrm{End}^{-1}(\wedge^\bullet X)$. (We refer to Sect. 1.2 for definitions and notations concerning symplectic structures.) In fact, on a compact symplectic manifold,[3] both the spectral sequences associated to the canonical double complex

---

[3]More in general, given a compact manifold X endowed with a Poisson bracket $\{\cdot, \cdot\cdot\}$, and denoted by G the Poisson tensor associated to $\{\cdot, \cdot\cdot\}$, by following J.-L. Koszul, [Kos85], one can define $\delta := [\iota_G, \mathrm{d}] \in \mathrm{End}^{-1}(\wedge^\bullet X)$. One has that $\delta^2 = 0$ and $[\mathrm{d}, \delta] = 0$, [Kos85, pp. 266, 265], see also [Bry88, Proposition 1.2.3, Theorem 1.3.1].

It holds that, on any compact Poisson manifold, the first spectral sequence $'E_r^{\bullet, \bullet}$ associated to the canonical double complex $(\mathrm{Doub}^{\bullet, \bullet} \wedge^\bullet X, \mathrm{d} \otimes_{\mathbb{R}} \mathrm{id}, \delta \otimes_{\mathbb{R}} \beta)$ degenerates at the first level, [FIdL98, Theorem 2.5].

On the other hand, an example of a compact Poisson manifold (more precisely, of a nilmanifold endowed with a co-symplectic structure) such that the second spectral sequence $''E_r^{\bullet, \bullet}(\mathrm{Doub}^{\bullet, \bullet} \wedge^\bullet X, \mathrm{d} \otimes_{\mathbb{R}} \mathrm{id}, \delta \otimes_{\mathbb{R}} \beta)$ does not degenerate at the first level has been provided by M. Fernández, R. Ibáñez, and M. de León, [FIdL98, Theorem 5.1].

In fact, on a compact $2n$-dimensional manifold X endowed with a symplectic structure ω, the symplectic-$\star$-operator $\star_\omega: \wedge^\bullet X \to \wedge^{2n-\bullet} X$ induces the isomorphism $\star_\omega: 'E_r^{\bullet_1, \bullet_2} \xrightarrow{\simeq} ''E_r^{\bullet_2, 2n + \bullet_1}$, [FIdL98, Theorem 2.9].

$\left(\mathrm{Doub}^{\bullet,\bullet}\wedge^{\bullet}X,\, \mathrm{d}\otimes_{\mathbb{R}}\mathrm{id},\, \mathrm{d}^{\Lambda}\otimes_{\mathbb{R}}\beta\right)$ degenerate at the first level, [Bry88, Theorem 2.3.1], [FIdL98, Theorem 2.5]; see also [FIdL98, Theorem 2.9], [Cav06, Theorem 5.2]. Hence, one gets the following result[4].

Theorem 2.16 ([AT13a, Theorem 4.4]). *Let X be a compact manifold endowed with a symplectic structure ω. The inequality*

$$\dim_{\mathbb{R}} H^{\bullet}_{\mathrm{d}+\mathrm{d}^{\Lambda}}(X;\mathbb{R}) + \dim_{\mathbb{R}} H^{\bullet}_{\mathrm{d}\mathrm{d}^{\Lambda}}(X;\mathbb{R}) \;\geq\; 2\dim_{\mathbb{R}} H^{\bullet}_{dR}(X;\mathbb{R}) \qquad (2.6)$$

holds. Furthermore, the equality in (2.6) holds if and only if X satisfies the Hard Lefschetz Condition.

Example 2.1 ([AT13a, Example 4.6]). Consider the Iwasawa manifold $\mathbb{I}_3 := \mathbb{Z}\,[\mathrm{i}]^3 \big\backslash$ $\left(\mathbb{C}^3,\, *\right)$, and take a $\left(\mathbb{C}^3,\, *\right)$-left-invariant co-frame $\left\{e^j\right\}_{j\in\{1,\ldots,6\}}$ of T^*X such that

$$\mathrm{d}\,e^1 = \mathrm{d}\,e^2 = \mathrm{d}\,e^3 = \mathrm{d}\,e^4 = 0\,, \qquad \mathrm{d}\,e^5 = -e^{13} + e^{24}\,, \qquad \mathrm{d}\,e^6 = -e^{14} - e^{23}$$

(in order to simplify notation, we shorten, e.g., $e^{12} := e^1 \wedge e^2$).

Consider the $\left(\mathbb{C}^3,\, *\right)$-left-invariant almost-Kähler structure (J, ω, g) on $\mathbb{I}_3$ defined by

$$Je^1 := -e^6\,, \quad Je^2 := -e^5\,, \quad Je^3 := -e^4\,, \quad \omega := e^{16} + e^{25} + e^{34}\,,$$

$$g := \omega\,(\cdot,\, J\cdot\cdot)\,;$$

(it has been studied in [ATZ12, Sect. 4] as an example of an almost-Kähler structure non-inducing a decomposition in cohomology according to the almost-complex structure, [ATZ12, Proposition 4.1]).

By using the theorem *à la* Nomizu for the symplectic cohomologies proven by M. Macrì, [Mac13, Theorem 3, Remark 4], see also [AK13b, Theorems 3.2 and 3.4], we can compute the symplectic cohomologies of the Iwasawa manifold $\mathbb{I}_3$ endowed with the $\left(\mathbb{C}^3,\, *\right)$-left-invariant symplectic structure ω by using just $\left(\mathbb{C}^3,\, *\right)$-left-invariant forms; we summarize the dimensions of the symplectic cohomologies in Table 2.2.

---

[4]We recall that the symplectic cohomologies, introduced by L.-S. Tseng and S.-T. Yau in [TY12a, Sect. 3], are defined as

$$H^{\bullet}_{\mathrm{d}+\mathrm{d}^{\Lambda}}(X;\mathbb{R}) := H^{\bullet}_{(\mathrm{d},\mathrm{d}^{\Lambda};\mathrm{d}\mathrm{d}^{\Lambda})}(\wedge^{\bullet}X) := \frac{\ker\left(\mathrm{d}+\mathrm{d}^{\Lambda}\right)}{\operatorname{im}\mathrm{d}\,\mathrm{d}^{\Lambda}}$$

and

$$H^{\bullet}_{\mathrm{d}\mathrm{d}^{\Lambda}}(X;\mathbb{R}) := H^{\bullet}_{(\mathrm{d}\mathrm{d}^{\Lambda};\mathrm{d},\mathrm{d}^{\Lambda})}(\wedge^{\bullet}X) := \frac{\ker\mathrm{d}\,\mathrm{d}^{\Lambda}}{\operatorname{im}\mathrm{d}+\operatorname{im}\mathrm{d}^{\Lambda}}\,.$$

Table 2.2 The symplectic cohomologies of the Iwasawa manifold $\mathbb{I}_3 := \mathbb{Z}[i]^3 \backslash (\mathbb{C}^3, *)$ endowed with the symplectic structure $\omega := e^1 \wedge e^6 + e^2 \wedge e^5 + e^3 \wedge e^4$

$\dim_{\mathbb{C}} H_\sharp^\bullet(\mathbb{I}_3)$	$(d; d)$	$(d^\Lambda; d^\Lambda)$	$(d, d^\Lambda; d d^\Lambda)$	$(d d^\Lambda; d, d^\Lambda)$
0	1	1	1	1
1	4	4	4	4
2	8	8	9	10
3	10	10	11	11
4	8	8	10	9
5	4	4	4	4
6	1	1	1	1

Cohomological Aspects of Generalized Complex Manifolds. Finally, consider a compact differentiable manifold X endowed with an H-twisted generalized complex structure $\mathcal{J}$, for a d-closed 3-form H on X. (We refer to Sect. 1.3 for definitions and notations on generalized complex structures.)

As an application of Theorem 2.15, we get the following result.

Theorem 2.17 ([AT13a, Theorems 4.10, 4.11]). *Let X be a compact differentiable manifold endowed with an H-twisted generalized complex structure $\mathcal{J}$. Then*

$$\dim_{\mathbb{C}} GH^\bullet_{BC_{\mathcal{J},H}}(X) + \dim_{\mathbb{C}} GH^\bullet_{A_{\mathcal{J},H}}(X) \geq \dim_{\mathbb{C}} GH^\bullet_{\bar{\partial}_{\mathcal{J},H}}(X) + \dim_{\mathbb{C}} GH^\bullet_{\partial_{\mathcal{J},H}}(X).$$
$$(2.7)$$

Furthermore, the following conditions are equivalent:

- *X satisfies the $\partial_{\mathcal{J},H}\bar{\partial}_{\mathcal{J},H}$-Lemma;*
- *the Hodge and Frölicher spectral sequences associated to the canonical double complex $\left(\mathrm{Doub}^{\bullet,\bullet} U_{\mathcal{J}}^\bullet, \partial_{\mathcal{J},H} \otimes_{\mathbb{C}} \mathrm{id}, \bar{\partial}_{\mathcal{J},H} \otimes_{\mathbb{C}} \beta\right)$ degenerate at the first level and the equality in (2.7),*

$$\dim_{\mathbb{C}} GH^\bullet_{BC_{\mathcal{J},H}}(X) + \dim_{\mathbb{C}} GH^\bullet_{A_{\mathcal{J},H}}(X) = \dim_{\mathbb{C}} GH^\bullet_{\bar{\partial}_{\mathcal{J},H}}(X) + \dim_{\mathbb{C}} GH^\bullet_{\partial_{\mathcal{J},H}}(X),$$

holds.

Note that, by viewing a complex structure as a generalized complex structure of maximum type, Theorem 2.17 gives the following result, to be compared with [AT13b, Theorems A and B], see Theorems 2.13 and 2.14.

Corollary 2.3 ([AT13a, Corollary 4.14]). *Let X be a compact complex manifold. Then the inequality*

$$\sum_{p-q=\bullet} \dim_{\mathbb{C}} H^{p,q}_{BC}(X) \geq \sum_{p-q=\bullet} \dim_{\mathbb{C}} H^{p,q}_{\bar{\partial}}(X)$$

holds. Furthermore, X satisfies the $\partial\bar\partial$-Lemma if and only if (i) the Hodge and Frölicher spectral sequence of X degenerates at the first level, namely,

$$\dim_{\mathbb{C}} H^\bullet_{dR}(X;\mathbb{C}) = \dim_{\mathbb{C}} \mathrm{Tot}^\bullet H^{\bullet,\bullet}_{\bar\partial}(X),$$

and (ii) the equality

$$\sum_{p-q=\bullet} \dim_{\mathbb{C}} H^{p,q}_{BC}(X) = \sum_{p-q=\bullet} \dim_{\mathbb{C}} H^{p,q}_{\bar\partial}(X)$$

holds.

Example 2.2 ([AT13a]). As an example, in Table 2.3, we summarize the dimensions of the generalized cohomologies of the Iwasawa manifold and of its small deformations, referring to Sect. 3.2.1 for the computations.

Table 2.3 Generalized complex cohomologies of the Iwasawa manifold and of its small deformations, [AT13a, Table 2]

$\mathbb{I}_3 := \mathbb{Z}[i]^3\backslash\mathbb{C}^3$	$\dim_\mathbb{C}\mathrm{Tot}^{-3}$ $H_\sharp^{\bullet,-\bullet}(X)$				$\dim_\mathbb{C}\mathrm{Tot}^{-2}$ $H_\sharp^{\bullet,-\bullet}(X)$				$\dim_\mathbb{C}\mathrm{Tot}^{-1}$ $H_\sharp^{\bullet,-\bullet}(X)$				$\dim_\mathbb{C}\mathrm{Tot}^{0}$ $H_\sharp^{\bullet,-\bullet}(X)$				$\dim_\mathbb{C}\mathrm{Tot}^{1}$ $H_\sharp^{\bullet,-\bullet}(X)$				$\dim_\mathbb{C}\mathrm{Tot}^{2}$ $H_\sharp^{\bullet,-\bullet}(X)$				$\dim_\mathbb{C}\mathrm{Tot}^{3}$ $H_\sharp^{\bullet,-\bullet}(X)$			
Classes	$\bar\partial$	∂	BC	A	$\bar\partial$	∂	BC	A	$\bar\partial$	∂	BC	A	$\bar\partial$	∂	BC	A	$\bar\partial$	∂	BC	A	$\bar\partial$	∂	BC	A	$\bar\partial$	∂	BC	A
(i)	1	1	1	1	5	5	5	5	11	11	11	11	12	12	12	12	11	11	11	11	5	5	5	5	1	1	1	1
(ii.a)	1	1	1	1	4	4	4	4	9	9	11	11	10	10	11	11	9	9	11	11	4	4	4	4	1	1	1	1
(ii.b)	1	1	1	1	4	4	4	4	9	9	11	11	10	10	10	10	9	9	11	11	4	4	4	4	1	1	1	1
(iii.a)	1	1	1	1	3	3	3	3	8	8	11	11	10	10	11	11	8	8	11	11	3	3	3	3	1	1	1	1
(iii.b)	1	1	1	1	3	3	3	3	8	8	11	11	10	10	10	10	8	8	11	11	3	3	3	3	1	1	1	1
	$b_0 = 1$				$b_1 = 4$				$b_2 = 8$				$b_3 = 10$				$b_4 = 8$				$b_5 = 4$				$b_6 = 1$			

Chapter 3
Cohomology of Nilmanifolds

Abstract Nilmanifolds and solvmanifolds appear as "toy-examples" in non-Kähler geometry: indeed, on the one hand, non-tori nilmanifolds admit no Kähler structure, (Benson and Gordon, Topology 27(4):513–518, 1988; Lupton and Oprea, J. Pure Appl. Algebra 91(1–3):193–207, 1994), and, more in general, solvmanifolds admitting a Kähler structure are characterized, (Hasegawa, Proc. Am. Math. Soc. 106(1):65–71, 1989); on the other hand, the geometry and cohomology of solvmanifolds can be often reduced to study left-invariant geometry.

In Sect. 3.1, it is shown that, for certain classes of complex structures on *nilmanifolds* (that is, compact quotients of connected simply-connected nilpotent Lie groups by co-compact discrete subgroups), the de Rham, Dolbeault, Bott-Chern, and Aeppli cohomologies are completely determined by the associated Lie algebra endowed with the induced linear complex structure, Theorem 3.6, giving a sort of result *à la* Nomizu for the Bott-Chern cohomology. This will allow us to explicitly study the Bott-Chern and Aeppli cohomologies of the *Iwasawa manifold* and of its small deformations, in Sect. 3.2, and of the complex structures on six-dimensional nilmanifolds in M. Ceballos, A. Otal, L. Ugarte, and R. Villacampa's classification, (Ceballos et al., Classification of complex structures on 6-dimensional nilpotent Lie algebras, arXiv:1111.5873v3 [math.DG], 2011), in Sect. 3.3. Finally, in Appendix: Cohomology of Solvmanifolds, we recall some facts concerning cohomologies of solvmanifolds.

3.1 Cohomology Computations for Special Nilmanifolds

We are now interested in studying the Bott-Chern and Aeppli cohomologies in the special case of left-invariant complex structures on nilmanifolds and solvmanifolds.

In this section, we firstly recall some results concerning the computation of the de Rham cohomology and of the Dolbeault cohomology, for nilmanifolds and solvmanifolds, endowed with left-invariant complex structures, Sect. 3.1.2, referring to [Nom54, Hat60], respectively [Sak76, CFGU00, CF01, Rol09a, Rol11a]; then, we state and prove the results obtained in [Ang11] about computation of the Bott-Chern and Aeppli cohomologies, Theorems 3.5 and 3.7. Using these tools, one can compute the de Rham, Dolbeault, Bott-Chern and Aeppli cohomologies for the Iwasawa manifold and for its small deformations, Sects. 3.2.2–3.2.4.

3.1.1 Left-Invariant Complex Structures on Solvmanifolds

We start by recalling some facts and notations concerning left-invariant complex structures on solvmanifolds.

Let $X = \Gamma \backslash G$ be a solvmanifold, that is, a compact quotient of a connected simply-connected solvable Lie group G by a discrete and co-compact subgroup Γ; the Lie algebra naturally associated to G will be denoted by $\mathfrak{g}$ and its complexification by $\mathfrak{g}_{\mathbb{C}} := \mathfrak{g} \otimes_{\mathbb{R}} \mathbb{C}$. We recall that, dealing with G-left-invariant objects on X, we mean objects on X obtained by objects on G that are invariant under the action of G on itself given by left-translations; note that G-left-invariant objects on X are uniquely determined by objects on $\mathfrak{g}$. In particular, a G-left-invariant complex structure J on X is uniquely determined by a linear complex structure J on $\mathfrak{g}$ satisfying the integrability condition $\mathrm{Nij}_J = 0$, [NN57, Theorem 1.1]; the set of G-left-invariant complex structures on X is denoted by

$$\mathcal{C}(\mathfrak{g}) := \left\{ J \in \mathrm{End}(\mathfrak{g}) \ : \ J^2 = -\mathrm{id}_{\mathfrak{g}} \text{ and } \mathrm{Nij}_J = 0 \right\} .$$

Recall that the exterior differential d on X can be written using only the action of $\Gamma(X; TX)$ on $\mathcal{C}^{\infty}(X)$ and the Lie bracket of the Lie algebra of vector fields on X: more precisely, recall that, if $\varphi \in \wedge^k X$ and $X_0, \ldots, X_k \in \mathcal{C}^{\infty}(X; TX)$, then

$$d\varphi(X_0, \ldots, X_k) = \sum_{j=0}^{k} (-1)^j \, X_j \, \varphi(X_0, \ldots, X_{j-1}, X_{j+1}, \ldots, X_k)$$

$$+ \sum_{0 \leq j < h \leq k} (-1)^{j+h-1} \, \varphi([X_j, X_h], X_0, \ldots, X_{j-1}, X_{j+1}, \ldots, X_{h-1}, X_{h+1}, \ldots, X_k) .$$

Hence one has a differential complex $(\wedge^\bullet \mathfrak{g}^*, d)$, which is isomorphic, as a differential complex, to the differential sub-complex $\left(\wedge^\bullet_{\mathrm{inv}} X, \, d\lfloor_{\wedge^\bullet_{\mathrm{inv}} X} \right)$ of $(\wedge^\bullet X, d)$ given by the G-left-invariant forms on X.

If a G-left-invariant complex structure on X is given, then one also has the double complex $\left(\wedge^{\bullet, \bullet} \mathfrak{g}_{\mathbb{C}}^*, \partial, \bar{\partial} \right)$, which is isomorphic, as a double complex, to the double sub-complex $\left(\wedge^{\bullet, \bullet}_{\mathrm{inv}} X, \, \partial\lfloor_{\wedge^{\bullet, \bullet}_{\mathrm{inv}} X}, \, \bar{\partial}\lfloor_{\wedge^{\bullet, \bullet}_{\mathrm{inv}} X} \right)$ of $\left(\wedge^{\bullet, \bullet} X, \partial, \bar{\partial} \right)$ given by the G-left-invariant forms on X.

Finally, given a G-left-invariant complex structure on G and fixed $(p, q) \in \mathbb{N}^2$, one also has the following complexes and the following maps of complexes:

$$
\begin{array}{ccccc}
\wedge^{p-1,q-1}\mathfrak{g}_{\mathbb{C}}^* & \xrightarrow{\partial\bar{\partial}} & \wedge^{p,q}\mathfrak{g}_{\mathbb{C}}^* & \xrightarrow{d} & \wedge^{p+q+1}\mathfrak{g}_{\mathbb{C}}^* \\
\downarrow{\simeq} & & \downarrow{\simeq} & & \downarrow{\simeq} \\
\wedge_{\mathrm{inv}}^{p-1,q-1}X & \xrightarrow{\partial\bar{\partial}} & \wedge_{\mathrm{inv}}^{p,q}X & \xrightarrow{d} & \wedge_{\mathrm{inv}}^{p+q+1}(X;\mathbb{C}) \\
\cap{\,i} & & \cap{\,i} & & \cap{\,i} \\
\wedge^{p-1,q-1}X & \xrightarrow{\partial\bar{\partial}} & \wedge^{p,q}X & \xrightarrow{d} & \wedge^{p+q+1}(X;\mathbb{C})
\end{array}
\tag{3.1}
$$

and

$$
\begin{array}{ccccc}
\wedge^{p-1,q}\mathfrak{g}_{\mathbb{C}}^* \oplus \wedge^{p,q-1}\mathfrak{g}_{\mathbb{C}}^* & \xrightarrow{\partial+\bar{\partial}} & \wedge^{p,q}\mathfrak{g}_{\mathbb{C}}^* & \xrightarrow{\partial\bar{\partial}} & \wedge^{p+1,q+1}\mathfrak{g}_{\mathbb{C}}^* \\
\downarrow{\simeq} & & \downarrow{\simeq} & & \downarrow{\simeq} \\
\wedge_{\mathrm{inv}}^{p-1,q}X \oplus \wedge_{\mathrm{inv}}^{p,q-1}X & \xrightarrow{\partial+\bar{\partial}} & \wedge_{\mathrm{inv}}^{p,q}X & \xrightarrow{\partial\bar{\partial}} & \wedge_{\mathrm{inv}}^{p+1,q+1}X \\
\cap{\,i} & & \cap{\,i} & & \cap{\,i} \\
\wedge^{p-1,q}X \oplus \wedge^{p,q-1}X & \xrightarrow{\partial+\bar{\partial}} & \wedge^{p,q}X & \xrightarrow{\partial\bar{\partial}} & \wedge^{p+1,q+1}X
\end{array}
\tag{3.2}
$$

For $\sharp \in \left\{\bar{\partial},\, \partial,\, BC,\, A\right\}$ and $\mathbb{K} \in \{\mathbb{R}, \mathbb{C}\}$, we will write $H_{dR}^{\bullet}(\mathfrak{g}; \mathbb{K})$ and $H_{\sharp}^{\bullet,\bullet}(\mathfrak{g}_{\mathbb{C}})$ to denote the cohomology groups of the corresponding complexes of forms on $\mathfrak{g}$, which are isomorphic to the cohomology groups of the corresponding complexes of G-left-invariant forms on X. The rest of this section is devoted to the problem whether these cohomologies are isomorphic to the corresponding cohomologies on X.

3.1.2 Classical Results on Computations of the de Rham and Dolbeault Cohomologies

In this section, we collect some results, by K. Nomizu [Nom54], A. Hattori [Hat60], S. Console and A.M. Fino [CF01], Y. Sakane [Sak76], L.A. Cordero, M. Fernández, A. Gray, and L. Ugarte [CFGU00], S. Rollenske [Rol09a, Rol11a, Rol09b], concerning the computation of the de Rham cohomology and the Dolbeault cohomology for nilmanifolds and solvmanifolds, endowed with left-invariant complex structures.

First of all, we recall the following result, concerning the de Rham cohomology: it was firstly proven by K. Nomizu for nilmanifolds, and then generalized by

A. Hattori to the case of completely-solvable solvmanifolds. (See also [Mat51, Theorems 5 and 6].)

Theorem 3.1 ([Nom54, Theorem 1], [Hat60, Corollary 4.2]). *Let $X = \Gamma \backslash G$ be a nilmanifold, or, more in general, a completely-solvable solvmanifold, and denote the Lie algebra naturally associated to G by $\mathfrak{g}$. The map of differential complexes $(\wedge^\bullet \mathfrak{g}^*, d) \to (\wedge^\bullet X, d)$ is a quasi-isomorphism:*

$$i\colon H^\bullet_{dR}(\mathfrak{g};\mathbb{R}) \overset{\simeq}{\to} H^\bullet_{dR}(X;\mathbb{R}) \ .$$

Proof (sketch). We just sketch the idea of the proof for nilmanifolds, as in [Rol11a, Sect. 2]. The detailed proof, for the more general case of completely-solvable solvmanifolds, can be found, for example, in [TO97, Sect. 3.2].

The idea is to view X as a torus-bundle series and to argue by induction.

More precisely, suppose to have a tower of bundles

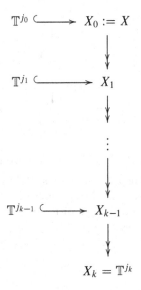

for some $k \in \mathbb{N} \backslash \{0\}$, where, for $\ell \in \{0, \ldots, k-1\}$, the fibre $\mathbb{T}^{j_\ell}$ is a j_ℓ-dimensional torus and the base $X_{\ell+1}$ is a nilmanifold with $\dim X_{\ell+1} < \dim X_\ell$, and with $X_k = \mathbb{T}^{j_k}$ a j_k-dimensional torus. By the Leray and Serre spectral sequence, [Ler50b, Ler50a, Ser51], see, e.g., [McC01, Theorem 5.2], and by noting that the de Rham cohomology of a torus is given by left-invariant forms, one can argue by induction.

The point now is to construct such a tower of bundles. Note that, for any $\ell \in \{0, \ldots, k-1\}$, the torus-bundle

$$\mathbb{T}^{j_\ell} \hookrightarrow X_\ell = \Gamma_\ell \backslash G_\ell \twoheadrightarrow X_{\ell+1} = \Gamma_{\ell+1} \backslash G_{\ell+1}$$

induces, at the level of the naturally associated Lie algebras (respectively, $\mathbb{R}^{j_\ell}$, and $\mathfrak{g}_\ell$, and $\mathfrak{g}_{\ell+1}$), the exact sequence

$$0 \to \mathbb{R}^{j_\ell} \to \mathfrak{g}_\ell \to \mathfrak{g}_{\ell+1} \to 0 \;.$$

Conversely, an exact sequence at the level of Lie algebras as the above induces a bundle at the level of the associated connected simply-connected Lie groups (respectively, $\mathbb{R}^{\ell_j}$, and G_ℓ, and $G_{\ell+1}$). Such bundle induces a bundle at the level of the corresponding nilmanifolds (respectively, $\mathbb{T}^{j_\ell}$, and $X_\ell = \Gamma_\ell \backslash G_\ell$, and $X_{\ell+1} = \Gamma_{\ell+1} \backslash G_{\ell+1}$) if it is compatible with the (rational structures associated to the) lattices induced by Γ. Hence it suffices to note that the descending central series of $\mathfrak{g}$ gives exact sequences that are compatible with the lattices induced by Γ.

□

Remark 3.1. In particular, for a nilmanifold $X = \Gamma \backslash G$, the dga $(\wedge^\bullet \mathfrak{g}^*, \mathrm{d})$ is the minimal model of $(\wedge^\bullet X, \mathrm{d})$, where $\mathfrak{g}$ denotes the Lie algebra naturally associated to G.

Remark 3.2. In fact, more in general, $(\wedge^\bullet \mathfrak{g}^*, \mathrm{d}) \to (\wedge^\bullet X, \mathrm{d})$ is a quasi-isomorphism also for solvmanifolds $X = \Gamma \backslash G$ satisfying the *Mostow condition*, namely, such that $\mathrm{Ad}\,\Gamma$ and $\mathrm{Ad}\,G$ have the same Zariski closure in $\mathrm{Aut}(\mathfrak{g}_\mathbb{C})$ (where $\mathfrak{g}$ is the Lie algebra naturally associated to G), as G.D. Mostow proved, [Mos54, Theorem 8.2, Corollary 8.1], see also [Rag72, Corollary 7.29]. In [CM12], six-dimensional non-completely-solvable almost-Abelian solvmanifolds for which Mostow condition does not hold are investigated, and their de Rham cohomology is computed; in particular, examples for which the de Rham cohomology is left-invariant even if the Mostow condition is not satisfies are provided, [CM12, Table 1]. See also [Gua07, CF11] for some results concerning the cohomology of solvmanifolds non-satisfying the Mostow condition.

Remark 3.3. In general, for a (possibly non-completely-solvable) solvmanifold, the inclusion $(\wedge^\bullet \mathfrak{g}^*, \mathrm{d}) \hookrightarrow (\wedge^\bullet X, \mathrm{d})$ induces just an injective map in cohomology, see [Rag72, Remark 7.30], [TO97, Theorem 3.2.10], compare also [CF01, Lemma 9] and Lemma 3.2. (In particular, as a consequence, one gets that any nilmanifold satisfies $b_1 \geq 2$, and any solvmanifold satisfies $b_1 \geq 1$, see, e.g., [Boc09, Corollaries 2.6 and 3.12].) An example of a non-completely-solvable solvmanifold for which such a map in cohomology is not surjective was provided by P. de Bartolomeis and A. Tomassini in [dBT06, Corollary 4.2, Remark 4.3], studying the Nakamura manifold, [Nak75, Sect. 2] (see also [CF11, Example 6.2]).

Remark 3.4. For further results about the de Rham cohomology of solvmanifolds, obtained by S. Console, by A.M. Fino, and by H. Kasuya, we refer to [Gua07, CF11, Kas13a, Kas12a, CFK13].

Similar results hold for the Dolbeault cohomology of nilmanifolds endowed with certain left-invariant complex structures; [Con06] and [Rol11a] are surveys on the known results.

First of all, we recall the following lemma by S. Console and A.M. Fino, [CF01]: the argument used in the proof can be generalized to Bott-Chern and Aeppli cohomologies, see Lemma 3.2.

Lemma 3.1 ([CF01, Lemma 9]). *Let $X = \Gamma \backslash G$ be a nilmanifold endowed with a G-left-invariant complex structure J, and denote the Lie algebra naturally associated to G by $\mathfrak{g}$. For any $p \in \mathbb{N}$, the map of complexes $\left(\wedge^{p,\bullet} \mathfrak{g}_{\mathbb{C}}^*, \overline{\partial} \right) \hookrightarrow \left(\wedge^{p,\bullet} X, \overline{\partial} \right)$ induces an injective homomorphism i in cohomology:*

$$ i \colon H_{\overline{\partial}}^{\bullet,\bullet} (\mathfrak{g}_{\mathbb{C}}) \hookrightarrow H_{\overline{\partial}}^{\bullet,\bullet}(X) \,. $$

Proof. We provide here a simple proof of the result, compare [AK13b, Corollary 1.2]. For a different argument, see [CF01, Lemma 9], see also [Ang11, Lemma 3.6], Lemma 3.2, [AK12, Sect. 1.2.2].

Consider the F.A. Belgun's symmetrization map, [Bel00, Theorem 7],

$$ \mu \colon \wedge^{\bullet} X \otimes \mathbb{C} \to \wedge^{\bullet} \mathfrak{g}_{\mathbb{C}}^*, \qquad \mu(\alpha) := \int_X \alpha \lfloor_m \eta(m) \,, $$

where η is a G-bi-invariant volume form on G such that $\int_X \eta = 1$, whose existence follows from J. Milnor's lemma [Mil76, Lemma 6.2]. One has that

$$ \mu \lfloor_{\wedge^{\bullet} \mathfrak{g}_{\mathbb{C}}^*} = \operatorname{id} \lfloor_{\wedge^{\bullet} \mathfrak{g}_{\mathbb{C}}^*} \,, $$

and that

$$ \operatorname{d} \circ \mu = \mu \circ \operatorname{d} \qquad \text{and} \qquad J \circ \mu = \mu \circ J \,, $$

since J is G-left-invariant, [Bel00, Theorem 7], [FG04, Theorem 2.1].

Hence one gets two maps of double complexes,

$$ i \colon \left(\wedge^{\bullet,\bullet} \mathfrak{g}_{\mathbb{C}}^*, \partial, \overline{\partial} \right) \to \left(\wedge^{\bullet,\bullet} X, \partial, \overline{\partial} \right) \quad \text{and} \quad \mu \colon \left(\wedge^{\bullet,\bullet} X, \partial, \overline{\partial} \right) \to \left(\wedge^{\bullet,\bullet} \mathfrak{g}_{\mathbb{C}}^*, \partial, \overline{\partial} \right) $$

such that

$$ \mu \circ i = \operatorname{id}_{\wedge^{\bullet,\bullet} \mathfrak{g}_{\mathbb{C}}^*} \,. $$

In particular, they induce the commutative diagram

$$ H_{\overline{\partial}}^{\bullet,\bullet} (\mathfrak{g}_{\mathbb{C}}) \xrightarrow{\ i \ } H_{\overline{\partial}}^{\bullet,\bullet}(X) \xrightarrow{\ \mu \ } H_{\overline{\partial}}^{\bullet,\bullet} (\mathfrak{g}_{\mathbb{C}}) $$

$$ \underset{\simeq}{\underbrace{\qquad\qquad\qquad}_{\operatorname{id}}} $$

from which it follows that the map $i \colon H_{\overline{\partial}}^{\bullet,\bullet} (\mathfrak{g}_{\mathbb{C}}) \to H_{\overline{\partial}}^{\bullet,\bullet}(X)$ is injective. □

For an arbitrary G-left-invariant complex structure on a nilmanifold $X = \Gamma \backslash G$, it is not known whether $i \colon H^{\bullet,\bullet}_{\bar{\partial}}(\mathfrak{g}_{\mathbb{C}}) \hookrightarrow H^{\bullet,\bullet}_{\bar{\partial}}(X)$ actually is an isomorphism, but some results are known for certain classes of G-left-invariant complex structures.

Theorem 3.2 ([Sak76, Theorem 1], [CFGU00, Main Theorem], [CF01, Theorem 2, Remark 4], [Rol09a, Theorem 1.10], [Rol11a, Corollary 3.10]). *Let $X = \Gamma \backslash G$ be a nilmanifold endowed with a G-left-invariant complex structure J, and denote the Lie algebra naturally associated to G by $\mathfrak{g}$. Then, for every $p \in \mathbb{N}$, the map of complexes*

$$\left(\wedge^{p,\bullet} \mathfrak{g}^*_{\mathbb{C}}, \bar{\partial} \right) \hookrightarrow \left(\wedge^{p,\bullet} X, \bar{\partial} \right) \tag{3.3}$$

is a quasi-isomorphism, namely,

$$i \colon H^{\bullet,\bullet}_{\bar{\partial}}(\mathfrak{g}_{\mathbb{C}}) \xrightarrow{\simeq} H^{\bullet,\bullet}_{\bar{\partial}}(X) ,$$

provided one of the following conditions holds:

- *X is holomorphically parallelizable;*
- *J is an Abelian complex structure;*
- *J is a nilpotent complex structure;*
- *J is a rational complex structure;*
- *$\mathfrak{g}$ admits a torus-bundle series compatible with J and with the rational structure induced by Γ;*
- *$\dim_{\mathbb{R}} \mathfrak{g} = 6$ and $\mathfrak{g}$ is not isomorphic to $\mathfrak{h}_7 := \left(0^3, 12, 13, 23 \right)$.*

Proof (sketch). We just sketch the idea of the proof, as in [Rol11a, Sect. 3]. For more details, see, e.g., [Sak76, CF01, Rol07, Rol09a].

The idea is to argue as in the proof of the Nomizu theorem, see Theorem 3.1, thanks to the Borel and Serre spectral sequence, [Hir95, Appendix Two].

The difference now is that the bundle series has to be compatible with the holomorphic structure. In terms of the Lie algebras exact sequences, they have to be compatible both with the rational structure induced by the lattice, and with the complex structure. One can prove that, whenever the complex structure is either holomorphically parallelizable, or Abelian, or nilpotent, or rational, one can construct a series of exact sequences of Lie algebras being compatible with both the rational and complex structures. For example, in the case J is Abelian, one just takes the descending central series $\{\mathfrak{g}^{[k]}\}_{k \in \mathbb{N}}$ of $\mathfrak{g}$. When J is rational, one takes $\{\mathfrak{g}^{[k]} + J \mathfrak{g}^{[k]}\}_{k \in \mathbb{N}}$. $\qquad\square$

We recall (see, e.g., [Rol09a, Definition 1.5]) that, given a nilpotent Lie algebra $\mathfrak{g}$, a *rational structure* for $\mathfrak{g}$ is a $\mathbb{Q}$-vector space $\mathfrak{g}_{\mathbb{Q}}$ such that $\mathfrak{g}_{\mathbb{Q}} \otimes_{\mathbb{Q}} \mathbb{R} = \mathfrak{g}$. A subalgebra $\mathfrak{h}$ of $\mathfrak{g}$ is said to be *rational with respect to a rational structure* $\mathfrak{g}_{\mathbb{Q}}$ if the $\mathbb{Q}$-vector space $\mathfrak{h} \cap \mathfrak{g}_{\mathbb{Q}}$ of $\mathfrak{h}$ is a rational structure for $\mathfrak{h}$. If G is the connected simply-connected Lie group associated to $\mathfrak{g}$, then any discrete co-compact subgroup Γ of G induces a rational structure for $\mathfrak{g}$, given by $\mathbb{Q} \log \Gamma$.

Consider a G-left-invariant complex structure on a nilmanifold $X = \Gamma \backslash G$ with associated Lie algebra $\mathfrak{g}$; we recall that:

- J is called *holomorphically parallelizable* if the holomorphic tangent bundle is holomorphically trivial, see, e.g., [Wan54, Nak75];
- J is called *Abelian* if $[Jx, Jy] = [x, y]$ for any $x, y \in \mathfrak{g}$, see, e.g., [BDMM95, ABDM11];
- J is called *nilpotent* if there exists a G-left-invariant co-frame $\{\omega^1, \ldots, \omega^n\}$ for $(T^{1,0}X)^*$ with respect to which the structure equations of X are of the form

$$\mathrm{d}\omega^j = \sum_{h<k<j} A_{hk}^j \, \omega^h \wedge \omega^k + \sum_{h,k<j} B_{hk}^j \, \omega^h \wedge \bar{\omega}^k$$

with $\left\{ A_{hk}^j, B_{hk}^j \right\}_{j,h,k} \subset \mathbb{C}$, see, e.g., [CFGU00];

- J is called *rational* if $J\left(\mathfrak{g}_{\mathbb{Q}}\right) \subseteq \mathfrak{g}_{\mathbb{Q}}$ where $\mathfrak{g}_{\mathbb{Q}}$ is the rational structure for $\mathfrak{g}$ induced by Γ, see, e.g., [CF01].

We recall also the following definitions, [Rol09a, Definition 1.8]. An ascending filtration $\left\{ S^j \mathfrak{g} \right\}_{j \in \{0,\ldots,k\}}$ on $\mathfrak{g}$ is called a *torus-bundle series compatible with a linear complex structure J on $\mathfrak{g}$ and a rational structure $\mathfrak{g}_{\mathbb{Q}}$ for $\mathfrak{g}$* if, for every $j \in \{1, \ldots, k\}$, it holds that *(i)* $S^j \mathfrak{g}$ is rational with respect to $\mathfrak{g}_{\mathbb{Q}}$ and an ideal in $S^{j+1}\mathfrak{g}$, *(ii)* $J S^j \mathfrak{g} = S^j \mathfrak{g}$, and *(iii)* $S^{j+1}\mathfrak{g}/S^j \mathfrak{g}$ is Abelian. If, in addition, it holds that *(iv)* $S^{j+1}\mathfrak{g}/S^j \mathfrak{g}$ is contained in the center of $\mathfrak{g}/S^j \mathfrak{g}$, then $\left\{ S^j \mathfrak{g} \right\}_{j \in \{0,\ldots,k\}}$ is called a *principal torus-bundle series compatible with J and $\mathfrak{g}_{\mathbb{Q}}$*. Finally, an ascending filtration $\left\{ S^j \mathfrak{g} \right\}_{j \in \{0,\ldots,k\}}$ on $\mathfrak{g}$ is called a *stable (principal) torus-bundle series* if it is a (principal) torus-bundle series compatible with J and $\mathfrak{g}_{\mathbb{Q}}$ for any complex structure J and for any rational structure $\mathfrak{g}_{\mathbb{Q}}$. By S. Rollenske's theorem [Rol09a, Theorem B], every six-dimensional nilpotent Lie algebra except $\mathfrak{h}_7 := \left(0^3, 12, 13, 23\right)$ admits a stable torus-bundle series.

Remark 3.5. For further results about the Dolbeault cohomology of solvmanifolds, obtained by H. Kasuya, we refer to [Kas13b, Kas12a, CFK13, AK13a] (see also [Kas11, Kas12b, Kas12c] for applications).

The property of computing the Dolbeault cohomology using just left-invariant forms turns out to be open along curves of left-invariant complex structures: this was proven by S. Console and A.M. Fino, [CF01].

Theorem 3.3 ([CF01, Theorem 1]). *Let $X = \Gamma \backslash G$ be a nilmanifold endowed with a G-left-invariant complex structure J, and denote the Lie algebra naturally associated to G by $\mathfrak{g}$. Let $\mathcal{U} \subseteq \mathcal{C}(\mathfrak{g})$ be the subset containing the G-left-invariant complex structures J on X such that the inclusion i is an isomorphism:*

$$\mathcal{U} := \left\{ J \in \mathcal{C}(\mathfrak{g}) \ : \ i \colon H_{\bar{\partial}}^{\bullet,\bullet}(\mathfrak{g}_{\mathbb{C}}) \xrightarrow{\cong} H_{\bar{\partial}}^{\bullet,\bullet}(X) \right\} \subseteq \mathcal{C}(\mathfrak{g}) \ .$$

Then $\mathcal{U}$ is an open set in $\mathcal{C}(\mathfrak{g})$.

The strategy of the proof consists in proving that the dimension of the orthogonal of $H_{\bar{\partial}}^{\bullet,\bullet}(\mathfrak{g}_{\mathbb{C}})$ in $H_{\bar{\partial}}^{\bullet,\bullet}(X)$ with respect to a given J-Hermitian G-left-invariant metric on $X = \Gamma \backslash G$ is an upper-semi-continuous function in $J \in \mathcal{C}(\mathfrak{g})$ and thus, if it is zero for a given $J \in \mathcal{C}(\mathfrak{g})$, then it remains equal to zero in an open neighbourhood of J in $\mathcal{C}(\mathfrak{g})$. We will use the same argument in proving Theorem 3.7, which is a slight modification of the previous result in the case of the Bott-Chern cohomology.

The aforementioned results suggest the following conjecture.

Conjecture 3.1 ([Rol11a, Conjecture 1]; see also [CFGU00, p. 5406], [CF01, p. 112]). Let $X = \Gamma \backslash G$ be a nilmanifold endowed with a G-left-invariant complex structure J, and denote the Lie algebra naturally associated to G by $\mathfrak{g}$. Then, for any $p \in \mathbb{N}$, the map of complexes (3.3) is a quasi-isomorphisms, that is,

$$i \colon H_{\bar{\partial}}^{\bullet,\bullet}(\mathfrak{g}_{\mathbb{C}}) \xrightarrow{\simeq} H_{\bar{\partial}}^{\bullet,\bullet}(X) .$$

Note that, since i is always injective by [CF01, Lemma 9], this is equivalent to asking that

$$\dim_{\mathbb{C}} \left(H_{\bar{\partial}}^{\bullet,\bullet}(\mathfrak{g}_{\mathbb{C}}) \right)^{\perp} = 0 ,$$

where the orthogonality is meant with respect to the inner product induced by a given J-Hermitian G-left-invariant metric g on X.

Finally, as an application of the previous results, we recall the following theorem by S. Rollenske, concerning the deformations of left-invariant complex structures on nilmanifolds.

Theorem 3.4 ([Rol09b, Theorem 2.6]). *Let $X = \Gamma \backslash G$ be a nilmanifold endowed with a G-left-invariant complex structure J, and denote the Lie algebra naturally associated to G by $\mathfrak{g}$. Suppose that, for $p = 1$, the map of complexes (3.3) is a quasi-isomorphism: $i \colon H_{\bar{\partial}}^{1,q}(\mathfrak{g}_{\mathbb{C}}) \xrightarrow{\simeq} H_{\bar{\partial}}^{1,q}(X)$ for every $q \in \mathbb{N}$. Then all small deformations of the complex structure J are again G-left-invariant complex structures. More precisely, the Kuranishi family of X contains only G-left-invariant complex structures.*

Remark 3.6. We refer to [Kas12c] by H. Kasuya for results concerning deformations of solvmanifolds.

3.1.3 The Bott-Chern Cohomology on Solvmanifolds

We recall here some results concerning the computation of the Bott-Chern cohomology for nilmanifolds and solvmanifolds, [Ang11].

Firstly, we prove a slight modification of [CF01, Lemma 9] proven by S. Console and A.M. Fino for the Dolbeault cohomology: we repeat here their argument for the case of the Bott-Chern cohomology.

Lemma 3.2 ([Ang11, Lemma 3.6]). *Let $X = \Gamma \backslash G$ be a solvmanifold endowed with a G-left-invariant complex structure J, and denote the Lie algebra naturally associated to G by $\mathfrak{g}$. The map of complexes (3.1) induces an injective homomorphism*

$$i \colon H_{BC}^{\bullet,\bullet}(\mathfrak{g}_{\mathbb{C}}) \hookrightarrow H_{BC}^{\bullet,\bullet}(X) \,.$$

Proof. Fix $(p,q) \in \mathbb{N}^2$. Let g be a J-Hermitian G-left-invariant metric on X and consider the induced inner product $\langle \cdot | \cdot \cdot \rangle$ on $\wedge^{\bullet,\bullet} X$. Hence, both ∂, $\bar{\partial}$, and their adjoints ∂^*, $\bar{\partial}^*$ preserve the G-left-invariant forms on X and therefore also $\tilde{\Delta}_{BC}$ does. In such a way, we get a Hodge decomposition also at the level of G-left-invariant forms:

$$\wedge^{p,q} \mathfrak{g}_{\mathbb{C}}^* = \ker \tilde{\Delta}_{BC}\lfloor_{\wedge^{p,q}\mathfrak{g}_{\mathbb{C}}^*} \oplus \operatorname{im} \partial\bar{\partial}\lfloor_{\wedge^{p-1,q-1}\mathfrak{g}_{\mathbb{C}}^*} \oplus \left(\operatorname{im} \partial^*\lfloor_{\wedge^{p+1,q}\mathfrak{g}_{\mathbb{C}}^*} + \operatorname{im} \bar{\partial}^*\lfloor_{\wedge^{p,q+1}\mathfrak{g}_{\mathbb{C}}^*} \right).$$

Now, take $[\omega] \in H_{BC}^{p,q}(\mathfrak{g}_{\mathbb{C}})$ such that $i[\omega] = 0$ in $H_{BC}^{p,q}(X)$, that is, ω is a G-left-invariant (p,q)-form on X and there exists a (possibly non-G-left-invariant) $(p-1,q-1)$-form η on X such that $\omega = \partial\bar{\partial}\eta$. Up to zero terms in $H_{BC}^{p,q}(\mathfrak{g}_{\mathbb{C}})$, we may assume that $\eta \in \left(i\left(\wedge^{p,q}\mathfrak{g}_{\mathbb{C}}^* \right) \right)^{\perp} \subseteq \wedge^{p,q} X$. Therefore, since $\bar{\partial}^*\partial^*\partial\bar{\partial}\eta$ is a G-left-invariant form (being $\partial\bar{\partial}\eta$ a G-left-invariant form), we have that

$$0 = \left\langle \bar{\partial}^*\partial^*\partial\bar{\partial}\eta \,\middle|\, \eta \right\rangle = \left\| \partial\bar{\partial}\eta \right\|^2 = \|\omega\|^2$$

and therefore $\omega = 0$. $\square$

The second general result says that, if the Dolbeault and de Rham cohomologies of a solvmanifold are computed using just left-invariant forms, then also the Bott-Chern cohomology is computed using just left-invariant forms. The idea of the proof is inspired by [Sch07, Sect. 1.c], where M. Schweitzer used a similar argument to explicitly compute the Bott-Chern cohomology in the special case of the Iwasawa manifold. (Compare also [AK12, Theorems 1.3 and 1.6, Sect. 2.3, Corollary 2.7].)

Theorem 3.5 ([Ang11, Theorem 3.7]). *Let $X = \Gamma \backslash G$ be a solvmanifold endowed with a G-left-invariant complex structure J, and denote the Lie algebra naturally associated to G by $\mathfrak{g}$. Suppose that*

$$i \colon H_{dR}^{\bullet}(\mathfrak{g}; \mathbb{C}) \xrightarrow{\simeq} H_{dR}^{\bullet}(X; \mathbb{C}) \qquad and \qquad i \colon H_{\bar{\partial}}^{\bullet,\bullet}(\mathfrak{g}_{\mathbb{C}}) \xrightarrow{\simeq} H_{\bar{\partial}}^{\bullet,\bullet}(X) \,.$$

Then also

$$i \colon H_{BC}^{\bullet,\bullet}(\mathfrak{g}_{\mathbb{C}}) \xrightarrow{\simeq} H_{BC}^{\bullet,\bullet}(X) \,.$$

Proof. Fix $(p,q) \in \mathbb{N}^2$. We prove the theorem as a consequence of the following claims.

Claim 1—*It suffices to prove that $\frac{\operatorname{im} d \cap \wedge^{p,q} X}{\operatorname{im} \partial \bar{\partial}}$ can be computed using just G-left-invariant forms.*

Indeed, in the hypothesis that $\frac{\operatorname{im} d \cap \wedge^{p,q} X}{\operatorname{im} \partial \bar{\partial}}$ can be computed using just G-left-invariant forms, the commutative diagram

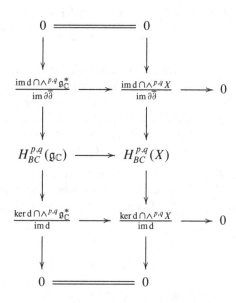

have exact rows and exact columns; hence, by using the Five Lemma, one gets that the map $H^{p,q}_{BC}(\mathfrak{g}\mathbb{C}) \to H^{p,q}_{BC}(X)$ is surjective, in fact, by Lemma 3.2, an isomorphism.

Claim 2—*Under the hypothesis that the Dolbeault cohomology is computed using just G-left-invariant forms, if ψ is a G-left-invariant $\bar{\partial}$-closed form, then every solution ϕ of $\bar{\partial}\phi = \psi$ is G-left-invariant up to $\bar{\partial}$-exact terms.*

Indeed, since $[\psi] = 0$ in $H^{\bullet,\bullet}_{\bar{\partial}}(X)$, there is a G-left-invariant form α such that $\psi = \bar{\partial}\alpha$. Hence, $\phi - \alpha$ defines a class in $H^{\bullet,\bullet}_{\bar{\partial}}(X)$ and hence $\phi - \alpha$ is G-left-invariant up to a $\bar{\partial}$-exact form, and so ϕ is.

Claim 3—*Under the hypothesis that the Dolbeault cohomology is computed using just G-left-invariant forms, the space $\frac{\operatorname{im} d \cap \wedge^{p,q} X}{\operatorname{im} \partial \bar{\partial}}$ can be computed using just G-left-invariant forms.*

Consider

$$\omega^{p,q} = d\eta \quad \operatorname{mod} \operatorname{im} \partial\bar{\partial} \in \frac{\operatorname{im} d \cap \wedge^{p,q} X}{\operatorname{im} \partial\bar{\partial}}. \tag{3.4}$$

Decomposing $\eta =: \sum_{p,q} \eta^{p,q}$ in pure-type components, the equality (3.4) is equivalent to the system

$$
\begin{cases}
\partial\eta^{p+q-1,0} = 0 & \text{mod im } \partial\bar\partial \\[4pt]
\bar\partial\eta^{p+q-\ell,\ell-1} + \partial\eta^{p+q-\ell-1,\ell} = 0 & \text{mod im } \partial\bar\partial \text{ for } \ell \in \{1,\dots,q-1\} \\[4pt]
\bar\partial\eta^{p,q-1} + \partial\eta^{p-1,q} = \omega^{p,q} & \text{mod im } \partial\bar\partial \\[4pt]
\bar\partial\eta^{\ell,p+q-\ell-1} + \partial\eta^{\ell-1,p+q-\ell} = 0 & \text{mod im } \partial\bar\partial \text{ for } \ell \in \{1,\dots,p-1\} \\[4pt]
\bar\partial\eta^{0,p+q-1} = 0 & \text{mod im } \partial\bar\partial
\end{cases}
.
$$

Applying several times `Claim` 2, we may suppose that the forms $\eta^{\ell,p+q-\ell-1}$, with $\ell \in \{0,\dots,p-1\}$, are G-left-invariant: indeed, they are G-left-invariant up to $\bar\partial$-exact terms, but $\bar\partial$-exact terms give no contribution in the system, since it is modulo im $\partial\bar\partial$. Analogously, using the conjugate version of `Claim` 2, we may suppose that the forms $\eta^{p+q-\ell-1,\ell}$, with $\ell \in \{0,\dots,q-1\}$, are G-left-invariant. Then we may suppose that $\omega^{p,q} = \bar\partial\eta^{p,q-1} + \partial\eta^{p-1,q}$ is G-left-invariant.   □

Remark 3.7. Let $X = \Gamma \backslash G$ be a solvmanifold endowed with a G-left-invariant complex structure J, and denote the Lie algebra naturally associated to G by $\mathfrak{g}$. Note that, if the map of complexes $i \colon \left(\wedge^{p,\bullet} \mathfrak{g}_{\mathbb{C}}^*, \bar\partial \right) \to \left(\wedge^{p,\bullet} X, \bar\partial \right)$ is a quasi-isomorphism for every $p \in \mathbb{N}$, that is,

$$
i \colon H_{\bar\partial}^{\bullet,\bullet}(\mathfrak{g}_{\mathbb{C}}) \xrightarrow{\ \simeq\ } H_{\bar\partial}^{\bullet,\bullet}(X) ,
$$

then also the map of complexes $i \colon (\wedge^\bullet \mathfrak{g}^*, \mathrm{d}) \to (\wedge^\bullet X, \mathrm{d})$ is a quasi-isomorphism, that is,

$$
i \colon H_{dR}^\bullet(\mathfrak{g}; \mathbb{C}) \xrightarrow{\ \simeq\ } H_{dR}^\bullet(X; \mathbb{C}) .
$$

Indeed, the map of double complexes $i \colon \left(\wedge^{\bullet,\bullet} \mathfrak{g}_{\mathbb{C}}^*, \partial, \bar\partial \right) \to \left(\wedge^{\bullet,\bullet} X, \partial, \bar\partial \right)$ induces a map between the corresponding Hodge and Frölicher spectral sequences:

$$
i \colon \left\{ \left(E_r^{\bullet,\bullet}(\mathfrak{g}_{\mathbb{C}}), \mathrm{d}_r \right) \right\}_{r \in \mathbb{N}} \to \left\{ \left(E_r^{\bullet,\bullet}(X), \mathrm{d}_r \right) \right\}_{r \in \mathbb{N}} .
$$

Since, see, e.g., [McC01, Theorem 2.15],

$$
E_1^{\bullet,\bullet}(\mathfrak{g}_{\mathbb{C}}) \simeq H_{\bar\partial}^{\bullet,\bullet}(\mathfrak{g}) \ \Rightarrow\ H_{dR}^\bullet(\mathfrak{g}_{\mathbb{C}}) \qquad \text{and}
$$

$$
E_1^{\bullet,\bullet}(X) \simeq H_{\bar\partial}^{\bullet,\bullet}(X) \ \Rightarrow\ H_{dR}^\bullet(X; \mathbb{C}) ,
$$

one gets that, if $i \colon E_1^{\bullet,\bullet}(\mathfrak{g}_{\mathbb{C}}) \to E_1^{\bullet,\bullet}(X)$ is an isomorphism, then also $i \colon H_{dR}^\bullet(\mathfrak{g}_{\mathbb{C}}) \to H_{dR}^\bullet(X; \mathbb{C})$ is an isomorphism, see, e.g., [McC01, Theorem 3.5].

Remark 3.8. The results in Lemma 3.2 and Theorem 3.5 can be generalized to an algebraic and more general context, see [AK12, Theorems 1.6 and 1.3]: as corollaries of such a generalization, one gets both results on the cohomologies of solvmanifolds endowed with left-invariant complex structures, see [AK12, Sect. 2.4], and another proof of a result concerning the symplectic cohomologies of solvmanifolds endowed with left-invariant symplectic structures, originally proven by M. Macrì, [Mac13, Theorem 3], see [AK13b, Theorems 3.2 and 3.4].

As a corollary of [Nom54, Theorem 1], [Sak76, Theorem 1], [CFGU00, Main Theorem], [CF01, Theorem 2, Remark 4], [Rol09a, Theorem 1.10], [Rol11a, Corollary 3.10], and Theorem 3.5, we get the following result.

Theorem 3.6 ([Ang11, Theorem 3.8]). *Let* $X = \Gamma \backslash G$ *be a nilmanifold endowed with a G-left-invariant complex structure J, and denote the Lie algebra naturally associated to G by $\mathfrak{g}$. Suppose that one of the following conditions holds:*

- *X is holomorphically parallelizable;*
- *J is an Abelian complex structure;*
- *J is a nilpotent complex structure;*
- *J is a rational complex structure;*
- *$\mathfrak{g}$ admits a torus-bundle series compatible with J and with the rational structure induced by Γ;*
- *$\dim_{\mathbb{R}} \mathfrak{g} = 6$ and $\mathfrak{g}$ is not isomorphic to $\mathfrak{h}_7 := (0^3, 12, 13, 23)$.*

Then the de Rham, Dolbeault, Bott-Chern and Aeppli cohomologies can be computed as the cohomologies of the corresponding sub-complexes given by the space of G-left-invariant forms on X; in other words, the inclusions of the several sub-complexes of G-left-invariant forms on X into the corresponding complexes of forms on X are quasi-isomorphisms:

$$i : H_{dR}^{\bullet}(\mathfrak{g}; \mathbb{R}) \xrightarrow{\simeq} H_{dR}^{\bullet}(X; \mathbb{R}) \qquad and \qquad i : H_{\sharp}^{\bullet,\bullet}(\mathfrak{g}_{\mathbb{C}}) \xrightarrow{\simeq} H_{\sharp}^{\bullet,\bullet}(X) ,$$

for $\sharp \in \{\partial, \bar{\partial}, BC, A\}$.

Remark 3.9. Note that Theorem 3.6, and [Hat60, Corollary 4.2], allow to straightforwardly compute the de Rham, Dolbeault, Bott-Chern, and Aeppli cohomologies of nilmanifolds, endowed with certain left-invariant complex structures, respectively the de Rham cohomology of completely-solvable solvmanifolds, just by computing the space of left-invariant (Δ, or $\bar{\square}$, or $\tilde{\Delta}_{BC}$, or $\tilde{\Delta}_A$-)harmonic forms with respect to a left-invariant Riemannian, or Hermitian, metric.

Indeed, suppose that X is a nilmanifold, endowed with a left-invariant complex structure, or a completely-solvable solvmanifold, satisfying $i : H_{dR}^{\bullet}(\mathfrak{g}; \mathbb{R}) \xrightarrow{\simeq} H_{dR}^{\bullet}(X; \mathbb{R})$, or $i : H_{\sharp}^{\bullet,\bullet}(\mathfrak{g}_{\mathbb{C}}) \xrightarrow{\simeq} H_{\sharp}^{\bullet,\bullet}(X)$, for some $\sharp \in \{\partial, \bar{\partial}, BC, A\}$. Let g be a left-invariant Riemannian, or Hermitian, metric on X. Hence, the operators $\Delta, \bar{\square}, \tilde{\Delta}_{BC}, \tilde{\Delta}_A$ send the subspace of left-invariant forms to the subspace of left-invariant forms, and induce the self-adjoint operators

$$\Delta \in \text{End}\left(\wedge^{\bullet}\mathfrak{g}^*\right) , \qquad \bar{\square} \in \text{End}\left(\wedge^{\bullet,\bullet}\mathfrak{g}_{\mathbb{C}}^*\right) , \qquad \tilde{\Delta}_{BC} \in \text{End}\left(\wedge^{\bullet,\bullet}\mathfrak{g}_{\mathbb{C}}^*\right) ,$$

$$\tilde{\Delta}_A \in \text{End}\left(\wedge^{\bullet,\bullet}\mathfrak{g}_{\mathbb{C}}^*\right) ,$$

with respect to the inner products $\langle \cdot, \cdots \rangle$ induced by g on the space $\wedge^\bullet \mathfrak{g}^*$ and on the space $\wedge^{\bullet,\bullet} \mathfrak{g}_\mathbb{C}^*$. Hence, one gets the orthogonal decompositions

$$\wedge^\bullet \mathfrak{g}^* = \ker \Delta \oplus \operatorname{im} \Delta , \qquad\qquad \wedge^{\bullet,\bullet} \mathfrak{g}_\mathbb{C}^* = \ker \overline{\square} \oplus \operatorname{im} \overline{\square} ,$$

$$\wedge^{\bullet,\bullet} \mathfrak{g}_\mathbb{C}^* = \ker \tilde{\Delta}_{BC} \oplus \operatorname{im} \tilde{\Delta}_{BC} , \qquad\qquad \wedge^{\bullet,\bullet} \mathfrak{g}_\mathbb{C}^* = \ker \tilde{\Delta}_A \oplus \operatorname{im} \tilde{\Delta}_A$$

(one could argue also by using the F.A. Belgun symmetrization trick [Bel00, Theorem 7]). It follows that

$$H_{dR}^\bullet (\mathfrak{g}; \mathbb{R}) \simeq \ker \Delta , \qquad H_{\overline{\partial}}^{\bullet,\bullet} (\mathfrak{g}_\mathbb{C}) \simeq \ker \overline{\square} , \qquad H_{BC}^{\bullet,\bullet} (\mathfrak{g}_\mathbb{C}) \simeq \ker \tilde{\Delta}_{BC} ,$$

$$H_A^{\bullet,\bullet} (\mathfrak{g}_\mathbb{C}) \simeq \ker \tilde{\Delta}_A .$$

Remark 3.10. Let $X = \Gamma \backslash G$ be a $2n$-dimensional solvmanifold endowed with a G-left-invariant complex structure J, and denote the Lie algebra naturally associated to G by $\mathfrak{g}$. The map of complexes (3.2) induces an injective homomorphism

$$i \colon H_A^{\bullet,\bullet} (\mathfrak{g}_\mathbb{C}) \hookrightarrow H_A^{\bullet,\bullet}(X) .$$

Furthermore, if $i \colon H_{BC}^{\bullet,\bullet} (\mathfrak{g}_\mathbb{C}) \xrightarrow{\simeq} H_{BC}^{\bullet,\bullet}(X)$, then the map of complexes (3.2) is a quasi-isomorphism, that is,

$$i \colon H_A^{\bullet,\bullet} (\mathfrak{g}_\mathbb{C}) \xrightarrow{\simeq} H_A^{\bullet,\bullet}(X) .$$

Indeed, fix a G-left-invariant Hermitian metric g on X. Recall that

$$* \colon H_A^{\bullet_1,\bullet_2}(X) \xrightarrow{\simeq} H_{BC}^{n-\bullet_2, n-\bullet_1}(X)$$

is an isomorphism, [Sch07, Sect. 2.c]. Analogously, note that, by Remark 3.9 and since g is G-left-invariant, the map $* \colon \wedge^{\bullet_1,\bullet_2} \mathfrak{g}_\mathbb{C}^* \xrightarrow{\simeq} \wedge^{n-\bullet_2, n-\bullet_1} \mathfrak{g}_\mathbb{C}^*$ induces an isomorphism

$$* \colon H_A^{\bullet_1,\bullet_2} (\mathfrak{g}_\mathbb{C}) \xrightarrow{\simeq} H_{BC}^{n-\bullet_2, n-\bullet_1} (\mathfrak{g}_\mathbb{C}) .$$

Note also that the diagram

$$
\begin{array}{ccc}
H_A^{\bullet_1,\bullet_2} (\mathfrak{g}_\mathbb{C}) & \xrightarrow{\ i\ } & H_A^{\bullet_1,\bullet_2}(X) \\[2mm]
{\scriptstyle *} \downarrow {\scriptstyle \simeq} & & {\scriptstyle \simeq} \downarrow {\scriptstyle *} \\[2mm]
H_{BC}^{n-\bullet_2, n-\bullet_1} (\mathfrak{g}_\mathbb{C}) & \xrightarrow[\ i\ ]{} & H_{BC}^{n-\bullet_2, n-\bullet_1}(X)
\end{array}
$$

commutes, since g is G-left-invariant. Since the map $i\colon H_{BC}^{n-\bullet_2,n-\bullet_1}(\mathfrak{g}_{\mathbb{C}}) \hookrightarrow$ $H_{BC}^{n-\bullet_2,n-\bullet_1}(X)$ is injective by Lemma 3.2, then also the map $H_A^{\bullet_1,\bullet_2}(\mathfrak{g}_{\mathbb{C}}) \to$ $H_A^{\bullet_1,\bullet_2}(X)$ is injective. If $i\colon H_{BC}^{n-\bullet_2,n-\bullet_1}(\mathfrak{g}_{\mathbb{C}}) \hookrightarrow H_{BC}^{n-\bullet_2,n-\bullet_1}(X)$ is actually an isomorphism, then also $i\colon H_A^{\bullet_1,\bullet_2}(\mathfrak{g}_{\mathbb{C}}) \to H_A^{\bullet_1,\bullet_2}(X)$ is an isomorphism.

Remark 3.11. In order to generalize Theorem 3.6 to solvmanifolds, in [AK12, AK13a], finite-dimensional complexes allowing to compute the Bott-Chern cohomology of certain classes of solvmanifolds endowed with left-invariant complex structures are constructed, yielding explicit computations for the completely-solvable Nakamura manifold and for the holomorphically parallelizable Nakamura manifold. See Appendix: Cohomology of Solvmanifolds for further details, for which we refer to [AK12, AK13a].

A slight modification of [CF01, Theorem 1] by S. Console and A.M. Fino gives the following result, which says that the property of computing the Bott-Chern cohomology using just left-invariant forms is open in the space of left-invariant complex structures on solvmanifolds.

Theorem 3.7 ([Ang11, Theorem 3.9]). *Let $X = \Gamma \backslash G$ be a solvmanifold endowed with a G-left-invariant complex structure J, and denote the Lie algebra naturally associated to G by $\mathfrak{g}$. Let $\sharp \in \{\partial, \bar{\partial}, BC, A\}$. Suppose that*

$$i\colon H_{\sharp_J}^{\bullet,\bullet}(\mathfrak{g}_{\mathbb{C}}) \xrightarrow{\simeq} H_{\sharp_J}^{\bullet,\bullet}(X) \ .$$

Then there exists an open neighbourhood $\mathcal{U}$ of J in $\mathcal{C}(\mathfrak{g})$ such that any $\tilde{J} \in \mathcal{U}$ still satisfies

$$i\colon H_{\sharp_{\tilde{J}}}^{\bullet,\bullet}(\mathfrak{g}_{\mathbb{C}}) \xrightarrow{\simeq} H_{\sharp_{\tilde{J}}}^{\bullet,\bullet}(X) \ .$$

In other words, the set

$$\mathcal{U} := \left\{ J \in \mathcal{C}(\mathfrak{g}) \ : \ i\colon H_{\sharp_{\tilde{J}}}^{\bullet,\bullet}(\mathfrak{g}_{\mathbb{C}}) \xrightarrow{\simeq} H_{\sharp_{\tilde{J}}}^{\bullet,\bullet}(X) \right\}$$

is open in $\mathcal{C}(\mathfrak{g})$.

Proof. As a matter of notation, for $\varepsilon > 0$ small enough, we consider

$$\{(X, J_t) \ : \ t \in \Delta(0, \varepsilon)\} \twoheadrightarrow \Delta(0, \varepsilon)$$

a complex-analytic family of G-left-invariant complex structures on X, where $\Delta(0, \varepsilon) := \{t \in \mathbb{C}^m \ : \ |t| < \varepsilon\}$ for some $m \in \mathbb{N} \setminus \{0\}$; moreover, let $\{g_t\}_{t \in \Delta(0,\varepsilon)}$ be a family of J_t-Hermitian G-left-invariant metrics on X depending smoothly on t. We will denote by $\bar{\partial}_t := \bar{\partial}_{J_t}$ and $\bar{\partial}_t^* := -*_{g_t} \partial_{J_t} *_{g_t}$ the $\bar{\partial}$ operator and its g_t-adjoint respectively for the Hermitian structure (J_t, g_t) and we set $\Delta_t := \Delta_{\sharp_{J_t}}$ one of the differential operators involved in the definition of the Dolbeault, conjugate Dolbeault, Bott-Chern or Aeppli cohomologies with respect to (J_t, g_t);

we remark that Δ_t is a self-adjoint elliptic differential operator for all the considered cohomologies.

By hypothesis, we have that $\left(H^{\bullet,\bullet}_{\sharp_{J_0}}(\mathfrak{g}_\mathbb{C})\right)^\perp = \{0\}$, where the orthogonality is meant with respect to the inner product induced by g_0, and we have to prove the same replacing 0 with $t \in \Delta(0,\varepsilon)$. Therefore, it will suffice to prove that

$$\Delta(0,\varepsilon) \ni t \mapsto \dim_\mathbb{C} \left(H^{\bullet,\bullet}_{\sharp_{J_t}}(\mathfrak{g}_\mathbb{C})\right)^\perp \in \mathbb{N}$$

is an upper-semi-continuous function at 0. For any $t \in \Delta(0,\varepsilon)$, being Δ_t a self-adjoint elliptic differential operator, there exists a complete orthonormal basis $\{e_i(t)\}_{i \in I}$ of eigen-forms for Δ_t spanning $\left(\wedge^{\bullet,\bullet}_{J_t} \mathfrak{g}^*_\mathbb{C}\right)^\perp$, the orthogonal complement of the space of G-left-invariant forms, see [KS60, Theorem 1]. For any $i \in I$ and $t \in \Delta(0,\varepsilon)$, let $a_i(t)$ be the eigen-value corresponding to $e_i(t)$; Δ_t depending differentiably on $t \in \Delta(0,\varepsilon)$, for any $i \in I$, the function $\Delta(0,\varepsilon) \ni t \mapsto a_i(t) \in \mathbb{C}$ is continuous, see [KS60, Theorem 2]. Therefore, for any $t_0 \in \Delta(0,\varepsilon)$, choosing a constant $c > 0$ such that $c \notin \overline{\{a_i(t_0) \; : \; i \in I\}}$, the function

$$\Psi_c \colon \Delta(0,\varepsilon) \to \mathbb{N}, \qquad t \mapsto \dim \operatorname{span}\{e_i(t) \; : \; a_i(t) < c\}$$

is locally constant at t_0; moreover, for any $t \in \Delta(0,\varepsilon)$ and for any $c > 0$, we have

$$\Psi_c(t) \geq \dim_\mathbb{C} \left(H^{\bullet,\bullet}_{\sharp_{J_t}}(\mathfrak{g}_\mathbb{C})\right)^\perp .$$

Since the spectrum of Δ_{t_0} has no accumulation point for any $t_0 \in \Delta(0,\varepsilon)$, see [KS60, Theorem 1], the theorem follows choosing $c > 0$ small enough so that $\Psi_c(0) = \dim_\mathbb{C} \left(H^{\bullet,\bullet}_{\sharp_{J_0}}(\mathfrak{g}_\mathbb{C})\right)^\perp$. $\square$

In particular, the left-invariant complex structures on nilmanifolds belonging to the classes of Theorem 3.6, and their small deformations satisfy the following conjecture, which generalizes Conjecture 3.1.

Conjecture 3.2 ([Rol11a, Conjecture 1], [Ang11, Conjecture 3.10]; see also [CFGU00, p. 5406], [CF01, p. 112]). Let $X = \Gamma \backslash G$ be a nilmanifold endowed with a G-left-invariant complex structure J, and denote the Lie algebra naturally associated to G by $\mathfrak{g}$. Then the de Rham, Dolbeault, Bott-Chern and Aeppli cohomologies can be computed as the cohomologies of the corresponding subcomplexes given by the space of G-left-invariant forms on X, that is,

$$\dim_\mathbb{R} \left(H^\bullet_{dR}(\mathfrak{g};\mathbb{R})\right)^\perp = 0 \qquad \text{and} \qquad \dim_\mathbb{C} \left(H^{\bullet,\bullet}_\sharp(\mathfrak{g}_\mathbb{C})\right)^\perp = 0 ,$$

where $\sharp \in \{\partial, \bar{\partial}, BC, A\}$, and the orthogonality is meant with respect to the inner product induced by a given J-Hermitian G-left-invariant metric g on X.

3.2 The Cohomologies of the Iwasawa Manifold and of Its Small Deformations

The Iwasawa manifold is one of the simplest example of non-Kähler complex manifold: as such, it has been studied by several authors, and it has turned out to be a fruitful source of interesting behaviours, see, e.g., [FG86, Nak75, AB90, Bas99, AGS97, KS04, Ye08, Sch07, AT11, Ang11, Fra11].

In this section, we recall the construction of the Iwasawa manifold Sect. 3.2.1.1, see, e.g., [FG86], [Nak75, Sect. 2], and of its Kuranishi space, Sect. 3.2.1.2, see [Nak75, Sect. 3]; then we write down the de Rham cohomology, Sect. 3.2.2, and the Dolbeault cohomology, Sect. 3.2.3 (using [Nom54, Theorem 1], and [Sak76, Theorem 1] and [CF01, Theorem 1]), and we compute the Bott-Chern and Aeppli cohomologies, Sect. 3.2.4 (using Theorems 3.6 and 3.7), of the Iwasawa manifold and of its small deformations.

3.2.1 The Iwasawa Manifold and Its Small Deformations

3.2.1.1 The Iwasawa Manifold

In order to answer to a question by S. Iitaka in [Iit72], I. Nakamura classified in [Nak75, Sect. 2] the three-dimensional holomorphically parallelizable solvmanifolds into four classes by numerical invariants, giving the Iwasawa manifold $\mathbb{I}_3$ as an example in the second class.

Let $\mathbb{H}(3; \mathbb{C})$ be the three-dimensional *Heisenberg group* over $\mathbb{C}$ defined by

$$\mathbb{H}(3; \mathbb{C}) := \left\{ \begin{pmatrix} 1 & z^1 & z^3 \\ 0 & 1 & z^2 \\ 0 & 0 & 1 \end{pmatrix} \in GL(3; \mathbb{C}) \ : \ z^1, z^2, z^3 \in \mathbb{C} \right\},$$

where the product is the one induced by matrix multiplication. (Equivalently, one can consider $\mathbb{H}(3; \mathbb{C})$ as isomorphic to $(\mathbb{C}^3, *)$, where the group structure $*$ on $\mathbb{C}^3$ is defined as

$$(z_1, z_2, z_3) * (w_1, w_2, w_3) := (z_1 + w_1, z_2 + w_2, z_3 + z_1 w_2 + w_3) \ .)$$

It is straightforward to prove that $\mathbb{H}(3; \mathbb{C})$ is a connected simply-connected complex 2-step nilpotent Lie group, that is, the Lie algebra $(\mathfrak{h}_3, [\cdot, \cdot\cdot])$ naturally associated to $\mathbb{H}(3; \mathbb{C})$ satisfies $[\mathfrak{h}_3, \mathfrak{h}_3] \neq 0$ and $[\mathfrak{h}_3, [\mathfrak{h}_3, \mathfrak{h}_3]] = 0$.

One finds that

$$\begin{cases} \varphi^1 := d z^1 \\ \varphi^2 := d z^2 \\ \varphi^3 := d z^3 - z^1 d z^2 \end{cases}$$

Fig. 3.1 The double complex $\left(\wedge^{\bullet,\bullet}\left(\mathfrak{h}_3 \otimes_{\mathbb{R}} \mathbb{C}\right)^*, \partial, \bar{\partial}\right)$

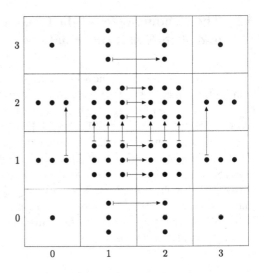

is a $\mathbb{H}(3; \mathbb{C})$-left-invariant co-frame for the space of $(1,0)$-forms on $\mathbb{H}(3; \mathbb{C})$, and that the structure equations with respect to this co-frame are

$$\begin{cases} \mathrm{d}\,\varphi^1 = 0 \\ \mathrm{d}\,\varphi^2 = 0 \\ \mathrm{d}\,\varphi^3 = -\varphi^1 \wedge \varphi^2 \end{cases}.$$

Consider the action on the left of $\mathbb{H}(3; \mathbb{Z}[\mathrm{i}]) := \mathbb{H}(3; \mathbb{C}) \cap \mathrm{GL}(3; \mathbb{Z}[\mathrm{i}])$ on $\mathbb{H}(3; \mathbb{C})$ and take the compact quotient

$$\mathbb{I}_3 := \mathbb{H}(3; \mathbb{Z}[\mathrm{i}]) \backslash \mathbb{H}(3; \mathbb{C}).$$

One gets that $\mathbb{I}_3$ is a three-dimensional complex nilmanifold, whose ($\mathbb{H}(3; \mathbb{C})$-left-invariant) complex structure J_0 is the one inherited by the standard complex structure on $\mathbb{C}^3$; $\mathbb{I}_3$ is called the *Iwasawa manifold*. (See also [KS04].)

The forms φ^1, φ^2 and φ^3, being $\mathbb{H}(3; \mathbb{C})$-left-invariant, define a co-frame also for $\left(T^{1,0}\mathbb{I}_3\right)^*$. Note that $\mathbb{I}_3$ is a holomorphically parallelizable manifold, that is, its holomorphic tangent bundle is holomorphically trivial. Since, for example, φ^3 is a non-closed holomorphic form, it follows that $\mathbb{I}_3$ admits no Kähler metric. In fact, one can show that $\mathbb{I}_3$ is not formal, having a non-zero Massey triple product, see [FG86, p. 158]; therefore the underlying differentiable manifold of $\mathbb{I}_3$ has no complex structure admitting Kähler metrics, see [DGMS75, Main Theorem], even though all the topological obstructions concerning the Betti numbers are satisfied. Nevertheless, $\mathbb{I}_3$ admits the balanced metric $\omega := \sum_{j=1}^3 \varphi^j \wedge \bar{\varphi}^j$.

We sketch in Fig. 3.1 the structure of the finite-dimensional double complex $\left(\wedge^{\bullet,\bullet}\left(\mathfrak{h}_3 \otimes_{\mathbb{R}} \mathbb{C}\right)^*, \partial, \bar{\partial}\right)$: the dots denote a basis of $\wedge^{\bullet,\bullet}\left(\mathfrak{h}_3 \otimes_{\mathbb{R}} \mathbb{C}\right)^*$, horizontal arrows are meant as ∂, vertical ones as $\bar{\partial}$ and zero arrows are not depicted.

Table 3.1 Dimensions of the cohomologies of the Iwasawa manifold

♯	$h_\sharp^{1,0}$	$h_\sharp^{0,1}$	$h_\sharp^{2,0}$	$h_\sharp^{1,1}$	$h_\sharp^{0,2}$	$h_\sharp^{3,0}$	$h_\sharp^{2,1}$	$h_\sharp^{1,2}$	$h_\sharp^{0,3}$	$h_\sharp^{3,1}$	$h_\sharp^{2,2}$	$h_\sharp^{1,3}$	$h_\sharp^{3,2}$	$h_\sharp^{2,3}$
$\bar\partial$	3	2	3	6	2	1	6	6	1	2	6	3	2	3
∂	2	3	2	6	3	1	6	6	1	3	6	2	3	2
BC	2	2	3	4	3	1	6	6	1	2	8	2	3	3
A	3	3	2	8	2	1	6	6	1	3	4	3	2	2
dR	$b_1 = 4$		$b_2 = 8$				$b_3 = 10$			$b_4 = 8$			$b_5 = 4$	

In Fig. 3.2, we highlight what generators of the double complex $\left(\wedge^{\bullet,\bullet}\left(\mathfrak{h}_3 \otimes_{\mathbb{R}} \mathbb{C}\right)^*, \partial, \bar\partial\right)$ give a non-trivial contribution in Dolbeault, respectively conjugate Dolbeault, respectively Bott-Chern, respectively Aeppli cohomology, in such a way to get immediately the corresponding cohomological diamonds. As a matter of notation, a filled point yields a non-trivial cohomology class, while empty points do not contribute to cohomology.

We summarize the dimensions of the cohomology groups for the Iwasawa manifold in Table 3.1.

Remark 3.12. The pictures in Fig. 3.1, as well as in Fig. 3.2, have been inspired by an answer by G. Kuperberg on "Dolbeault cohomology of Hopf manifolds" on MathOverflow, see http://mathoverflow.net/questions/25723/dolbeault-cohomology-of-hopf-manifolds. Such pictures have inspired and motivate Theorems 2.13 and 2.14.

3.2.1.2 Small Deformations of the Iwasawa Manifold

Still in [Nak75], I. Nakamura explicitly constructed the Kuranishi family of deformations of $\mathbb{I}_3$, showing that it is smooth and depends on six effective parameters, [Nak75, pp. 94–95], compare also [Rol11b, Corollary 4.9]. In particular, he computed the Hodge numbers of the small deformations of $\mathbb{I}_3$ proving that they have not to remain invariant along a complex-analytic family of complex structures, [Nak75, Theorem 2], compare also [Ye08, Sect. 4]; moreover, he proved in this way that the property of being holomorphically parallelizable is not stable under small deformations, [Nak75, p. 86], compare also [Rol11b, Theorem 5.1, Corollary 5.2].

Firstly, we recall in the following theorem the results by I. Nakamura concerning the Kuranishi space of the Iwasawa manifold.

Theorem 3.8 ([Nak75, pp. 94–96]). *Consider the Iwasawa manifold* $\mathbb{I}_3 := \mathbb{H}(3; \mathbb{Z}[i]) \backslash \mathbb{H}(3; \mathbb{C})$*. There exists a locally complete complex-analytic family of complex structures* $\{X_t = (\mathbb{I}_3, J_t)\}_{t \in \Delta(0,\varepsilon)}$*, deformations of* $\mathbb{I}_3$*, depending on six parameters*

$$\mathbf{t} = (t_{11}, t_{12}, t_{21}, t_{22}, t_{31}, t_{32}) \in \Delta(\mathbf{0}, \varepsilon) \subset \mathbb{C}^6,$$

where $\varepsilon > 0$ *is small enough,* $\Delta(\mathbf{0}, \varepsilon) := \{s \in \mathbb{C}^6 : |s| < \varepsilon\}$*, and* $X_0 = \mathbb{I}_3$*.*

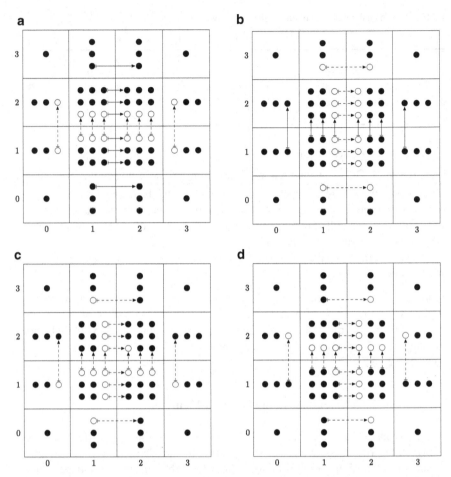

Fig. 3.2 Cohomologies of the Iwasawa manifold. (**a**) The diagram for the Dolbeault cohomology $H_{\bar{\partial}}^{\bullet,\bullet}(\mathbb{I}_3)$ of the Iwasawa manifold $\mathbb{I}_3$. (**b**) The diagram for the conjugate Dolbeault cohomology $H_{\partial}^{\bullet,\bullet}(\mathbb{I}_3)$ of the Iwasawa manifold $\mathbb{I}_3$. (**c**) The diagram for the Bott-Chern cohomology $H_{BC}^{\bullet,\bullet}(\mathbb{I}_3)$ of the Iwasawa manifold $\mathbb{I}_3$. (**d**) The diagram for the Aeppli cohomology $H_{A}^{\bullet,\bullet}(\mathbb{I}_3)$ of the Iwasawa manifold $\mathbb{I}_3$

A set of holomorphic coordinates for $X_{\mathbf{t}}$ is given by

$$\begin{cases} \zeta^1 := \zeta^1(\mathbf{t}) := z^1 + \sum_{k=1}^{2} t_{1k}\, \bar{z}^k \\ \zeta^2 := \zeta^2(\mathbf{t}) := z^2 + \sum_{k=1}^{2} t_{2k}\, \bar{z}^k \\ \zeta^3 := \zeta^3(\mathbf{t}) := z^3 + \sum_{k=1}^{2} \left(t_{3k} + t_{2k}\, z^1 \right) \bar{z}^k + A\left(\bar{z}^1, \bar{z}^2 \right) - D\left(\mathbf{t} \right) \bar{z}^3 \end{cases}$$

where

$$D\left(\mathbf{t}\right) := \det \begin{pmatrix} t_{11} & t_{12} \\ t_{21} & t_{22} \end{pmatrix}$$

and

$$A\left(\bar{z}^1, \bar{z}^2\right) := \frac{1}{2}\left(t_{11}\, t_{21}\left(\bar{z}^1\right)^2 + 2\, t_{11}\, t_{22}\, \bar{z}^1\, \bar{z}^2 + t_{12}\, t_{22}\left(\bar{z}^2\right)^2\right) .$$

For every $\mathbf{t} \in \Delta(\mathbf{0}, \varepsilon)$, *the universal covering of* $X_{\mathbf{t}}$ *is* $\mathbb{C}^3$; *more precisely,*

$$X_{\mathbf{t}} = \Gamma_{\mathbf{t}} \backslash \mathbb{C}^3 ,$$

where $\Gamma_{\mathbf{t}}$ *is the subgroup generated by the transformations*

$$\left(\zeta^1, \zeta^2, \zeta^3\right) \overset{\left(\omega^1, \omega^2, \omega^3\right)}{\longmapsto} \left(\tilde{\zeta}^1, \tilde{\zeta}^2, \tilde{\zeta}^3\right) ,$$

varying $\left(\omega^1, \omega^2, \omega^3\right) \in (\mathbb{Z}\,[\mathrm{i}])^3$, *where*

$$\begin{cases} \tilde{\zeta}^1 := \zeta^1 + \left(\omega^1 + t_{11}\,\bar{\omega}^1 + t_{12}\,\bar{\omega}^2\right) \\[4pt] \tilde{\zeta}^2 := \zeta^2 + \left(\omega^2 + t_{21}\,\bar{\omega}^1 + t_{22}\,\bar{\omega}^2\right) \\[4pt] \tilde{\zeta}^3 := \zeta^3 + \left(\omega^3 + t_{31}\,\bar{\omega}^1 + t_{32}\,\bar{\omega}^2\right) + \omega^1\,\zeta^2 \\[4pt] \qquad + \left(t_{21}\,\bar{\omega}^1 + t_{22}\,\bar{\omega}^2\right)\left(\zeta^1 + \omega^1\right) + A\left(\bar{\omega}^1, \bar{\omega}^2\right) - D\left(\mathbf{t}\right)\,\bar{\omega}^3 \end{cases}$$

Proof (sketch). We just briefly sketch the idea of the proof, referring to [Nak75, pp. 94–96], see also [Uen75, pp. 217–219], for further details.

With the above notations, consider the $\mathbb{H}(3; \mathbb{C})$-left-invariant co-frame $\{\varphi^1, \varphi^2, \varphi^3\}$ of $\left(T^{1,0}\mathbb{I}_3\right)^*$, where

$$\varphi^1 := \mathrm{d}z^1 , \qquad \varphi^2 := \mathrm{d}z^2 , \qquad \varphi^3 := \mathrm{d}z^3 - z^1\,\mathrm{d}z^2 ,$$

and its dual frame $\{\vartheta_1, \vartheta_2, \vartheta_3\}$ of $T^{1,0}\mathbb{I}_3$, where

$$\vartheta_1 := \frac{\partial}{\partial z^1} , \qquad \vartheta_2 := \frac{\partial}{\partial z^2} + z^1\,\frac{\partial}{\partial z^3} , \qquad \vartheta_3 := \frac{\partial}{\partial z^3} .$$

One has

$$H^{0,1}\left(\mathbb{I}_3; \Theta_{\mathbb{I}_3}\right) = \mathbb{C}\left\{\vartheta_1 \otimes \bar{\varphi}^1,\, \vartheta_2 \otimes \bar{\varphi}^1,\, \vartheta_3 \otimes \bar{\varphi}^1,\, \vartheta_1 \otimes \bar{\varphi}^2,\, \vartheta_2 \otimes \bar{\varphi}^2,\, \vartheta_3 \otimes \bar{\varphi}^2\right\} .$$

The vector-valued $(0, 1)$-forms solving the Maurer and Cartan equation,

$$\begin{cases} \bar{\partial}\psi\left(\mathbf{t}\right) + \frac{1}{2}\left[\psi\left(\mathbf{t}\right), \psi\left(\mathbf{t}\right)\right] = 0 \\[4pt] \psi\left(\mathbf{0}\right) = 0 \\[4pt] \left.\dfrac{\partial\psi}{\partial t_{h\lambda}}\right|_{\mathbf{t}=0} = \vartheta_h \otimes \bar{\varphi}^\lambda \quad \text{for } h \in \{1, 2, 3\} \text{ and } \lambda \in \{1, 2\} \end{cases} ,$$

are

$$\psi(\mathbf{t}) = \sum_{h=1}^{3} \sum_{\lambda=1}^{2} t_{h\lambda} \vartheta_h \otimes \bar{\varphi}^{\lambda} - (t_{11}t_{22} - t_{21}t_{12}) \vartheta_3 \otimes \bar{\varphi}^3$$

Finally, a set of holomorphic coordinates for $X_{\mathbf{t}}$ is obtained by solving, for $\nu \in \{1, 2, 3\}$,

$$\begin{cases} \bar{\partial}\zeta^{\nu}(\mathbf{t}) - \psi(\mathbf{t})\zeta^{\nu}(\mathbf{t}) = 0 \\ \zeta^{\nu}(0) = z^{\nu} \end{cases}.$$

For $\mathbf{t}$ small enough, the map $\Phi_{\mathbf{t}} \colon (\zeta^1(\mathbf{t}), \zeta^2(\mathbf{t}), \zeta^3(\mathbf{t})) \mapsto (z^1, z^2, z^3)$ is a diffeomorphism of $\mathbb{C}^3$. Hence one gets the commutative diagram

$$\begin{array}{ccc} \mathbb{C}^3 & \xrightarrow{\Phi_{\mathbf{t}}} & \mathbb{C}^3 \\ \downarrow & \simeq & \downarrow \\ \mathbb{I}_3 & \xrightarrow{\simeq} & X_{\mathbf{t}} \end{array}$$

which induces a covering map $\mathbb{C}^3 \to X_{\mathbf{t}}$. Hence $X_{\mathbf{t}} = \Gamma_{\mathbf{t}} \backslash \mathbb{C}^3$ for a group $\Gamma_{\mathbf{t}}$ of analytic automorphisms of $\mathbb{C}^3$.   □

Remark 3.13. Note that, by [Rol11b, Theorem 4.5], if $X = \Gamma \backslash G$ is a holomorphically parallelizable nilmanifold and G is ν-step nilpotent, then $\mathrm{Kur}(X)$ is cut out by polynomial equations of degree at most ν; furthermore, by [Rol11b, Corollary 4.9], the Kuranishi space of X is smooth if and only if the associated Lie algebra $\mathfrak{g}$ to G is a free 2-step nilpotent Lie algebra, i.e., $\mathfrak{g} \simeq \mathfrak{b}_m$ with $m = \dim_{\mathbb{C}} H^{0,1}_{\bar{\partial}}(X)$, where $\mathfrak{b}_m := \mathbb{C}^m \oplus \wedge^2\mathbb{C}^m$ with Lie bracket $[a_1 + b_1 \wedge c_1, a_2 + b_2 \wedge c_2] := a_1 \wedge a_2$ for $a_1, b_1, c_1, a_2, b_2, c_2 \in \mathbb{C}^m$.

According to the classification by I. Nakamura, the small deformations of $\mathbb{I}_3$ are divided into three classes, *(i)–(iii)*, in terms of their Hodge numbers: such classes are explicitly described by means of polynomial relations in the parameters, see [Nak75, Sect. 3]. As we will see in Sect. 3.2.4, it turns out that the Bott-Chern cohomology yields a finer classification of the Kuranishi space of $\mathbb{I}_3$; more precisely, $h^{2,2}_{BC}$ assumes different values within class *(ii)*, respectively class *(iii)*, according to the rank of a certain matrix whose entries are related to the complex structure equations with respect to a suitable co-frame, whereas the numbers corresponding to class *(i)* coincide with those for $\mathbb{I}_3$: this allows a further subdivision of classes *(ii)* and *(iii)* into subclasses *(ii.a)*, *(ii.b)*, and *(iii.a)*, *(iii.b)*.

More precisely, the classes and subclasses of this classification are characterized by the following values of the parameters:

class *(i)* $t_{11} = t_{12} = t_{21} = t_{22} = 0$;
class *(ii)* $D(\mathbf{t}) = 0$ and $(t_{11}, t_{12}, t_{21}, t_{22}) \neq (0, 0, 0, 0)$:

  subclass *(ii.a)* $D(\mathbf{t}) = 0$ and rk $S = 1$;
  subclass *(ii.b)* $D(\mathbf{t}) = 0$ and rk $S = 2$;

class *(iii)* $D(\mathbf{t}) \neq 0$:

  subclass *(iii.a)* $D(\mathbf{t}) \neq 0$ and rk $S = 1$;
  subclass *(iii.b)* $D(\mathbf{t}) \neq 0$ and rk $S = 2$.

The matrix S is defined by

$$S := \begin{pmatrix} \overline{\sigma_{1\bar{1}}} & \overline{\sigma_{2\bar{2}}} & \overline{\sigma_{1\bar{2}}} & \overline{\sigma_{2\bar{1}}} \\ \sigma_{1\bar{1}} & \sigma_{2\bar{3}} & \sigma_{2\bar{1}} & \sigma_{1\bar{3}} \end{pmatrix}$$

where $\sigma_{1\bar{1}}, \sigma_{1\bar{3}}, \sigma_{2\bar{1}}, \sigma_{2\bar{3}} \in \mathbb{C}$ and $\sigma_{12} \in \mathbb{C}$ are complex numbers depending only on **t** such that

$$d\varphi_t^3 =: \sigma_{12}\, \varphi_t^1 \wedge \varphi_t^2 + \sigma_{1\bar{1}}\, \varphi_t^1 \wedge \bar{\varphi}_t^1 + \sigma_{1\bar{3}}\, \varphi_t^1 \wedge \bar{\varphi}_t^2 + \sigma_{2\bar{1}}\, \varphi_t^2 \wedge \bar{\varphi}_t^1 + \sigma_{2\bar{3}}\, \varphi_t^2 \wedge \bar{\varphi}_t^2 ,$$

being

$$\varphi_t^1 := d\zeta_t^1 , \qquad \varphi_t^2 := d\zeta_t^2 , \qquad \varphi_t^3 := d\zeta_t^3 - z_1\, d\zeta_t^2 - \left(t_{21}\, \bar{z}^1 + t_{22}\, \bar{z}^2 \right) d\zeta_t^1 ,$$

see Sect. 3.2.1.3. As we will show, see Sect. 3.2.1.3, the first order asymptotic behaviour of $\sigma_{12}, \sigma_{1\bar{1}}, \sigma_{1\bar{3}}, \sigma_{2\bar{1}}, \sigma_{2\bar{3}}$ for **t** near 0 is the following:

$$\begin{cases} \sigma_{12} = -1 + o\,(|\mathbf{t}|) \\ \sigma_{1\bar{1}} = t_{21} + o\,(|\mathbf{t}|) \\ \sigma_{1\bar{3}} = t_{22} + o\,(|\mathbf{t}|) \\ \sigma_{2\bar{1}} = -t_{11} + o\,(|\mathbf{t}|) \\ \sigma_{2\bar{3}} = -t_{12} + o\,(|\mathbf{t}|) \end{cases} \qquad \text{for} \qquad \mathbf{t} \in \text{classes } \textit{(i), (ii)} \text{ and } \textit{(iii)} , \qquad (3.5)$$

and, more precisely, for deformations in class *(ii)* we actually have that

$$\begin{cases} \sigma_{12} = -1 + o\,(|\mathbf{t}|) \\ \sigma_{1\bar{1}} = t_{21}\,(1 + o\,(1)) \\ \sigma_{1\bar{3}} = t_{22}\,(1 + o\,(1)) \\ \sigma_{2\bar{1}} = -t_{11}\,(1 + o\,(1)) \\ \sigma_{2\bar{3}} = -t_{12}\,(1 + o\,(1)) \end{cases} \qquad \text{for} \qquad \mathbf{t} \in \text{class } \textit{(ii)} . \qquad (3.6)$$

The complex manifold X_t is endowed with the J_t-Hermitian $\mathbb{H}(3;\mathbb{C})$-left-invariant metric g_t, which is defined as follows:

$$g_t := \sum_{j=1}^{3} \varphi_t^j \odot \bar{\varphi}_t^j .$$

3.2.1.3 Structure Equations for Small Deformations of the Iwasawa Manifold

In this section, we give the structure equations for the small deformations of the Iwasawa manifold; we will use these computations in Sects. 3.2.3 and 3.2.4 to write the Bott-Chern cohomology of X_t, and in Theorem 4.8 to prove that the cohomological property of being C^∞-pure-and-full is not stable under small deformations of the complex structure.

Fix $t \in \Delta(\mathbf{0}, \varepsilon) \subset \mathbb{C}^6$, and consider the small deformation X_t of the Iwasawa manifold $\mathbb{I}_3$. Consider the system of complex coordinates on X_t given by

$$\begin{cases} \zeta_t^1 := z^1 + \sum_{\lambda=1}^2 t_{1\lambda} \bar{z}^\lambda \\[2mm] \zeta_t^2 := z^2 + \sum_{\lambda=1}^2 t_{2\lambda} \bar{z}^\lambda \\[2mm] \zeta_t^3 := z^3 + \sum_{\lambda=1}^2 (t_{3\lambda} + t_{2\lambda} z^1) \bar{z}^\lambda + A(\bar{z}) \end{cases} .$$

Consider

$$\begin{cases} \varphi_t^1 := \mathrm{d}\,\zeta_t^1 \\[2mm] \varphi_t^2 := \mathrm{d}\,\zeta_t^2 \\[2mm] \varphi_t^3 := \mathrm{d}\,\zeta_t^3 - z_1 \,\mathrm{d}\,\zeta_t^2 - \left(t_{21}\,\bar{z}^1 + t_{22}\,\bar{z}^2 \right) \mathrm{d}\,\zeta_t^1 \end{cases}$$

as a co-frame of $(1,0)$-forms on X_t (that is, as a Γ_t-invariant co-frame of $(1,0)$-forms on $\mathbb{C}^3$). We want to write the structure equations for X_t with respect to this co-frame.

A straightforward computation gives

$$\begin{cases} z^1 = \gamma \left(\zeta_t^1 + \lambda_1\,\bar{\zeta}_t^1 + \lambda_2\,\zeta_t^2 + \lambda_3\,\bar{\zeta}_t^2 \right) \\[2mm] z^2 = \alpha \left(\mu_0\,\zeta_t^1 + \mu_1\,\bar{\zeta}_t^1 + \mu_2\,\zeta_t^2 + \mu_3\,\bar{\zeta}_t^2 \right) \end{cases} ,$$

where α, β, γ, λ_i (for $i \in \{1,2,3\}$), μ_j (for $j \in \{0,1,2,3\}$) are complex numbers depending just on $\mathbf{t}$, and defined as follows:

$$
\begin{cases}
\alpha := \dfrac{1}{1 - |t_{22}|^2 - t_{21}\,\bar{t}_{12}} \\[2mm]
\beta := t_{21}\,\bar{t}_{11} + t_{22}\,\bar{t}_{21} \\[3mm]
\gamma := \dfrac{1}{1 - |t_{11}|^2 - \alpha\,\beta\,(t_{11}\,\bar{t}_{12} + t_{12}\,\bar{t}_{22}) - t_{12}\,\bar{t}_{21}} \\[3mm]
\lambda_1 := -t_{11}\left(1 + \alpha\,\bar{t}_{12}\,t_{21} + \alpha\,|t_{22}|^2\right) \\[2mm]
\lambda_2 := \alpha\,(t_{11}\,\bar{t}_{12} + t_{12}\,\bar{t}_{22}) \\[2mm]
\lambda_3 := -t_{12}\left(1 + \alpha\,\bar{t}_{12}\,t_{21} + \alpha\,|t_{22}|^2\right) \\[2mm]
\mu_0 := \beta\,\gamma \\[2mm]
\mu_1 := \lambda_1\,\beta\,\gamma - t_{21} \\[2mm]
\mu_2 := 1 + \lambda_2\,\beta\,\gamma \\[2mm]
\mu_3 := \lambda_3\,\beta\,\gamma - t_{22}
\end{cases}
$$

For the complex structures in the class *(i)*, one checks that the structure equations (with respect to the co-frame $\{\varphi_t^1, \varphi_t^2, \varphi_t^3\}$) are the same as the ones for $\mathbb{I}_3$, that is,

$$
\begin{cases}
d\varphi_t^1 = 0 \\[1mm]
d\varphi_t^2 = 0 \qquad\qquad \text{for} \quad t \in \text{class } (i)\,. \\[1mm]
d\varphi_t^3 = -\varphi_t^1 \wedge \varphi_t^2
\end{cases}
$$

For small deformations in classes *(ii)* and *(iii)*, we have that

$$
\begin{cases}
d\varphi_t^1 = 0 \\[1mm]
d\varphi_t^2 = 0 \\[1mm]
d\varphi_t^3 = \sigma_{12}\,\varphi_t^1 \wedge \varphi_t^2 \\
\qquad\quad + \sigma_{1\bar{1}}\,\varphi_t^1 \wedge \bar{\varphi}_t^1 + \sigma_{1\bar{2}}\,\varphi_t^1 \wedge \bar{\varphi}_t^2 \\
\qquad\quad + \sigma_{2\bar{1}}\,\varphi_t^2 \wedge \bar{\varphi}_t^1 + \sigma_{2\bar{2}}\,\varphi_t^2 \wedge \bar{\varphi}_t^2
\end{cases}
\qquad \text{for} \qquad t \in \text{classes } (ii) \text{ and } (iii)\,,
$$

where $\sigma_{12}, \sigma_{1\bar{1}}, \sigma_{1\bar{2}}, \sigma_{2\bar{1}}, \sigma_{2\bar{2}} \in \mathbb{C}$ are complex numbers depending just on t. The asymptotic behaviour of $\sigma_{12}, \sigma_{1\bar{1}}, \sigma_{1\bar{2}}, \sigma_{2\bar{1}}, \sigma_{2\bar{2}} \in \mathbb{C}$ is the following:

$$\begin{cases} \sigma_{12} = -1 + o\left(|\mathbf{t}|\right) \\ \sigma_{1\bar{1}} = t_{21} + o\left(|\mathbf{t}|\right) \\ \sigma_{1\bar{2}} = t_{22} + o\left(|\mathbf{t}|\right) \\ \sigma_{2\bar{1}} = -t_{11} + o\left(|\mathbf{t}|\right) \\ \sigma_{2\bar{2}} = -t_{12} + o\left(|\mathbf{t}|\right) \end{cases} \qquad \text{for} \qquad \mathbf{t} \in \text{classes } (i),\ (ii) \text{ and } (iii)\ , \qquad (3.7)$$

more precisely, for deformations in class *(ii)* we actually have that

$$\begin{cases} \sigma_{12} = -1 + o\left(|\mathbf{t}|\right) \\ \sigma_{1\bar{1}} = t_{21}\left(1 + o\left(1\right)\right) \\ \sigma_{1\bar{2}} = t_{22}\left(1 + o\left(1\right)\right) \\ \sigma_{2\bar{1}} = -t_{11}\left(1 + o\left(1\right)\right) \\ \sigma_{2\bar{2}} = -t_{12}\left(1 + o\left(1\right)\right) \end{cases} \qquad \text{for} \qquad \mathbf{t} \in \text{class } (ii)\ . \qquad (3.8)$$

The explicit values of $\sigma_{12},\ \sigma_{1\bar{1}},\ \sigma_{1\bar{2}},\ \sigma_{2\bar{1}},\ \sigma_{2\bar{2}} \in \mathbb{C}$ in the case of class *(ii)* are the following, [AT11, p. 416]:

$$\begin{cases} \sigma_{12} := -\gamma + t_{21}\bar{\lambda}_3\bar{\gamma} + t_{22}\bar{\alpha}\bar{\mu}_3 \\[2mm] \sigma_{1\bar{1}} := t_{21}\,\overline{\gamma\left(1 + t_{21}\bar{t}_{12}\alpha + |t_{22}|^2\alpha\right)} \\[2mm] \sigma_{1\bar{2}} := t_{22}\,\overline{\gamma\left(1 + t_{21}\bar{t}_{12}\alpha + |t_{22}|^2\alpha\right)} \qquad \text{for} \qquad \mathbf{t} \in \text{class } (ii)\ . \\[2mm] \sigma_{2\bar{1}} := -t_{11}\,\gamma\left(1 + t_{21}\bar{t}_{12}\alpha + |t_{22}|^2\alpha\right) \\[2mm] \sigma_{2\bar{2}} := -t_{12}\,\gamma\left(1 + t_{21}\bar{t}_{12}\alpha + |t_{22}|^2\alpha\right) \end{cases}$$

Note that, for small deformations in class *(ii)*, one has $\sigma_{12} \neq 0$ and $\left(\sigma_{1\bar{1}},\ \sigma_{1\bar{2}},\ \sigma_{2\bar{1}},\ \sigma_{2\bar{2}}\right) \neq (0,\ 0,\ 0,\ 0)$.

3.2.2 The de Rham Cohomology of the Iwasawa Manifold and of Its Small Deformations

Recall that, by Ch. Ehresmann's theorem, every complex-analytic family of compact complex manifolds is locally trivial as a differentiable family of compact differentiable manifolds, see, e.g., [MK06, Theorem 4.1]. Therefore the de Rham cohomology of small deformations of the Iwasawa manifold is the same as the de

Table 3.2 A basis of harmonic representatives for the de Rham cohomology of the Iwasawa manifold with respect to the metric $g_0 := \sum_{j=1}^{3} \varphi^j \odot \bar{\varphi}^j$

$H_{dR}^k(\mathbb{I}_3; \mathbb{C})$	g_0-Harmonic representatives	$\dim_{\mathbb{C}} H_{dR}^k(\mathbb{I}_3; \mathbb{C})$
$k = 1$	$\varphi^1, \varphi^2, \bar{\varphi}^1, \bar{\varphi}^2$	4
$k = 2$	$\varphi^{13}, \varphi^{23}, \varphi^{1\bar{1}}, \varphi^{1\bar{2}}, \varphi^{2\bar{1}}, \varphi^{2\bar{2}}, \varphi^{\bar{1}\bar{3}}, \varphi^{\bar{2}\bar{3}}$	8
$k = 3$	$\varphi^{123}, \varphi^{13\bar{1}}, \varphi^{13\bar{2}}, \varphi^{23\bar{1}}, \varphi^{23\bar{2}}, \varphi^{1\bar{1}\bar{3}}, \varphi^{1\bar{2}\bar{3}}, \varphi^{2\bar{1}\bar{3}}, \varphi^{2\bar{2}\bar{3}}, \varphi^{\bar{1}\bar{2}\bar{3}}$	10
$k = 4$	$\varphi^{123\bar{1}}, \varphi^{123\bar{2}}, \varphi^{13\bar{1}\bar{3}}, \varphi^{13\bar{2}\bar{3}}, \varphi^{23\bar{1}\bar{3}}, \varphi^{23\bar{2}\bar{3}}, \varphi^{1\bar{1}\bar{2}\bar{3}}, \varphi^{2\bar{1}\bar{2}\bar{3}}$	8
$k = 5$	$\varphi^{123\bar{1}\bar{3}}, \varphi^{123\bar{2}\bar{3}}, \varphi^{13\bar{1}\bar{2}\bar{3}}, \varphi^{23\bar{1}\bar{2}\bar{3}}$	4

Fig. 3.3 Graph associated to the Iwasawa manifold

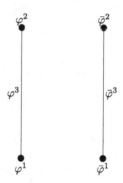

Rham cohomology of $\mathbb{I}_3$, which can be computed by using K. Nomizu's theorem [Nom54, Theorem 1].

In Table 3.2, we list the harmonic representatives with respect to the metric g_0 instead of their classes and, as usually, we shorten the notation writing, for example, $\varphi^{A\bar{B}} := \varphi^A \wedge \bar{\varphi}^B$.

Remark 3.14. Note that all the g_0-harmonic representatives of $H_{dR}^{\bullet}(\mathbb{I}_3; \mathbb{R})$ are of pure type with respect to J_0, that is, they are in $(\wedge^{p,q}\mathbb{I}_3 \oplus \wedge^{q,p}\mathbb{I}_3) \cap \wedge^{p+q}\mathbb{I}_3$ for some $p, q \in \{0, 1, 2, 3\}$; this is no more true for J_t with $\mathbf{t} \neq \mathbf{0}$ small enough, see Theorem 4.8.

Remark 3.15. In [PT09], H. Pouseele and P. Tirao studied nilmanifolds associated with graphs, [DM05], providing a combinatorial way for computing the second de Rham cohomology group, [PT09, Theorem 1], and characterizing the graphs giving rise to a symplectic or contact nilmanifold, [PT09, Theorem 3, Proposition 4].

The Iwasawa manifold can be seen as associated to the graph in Fig. 3.3.

3.2.3   The Dolbeault Cohomology of the Iwasawa Manifold and of Its Small Deformations

The Hodge numbers of the Iwasawa manifold and of its small deformations have been computed by I. Nakamura in [Nak75, p. 96]. The g_t-harmonic representatives for $H_{\bar\partial}^{\bullet,\bullet}(X_t)$, for $\mathbf{t}$ small enough, can be computed using the considerations in Sect. 3.1.2 and the structure equations given in Sect. 3.2.1.3. We collect here the results of the computations.

In order to reduce the number of cases under consideration, recall that, on a compact complex Hermitian manifold X of complex dimension n, for any $(p,q) \in \mathbb{N}^2$, the Hodge-$*$-operator and the conjugation induce an isomorphism

$$H_{\bar\partial}^{p,q}(X) \xrightarrow{\simeq} H_{\partial}^{n-q,n-p}(X) \xrightarrow{\simeq} \overline{H_{\bar\partial}^{n-p,n-q}(X)} \ .$$

- **1-forms.** It is straightforward to check that

$$H_{\bar\partial}^{1,0}(X_t) = \mathbb{C}\langle \varphi_t^1, \varphi_t^2, \varphi_t^3 \rangle \quad \text{for} \quad \mathbf{t} \in \text{class } (i)$$

and

$$H_{\bar\partial}^{0,1}(X_t) = \mathbb{C}\langle \bar\varphi_t^1, \bar\varphi_t^2 \rangle \quad \text{for} \quad \mathbf{t} \in \text{classes } (i), (ii), \text{ and } (iii) \ .$$

Since $\bar\partial \varphi_t^3 \neq 0$ for X_t in class (ii) or in class (iii), one has

$$H_{\bar\partial}^{1,0}(X_t) = \mathbb{C}\langle \varphi_t^1, \varphi_t^2 \rangle \quad \text{for} \quad \mathbf{t} \in \text{classes } (ii) \text{ and } (iii) \ :$$

this means in particular that X_t is not holomorphically parallelizable for $\mathbf{t}$ in classes (ii) and (iii), [Nak75, pp. 86, 96].

Summarizing,

$$\dim_{\mathbb{C}} H_{\bar\partial}^{1,0}(X_t) = \begin{cases} 3 & \text{for } \mathbf{t} \in \text{class } (i) \\ 2 & \text{for } \mathbf{t} \in \text{classes } (ii) \text{ and } (iii) \end{cases} \ ,$$

and

$$\dim_{\mathbb{C}} H_{\bar\partial}^{0,1}(X_t) = 2 \quad \text{for} \quad \mathbf{t} \in \text{classes } (i), (ii), \text{ and } (iii) \ .$$

- **2-forms.** A straightforward computation yields

$$H_{\bar\partial}^{2,0}(X_t) = \mathbb{C}\langle \varphi_t^{12}, \varphi_t^{13}, \varphi_t^{23} \rangle \quad \text{for} \quad \mathbf{t} \in \text{class } (i) \ ,$$

$$H_{\bar\partial}^{1,1}(X_t) = \mathbb{C}\langle \varphi_t^{1\bar1}, \varphi_t^{1\bar2}, \varphi_t^{2\bar1}, \varphi_t^{2\bar2}, \varphi_t^{3\bar1}, \varphi_t^{3\bar2} \rangle \quad \text{for} \quad \mathbf{t} \in \text{class } (i) \ ,$$

and

$$H_{\bar\partial}^{0,2}(X_t) = \mathbb{C}\left\langle \varphi_t^{\bar{1}\bar{3}}, \varphi_t^{\bar{2}\bar{3}} \right\rangle \quad \text{for} \quad \mathbf{t} \in \text{classes } (i), (ii), \text{ and } (iii).$$

We now compute $H_{\bar\partial}^{2,0}(X_t)$ for $\mathbf{t} \in$ classes (ii) and (iii). The $\mathbb{H}(3; \mathbb{C})$-left-invariant (2, 0)-forms are of the type $A\,\varphi_t^{12} + B\,\varphi_t^{13} + C\,\varphi_t^{23}$ with $A, B, C \in \mathbb{C}$, so one has to solve the linear system

$$\begin{pmatrix} 0 & 0 & 0 \\ 0 & -\sigma_{2\bar{1}} & \sigma_{1\bar{1}} \\ 0 & -\sigma_{2\bar{2}} & \sigma_{1\bar{2}} \end{pmatrix} \cdot \begin{pmatrix} A \\ B \\ C \end{pmatrix} = \begin{pmatrix} 0 \\ 0 \\ 0 \end{pmatrix};$$

since the associated matrix to the system has rank 0 for $\mathbf{t} \in$ class (i), rank 1 for $\mathbf{t} \in$ class (ii), and rank 2 for $\mathbf{t} \in$ class (iii), one concludes that

$$\dim_{\mathbb{C}} H_{\bar\partial}^{2,0}(X_t) = 2 \quad \text{for} \quad \mathbf{t} \in \text{class } (ii)$$

(the generators being φ_t^{12} and a linear combination of φ_t^{13} and φ_t^{23}) and

$$\dim_{\mathbb{C}} H_{\bar\partial}^{2,0}(X_t) = 1 \quad \text{for} \quad \mathbf{t} \in \text{class } (iii)$$

(the generator being φ_t^{12}).

It remains to compute $H_{\bar\partial}^{1,1}(X_t)$ for $\mathbf{t} \in$ classes (ii) and (iii). For such $\mathbf{t}$, one has that: three independent $\square_{J_t}$-harmonic (1, 1)-forms are of the type $\psi_1 = A\,\varphi_t^{1\bar{1}} + B\,\varphi_t^{1\bar{2}} + C\,\varphi_t^{2\bar{1}} + D\,\varphi_t^{2\bar{2}}$ where $A, B, C, D \in \mathbb{C}$ satisfy the equation

$$\left(\overline{\sigma_{1\bar{1}}} \ -\overline{\sigma_{1\bar{2}}} \ -\overline{\sigma_{2\bar{1}}} \ \overline{\sigma_{2\bar{2}}} \right) \cdot \begin{pmatrix} A \\ B \\ C \\ D \end{pmatrix} = 0,$$

whose matrix has rank 1 for $\mathbf{t} \in$ classes (ii) and (iii) (while its rank is 0 for $\mathbf{t} \in$ class (i)); two other independent $\square_{J_t}$-harmonic (1, 1)-forms are of the type $\psi_2 = E\,\varphi_t^{1\bar{3}} + F\,\varphi_t^{2\bar{3}} + G\,\varphi_t^{3\bar{1}} + H\,\varphi_t^{3\bar{2}}$ where $E, F, G, H \in \mathbb{C}$ are solution of the system

$$\begin{pmatrix} -\overline{\sigma_{12}} & 0 & -\overline{\sigma_{1\bar{2}}} & \overline{\sigma_{1\bar{1}}} \\ 0 & -\overline{\sigma_{12}} & -\overline{\sigma_{2\bar{1}}} & \overline{\sigma_{2\bar{2}}} \end{pmatrix} \cdot \begin{pmatrix} E \\ F \\ G \\ H \end{pmatrix} = \begin{pmatrix} 0 \\ 0 \end{pmatrix},$$

whose matrix has rank 2 for $\mathbf{t} \in$ classes *(i)*, *(ii)* and *(iii)*; note also that no $(1, 1)$-form with a non-zero component in $\varphi_t^{3\bar{3}}$ can be $\overline{\square}_{J_t}$-harmonic. Hence, one can conclude that

$$\dim_{\mathbb{C}} H_{\bar{\partial}}^{1,1}(X_t) = 5 \quad \text{for} \quad \mathbf{t} \in \text{classes } (ii) \text{ and } (iii) .$$

Summarizing,

$$\dim_{\mathbb{C}} H_{\bar{\partial}}^{2,0}(X_t) = \begin{cases} 3 & \text{for } \mathbf{t} \in \text{class } (i) \\ 2 & \text{for } \mathbf{t} \in \text{class } (ii) \\ 1 & \text{for } \mathbf{t} \in \text{class } (iii) \end{cases} ,$$

and

$$\dim_{\mathbb{C}} H_{\bar{\partial}}^{1,1}(X_t) = \begin{cases} 6 & \text{for } \mathbf{t} \in \text{class } (i) \\ 5 & \text{for } \mathbf{t} \in \text{classes } (ii) \text{ and } (iii) \end{cases} ,$$

and

$$\dim_{\mathbb{C}} H_{\bar{\partial}}^{0,2}(X_t) = 2 \quad \text{for} \quad \mathbf{t} \in \text{classes } (i), (ii), \text{ and } (iii) .$$

- **3-forms.** Finally, we have to compute $H_{\bar{\partial}}^{3,0}(X_t)$ and $H_{\bar{\partial}}^{2,1}(X_t)$. A straightforward linear algebra computation yields to

$$H_{\bar{\partial}}^{3,0}(X_t) = \mathbb{C} \langle \varphi_t^{123} \rangle \quad \text{for} \quad \mathbf{t} \in \text{classes } (i), (ii), \text{ and } (iii)$$

and

$$H_{\bar{\partial}}^{2,1}(X_t) = \mathbb{C} \langle \varphi_t^{12\bar{1}}, \varphi_t^{12\bar{2}}, \varphi_t^{13\bar{1}}, \varphi_t^{13\bar{2}}, \varphi_t^{23\bar{1}}, \varphi_t^{23\bar{2}} \rangle \quad \text{for} \quad \mathbf{t} \in \text{class } (i) .$$

It remains to compute $H_{\bar{\partial}}^{2,1}(X_t)$ for $\mathbf{t} \in$ classes *(ii)* and *(iii)*. Firstly, one notes that four of the six generators of the space of $\mathbb{H}(3; \mathbb{C})$-left-invariant $(2, 1)$-forms that are $\overline{\square}_{J_t}$-harmonic for $\mathbf{t} \in$ class *(i)* can be slightly modified to get four $\bar{\partial}_{J_t}$-holomorphic $(2, 1)$-forms for $\mathbf{t} \in$ class *(ii)* or class *(iii)*: more precisely, one has

$$H_{\bar{\partial}}^{2,1}(X_t)$$

$$\supseteq \mathbb{C} \left\langle \varphi_t^{13\bar{1}} - \frac{\sigma_{2\bar{2}}}{\overline{\sigma_{12}}} \varphi_t^{12\bar{3}}, \ \varphi_t^{13\bar{2}} - \frac{\sigma_{2\bar{1}}}{\overline{\sigma_{12}}} \varphi_t^{12\bar{3}}, \ \varphi_t^{23\bar{1}} - \frac{\sigma_{1\bar{2}}}{\overline{\sigma_{12}}} \varphi_t^{12\bar{3}}, \ \varphi_t^{23\bar{2}} - \frac{\sigma_{1\bar{1}}}{\overline{\sigma_{12}}} \varphi_t^{12\bar{3}} \right\rangle ;$$

in other words, four independent $\overline{\square}_{J_t}$-harmonic $(2, 1)$-forms are of the type $\psi_2 = C\,\varphi_t^{12\bar{3}} + D\,\varphi_t^{13\bar{1}} + E\,\varphi_t^{13\bar{2}} + F\,\varphi_t^{23\bar{1}} + G\,\varphi_t^{23\bar{2}}$, where $C, D, E, F, G \in \mathbb{C}$ are solution of the linear system

$$\begin{pmatrix} \overline{\sigma_{12}} & \sigma_{2\bar{2}} & -\sigma_{2\bar{1}} & -\sigma_{1\bar{2}} & \sigma_{1\bar{1}} \end{pmatrix} \cdot \begin{pmatrix} C \\ D \\ E \\ F \\ G \end{pmatrix} = 0 \,,$$

whose matrix has rank 1 for every $\mathbf{t} \in$ classes *(i)*, *(ii)*, and *(iii)*. Note that one can reduce to study the $\overline{\square}$-harmonicity of the $(2, 1)$-forms of the type $\psi_1 = A\,\varphi_t^{12\bar{1}} + B\,\varphi_t^{12\bar{2}}$: indeed, a $(2, 1)$-form $\psi = \psi_1 + \psi_2 + H\,\varphi_t^{13\bar{3}} + L\,\varphi_t^{23\bar{3}}$, where $H, L \in \mathbb{C}$, is $\overline{\square}$-harmonic if and only if $H = 0 = L$ and both ψ_1 and ψ_2 are $\overline{\square}$-harmonic. A $(2, 1)$-form of the type ψ_1 is $\overline{\square}$-harmonic if and only if $A, B \in \mathbb{C}$ solve the linear system

$$\begin{pmatrix} -\overline{\sigma_{1\bar{1}}} & \overline{\sigma_{1\bar{2}}} \\ -\overline{\sigma_{2\bar{1}}} & \overline{\sigma_{2\bar{2}}} \end{pmatrix} \cdot \begin{pmatrix} A \\ B \end{pmatrix} = \begin{pmatrix} 0 \\ 0 \end{pmatrix} \,,$$

whose matrix has rank 0 for $\mathbf{t} \in$ class *(i)*, rank 1 for $\mathbf{t} \in$ class *(ii)* and rank 2 for $\mathbf{t} \in$ class *(iii)*. In particular, one gets that

$$\dim_{\mathbb{C}} H_{\bar{\partial}}^{2,1}(X_t) = 5 \quad \text{for} \quad \mathbf{t} \in \text{class } (ii)$$

and

$$\dim_{\mathbb{C}} H_{\bar{\partial}}^{2,1}(X_t) = 4 \quad \text{for} \quad \mathbf{t} \in \text{class } (iii) \,.$$

Summarizing,

$$\dim_{\mathbb{C}} H_{\bar{\partial}}^{3,0}(X_t) = 1 \quad \text{for} \quad \mathbf{t} \in \text{classes } (i), (ii), \text{ and } (iii) \,,$$

and

$$\dim_{\mathbb{C}} H_{\bar{\partial}}^{2,1}(X_t) = \begin{cases} 6 & \text{for } \mathbf{t} \in \text{class } (i) \\ 5 & \text{for } \mathbf{t} \in \text{class } (ii) \\ 4 & \text{for } \mathbf{t} \in \text{class } (iii) \end{cases} \,.$$

We summarize the Hodge numbers of the Iwasawa manifold and of its small deformations in Table 3.3, see also [Nak75, p. 96], and the dimensions of the conjugate Dolbeault cohomology of the Iwasawa manifold and of its small deformations in Table 3.4.

Table 3.3 Dimensions of the Dolbeault cohomology of the Iwasawa manifold and of its small deformations, see also [Nak75, p. 96]

$H^{\bullet,\bullet}_{\bar\partial}$	$h^{1,0}_{\bar\partial}$	$h^{0,1}_{\bar\partial}$	$h^{2,0}_{\bar\partial}$	$h^{1,1}_{\bar\partial}$	$h^{0,2}_{\bar\partial}$	$h^{3,0}_{\bar\partial}$	$h^{2,1}_{\bar\partial}$	$h^{1,2}_{\bar\partial}$	$h^{0,3}_{\bar\partial}$	$h^{3,1}_{\bar\partial}$	$h^{2,2}_{\bar\partial}$	$h^{1,3}_{\bar\partial}$	$h^{3,2}_{\bar\partial}$	$h^{2,3}_{\bar\partial}$
$\mathbb{I}_3$ and (i)	3	2	3	6	2	1	6	6	1	2	6	3	2	3
(ii)	2	2	2	5	2	1	5	5	1	2	5	2	2	2
(iii)	2	2	1	5	2	1	4	4	1	2	5	1	2	2

Table 3.4 Dimensions of the conjugate Dolbeault cohomology of the Iwasawa manifold and of its small deformations

$H^{\bullet,\bullet}_{\partial}$	$h^{1,0}_{\partial}$	$h^{0,1}_{\partial}$	$h^{2,0}_{\partial}$	$h^{1,1}_{\partial}$	$h^{0,2}_{\partial}$	$h^{3,0}_{\partial}$	$h^{2,1}_{\partial}$	$h^{1,2}_{\partial}$	$h^{0,3}_{\partial}$	$h^{3,1}_{\partial}$	$h^{2,2}_{\partial}$	$h^{1,3}_{\partial}$	$h^{3,2}_{\partial}$	$h^{2,3}_{\partial}$
$\mathbb{I}_3$ and (i)	2	3	2	6	3	1	6	6	1	3	6	2	3	2
(ii)	2	2	2	5	2	1	5	5	1	2	5	2	2	2
(iii)	2	2	2	5	1	1	4	4	1	1	5	2	2	2

3.2.4 The Bott-Chern and Aeppli Cohomologies of the Iwasawa Manifold and of Its Small Deformations

In this section, using Theorems 3.6 and 3.7, we explicitly compute the dimensions of $H^{\bullet,\bullet}_{BC}(X_\mathbf{t})$, for $\mathbf{t}$ small enough, [Ang11, Sect. 5.3]: such numbers are summarized in Table 3.5.

In order to reduce the number of cases under consideration, recall that, on a compact complex Hermitian manifold X of complex dimension n, for every $(p, q) \in \mathbb{N}^2$, the conjugation induces an isomorphism $H^{p,q}_{BC}(X) \xrightarrow{\simeq} H^{q,p}_{BC}(X)$, and the Hodge-$*$-operator induces an isomorphism $H^{p,q}_{BC}(X) \xrightarrow{\simeq} H^{n-q,n-p}_{A}(X)$; furthermore, note that

$$H^{p,0}_{BC}(X) \simeq \ker\left(d\colon \wedge^{p,0}X \to \wedge^{p+1}(X;\mathbb{C})\right)$$

and that

$$H^{n,0}_{BC}(X) \simeq H^{n,0}_{\bar\partial}(X) \,.$$

- **1-forms** It is straightforward to check that

$$H^{1,0}_{BC}(X_\mathbf{t}) = \mathbb{C}\left\langle \varphi^1_\mathbf{t}, \varphi^2_\mathbf{t} \right\rangle \quad \text{for} \quad \mathbf{t} \in \text{classes } (i), (ii), \text{ and } (iii) \,.$$

- **2-forms** It is straightforward to compute

$$H^{2,0}_{BC}(X_\mathbf{t}) = \mathbb{C}\left\langle \varphi^{12}_\mathbf{t}, \varphi^{13}_\mathbf{t}, \varphi^{23}_\mathbf{t} \right\rangle \quad \text{for} \quad \mathbf{t} \in \text{class } (i) \,.$$

Table 3.5 Dimensions of the de Rham, Dolbeault, Bott-Chern, and Aeppli cohomologies of the Iwasawa manifold and of its small deformations

$H^\bullet_{dR}$	b_1	b_2	b_3	b_4	b_5
$\mathbb{I}_3$ and (i), (ii), (iii)	4	8	10	8	4

$H^{\bullet,\bullet}_{\bar{\partial}}$	$h^{1,0}_{\bar{\partial}}$	$h^{0,1}_{\bar{\partial}}$	$h^{2,0}_{\bar{\partial}}$	$h^{1,1}_{\bar{\partial}}$	$h^{0,2}_{\bar{\partial}}$	$h^{3,0}_{\bar{\partial}}$	$h^{2,1}_{\bar{\partial}}$	$h^{1,2}_{\bar{\partial}}$	$h^{0,3}_{\bar{\partial}}$	$h^{3,1}_{\bar{\partial}}$	$h^{2,2}_{\bar{\partial}}$	$h^{1,3}_{\bar{\partial}}$	$h^{3,2}_{\bar{\partial}}$	$h^{2,3}_{\bar{\partial}}$
$\mathbb{I}_3$ and (i)	3	2	3	6	2	1	6	6	1	2	6	3	2	3
(ii)	2	2	2	5	2	1	5	5	1	2	5	2	2	2
(iii)	2	2	1	5	2	1	4	4	1	2	5	1	2	2

$H^{\bullet,\bullet}_{BC}$	$h^{1,0}_{BC}$	$h^{0,1}_{BC}$	$h^{2,0}_{BC}$	$h^{1,1}_{BC}$	$h^{0,2}_{BC}$	$h^{3,0}_{BC}$	$h^{2,1}_{BC}$	$h^{1,2}_{BC}$	$h^{0,3}_{BC}$	$h^{3,1}_{BC}$	$h^{2,2}_{BC}$	$h^{1,3}_{BC}$	$h^{3,2}_{BC}$	$h^{2,3}_{BC}$
$\mathbb{I}_3$ and (i)	2	2	3	4	3	1	6	6	1	2	8	2	3	3
(ii.a)	2	2	2	4	2	1	6	6	1	2	7	2	3	3
(ii.b)	2	2	2	4	2	1	6	6	1	2	6	2	3	3
(iii.a)	2	2	1	4	1	1	6	6	1	2	7	2	3	3
(iii.b)	2	2	1	4	1	1	6	6	1	2	6	2	3	3

$H^{\bullet,\bullet}_{A}$	$h^{1,0}_{A}$	$h^{0,1}_{A}$	$h^{2,0}_{A}$	$h^{1,1}_{A}$	$h^{0,2}_{A}$	$h^{3,0}_{A}$	$h^{2,1}_{A}$	$h^{1,2}_{A}$	$h^{0,3}_{A}$	$h^{3,1}_{A}$	$h^{2,2}_{A}$	$h^{1,3}_{A}$	$h^{3,2}_{A}$	$h^{2,3}_{A}$
$\mathbb{I}_3$ and (i)	3	3	2	8	2	1	6	6	1	3	4	3	2	2
(ii.a)	3	3	2	7	2	1	6	6	1	2	4	2	2	2
(ii.b)	3	3	2	6	2	1	6	6	1	2	4	2	2	2
(iii.a)	3	3	2	7	2	1	6	6	1	1	4	1	2	2
(iii.b)	3	3	2	6	2	1	6	6	1	1	4	1	2	2

The computations for $H^{2,0}_{BC}(X_t)$ reduce to find $\psi = A\,\varphi_t^{12} + B\,\varphi_t^{13} + C\,\varphi_t^{23}$ where $A, B, C \in \mathbb{C}$ satisfy the linear system

$$
\begin{pmatrix} 0 & 0 & 0 \\ 0 & -\sigma_{2\bar{1}} & \sigma_{1\bar{1}} \\ 0 & -\sigma_{2\bar{2}} & \sigma_{1\bar{2}} \end{pmatrix} \cdot \begin{pmatrix} A \\ B \\ C \end{pmatrix} = \begin{pmatrix} 0 \\ 0 \\ 0 \end{pmatrix} ,
$$

whose matrix has rank 0 for $\mathbf{t} \in$ class (i), rank 1 for $\mathbf{t} \in$ class (ii), and rank 2 for $\mathbf{t} \in$ class (iii); so, in particular, we get that

$$
\dim_{\mathbb{C}} H^{2,0}_{BC}(X_t) = 2 \quad \text{for} \quad \mathbf{t} \in \text{class } (ii)
$$

and

$$
\dim_{\mathbb{C}} H^{2,0}_{BC}(X_t) = 1 \quad \text{for} \quad \mathbf{t} \in \text{class } (iii)
$$

(more precisely, for $\mathbf{t} \in$ class (iii) we have $H^{2,0}_{BC}(X_t) = \mathbb{C}\langle\varphi_t^{12}\rangle$).

It remains to compute $H^{1,1}_{BC}(X_t)$ for $\mathbf{t} \in$ classes (i), (ii), and (iii). First of all, it is easy to check that

$$
H^{1,1}_{BC}(X_t) \supseteq \mathbb{C}\langle\varphi_t^{1\bar{1}}, \varphi_t^{1\bar{2}}, \varphi_t^{2\bar{1}}, \varphi_t^{2\bar{2}}\rangle \quad \text{for} \quad \mathbf{t} \in \text{classes } (i), (ii), \text{ and } (iii) ,
$$

and equality holds if $\mathbf{t} \in$ class (i), hence, in particular, if $\mathbf{t} = \mathbf{0}$. This immediately implies that

$$H^{1,1}_{BC}(X_{\mathbf{t}}) = \mathbb{C}\left\langle \varphi_t^{1\bar{1}}, \varphi_t^{1\bar{2}}, \varphi_t^{2\bar{1}}, \varphi_t^{2\bar{2}} \right\rangle \quad \text{for} \quad \mathbf{t} \in \text{classes } (i), (ii), \text{ and } (iii) ;$$

indeed, the function $\mathbf{t} \mapsto \dim_{\mathbb{C}} H^{1,1}_{BC}(X_{\mathbf{t}})$ is upper-semi-continuous at 0, since $H^{1,1}_{BC}(X_{\mathbf{t}})$ is isomorphic to the kernel of the self-adjoint elliptic differential operator $\tilde{\Delta}_{BC J_{\mathbf{t}}} \lfloor_{\wedge^{1,1} X_{\mathbf{t}}}$. (One can explain this argument saying that the new parts appearing in the computations for $\mathbf{t} \neq \mathbf{0}$ are "too small" to balance out the lack for the ∂-closure or the $\bar{\partial}$-closure.) From another point of view, we can note that $(1,1)$-forms of the type $\psi = A \varphi_t^{1\bar{3}} + B \varphi_t^{2\bar{3}} + C \varphi_t^{3\bar{1}} + D \varphi_t^{3\bar{2}} + E \varphi_t^{3\bar{3}}$ are $\tilde{\Delta}_{BC J_{\mathbf{t}}}$-harmonic if and only if $E = 0$ and $A, B, C, D \in \mathbb{C}$ satisfy the linear system

$$\begin{pmatrix} -\bar{\sigma}_{12} & 0 & -\sigma_{1\bar{2}} & -\sigma_{1\bar{1}} \\ 0 & -\bar{\sigma}_{12} & -\sigma_{2\bar{2}} & -\sigma_{2\bar{1}} \\ \hline \overline{\sigma_{1\bar{2}}} & -\overline{\sigma_{1\bar{1}}} & \sigma_{12} & 0 \\ \overline{\sigma_{2\bar{2}}} & -\overline{\sigma_{2\bar{1}}} & 0 & \sigma_{12} \end{pmatrix} \cdot \begin{pmatrix} A \\ B \\ C \\ D \end{pmatrix} = \begin{pmatrix} 0 \\ 0 \\ 0 \\ 0 \end{pmatrix},$$

whose matrix has rank 4 for every $\mathbf{t} \in$ classes $(i), (ii)$, and (iii).

• **3-forms** It is straightforward to compute

$$H^{3,0}_{BC}(X_{\mathbf{t}}) = \mathbb{C}\left\langle \varphi_t^{123} \right\rangle \quad \text{for} \quad \mathbf{t} \in \text{classes } (i), (ii), \text{ and } (iii) .$$

Moreover,

$$H^{2,1}_{BC}(X_{\mathbf{t}}) = \mathbb{C}\left\langle \varphi_t^{12\bar{1}}, \varphi_t^{12\bar{2}}, \varphi_t^{13\bar{1}} - \frac{\sigma_{2\bar{2}}}{\overline{\sigma_{12}}} \varphi_t^{12\bar{3}}, \varphi_t^{13\bar{2}} + \frac{\sigma_{2\bar{1}}}{\overline{\sigma_{12}}} \varphi_t^{12\bar{3}}, \varphi_t^{23\bar{1}} \right.$$

$$\left. + \frac{\sigma_{1\bar{2}}}{\overline{\sigma_{12}}} \varphi_t^{12\bar{3}}, \varphi_t^{23\bar{2}} - \frac{\sigma_{1\bar{1}}}{\overline{\sigma_{12}}} \varphi_t^{12\bar{3}} \right\rangle$$

for $\mathbf{t} \in$ classes $(i), (ii)$, and (iii) ;

in particular,

$$H^{2,1}_{BC}(X_{\mathbf{t}}) = \mathbb{C}\left\langle \varphi_t^{12\bar{1}}, \varphi_t^{12\bar{2}}, \varphi_t^{13\bar{1}}, \varphi_t^{13\bar{2}}, \varphi_t^{23\bar{1}}, \varphi_t^{23\bar{2}} \right\rangle \quad \text{for} \quad \mathbf{t} \in \text{class } (i) .$$

From another point of view, one can easily check that

$$H^{2,1}_{BC}(X_{\mathbf{t}}) \supseteq \mathbb{C}\left\langle \varphi_t^{12\bar{1}}, \varphi_t^{12\bar{2}} \right\rangle \quad \text{for} \quad \mathbf{t} \in \text{classes } (i), (ii), \text{ and } (iii) ,$$

and that the $(2,1)$-forms of the type $\psi = A \varphi_t^{12\bar{3}} + B \varphi_t^{13\bar{1}} + C \varphi_t^{13\bar{2}} + D \varphi_t^{23\bar{1}} + E \varphi_t^{23\bar{2}} + F \varphi_t^{13\bar{3}} + G \varphi_t^{23\bar{3}}$ are $\tilde{\Delta}_{BC J_{\mathbf{t}}}$-harmonic if and only if $F = 0 = G$ and $A, B, C, D, E \in \mathbb{C}$ satisfy the equation

$$
\left(\overline{\sigma_{12}} \ \sigma_{2\bar{2}} \ -\sigma_{2\bar{1}} \ \sigma_{1\bar{2}} \ \sigma_{1\bar{1}}\right) \cdot
\begin{pmatrix} A \\ B \\ C \\ D \\ E \end{pmatrix} = 0 \, ,
$$

whose matrix has rank 1 for every $\mathbf{t} \in$ classes *(i)*, *(ii)*, and *(iii)*. Note in particular that the dimensions of $H^{3,0}_{BC}(X_t)$ and of $H^{2,1}_{BC}(X_t)$ do not depend on $\mathbf{t}$.

- **4-forms** It is straightforward to compute

$$
H^{3,1}_{BC}(X_t) = \mathbb{C}\left\langle \varphi_t^{123\bar{1}}, \varphi_t^{123\bar{2}} \right\rangle \quad \text{for} \quad \mathbf{t} \in \text{classes } (i), (ii), \text{ and } (iii)
$$

and

$$
H^{2,2}_{BC}(X_t) = \mathbb{C}\left\langle \varphi_t^{12\bar{1}\bar{3}}, \varphi_t^{12\bar{2}\bar{3}}, \varphi_t^{13\bar{1}\bar{2}}, \varphi_t^{13\bar{1}\bar{3}}, \varphi_t^{13\bar{2}\bar{3}}, \varphi_t^{23\bar{1}\bar{2}}, \varphi_t^{23\bar{1}\bar{3}}, \varphi_t^{23\bar{2}\bar{3}} \right\rangle
$$

$$
\text{for} \quad \mathbf{t} \in \text{class } (i) \, .
$$

Moreover, one can check that

$$
H^{2,2}_{BC}(X_t) \supseteq \mathbb{C}\left\langle \varphi_t^{12\bar{1}\bar{3}}, \varphi_t^{12\bar{2}\bar{3}}, \varphi_t^{13\bar{1}\bar{2}}, \varphi_t^{23\bar{1}\bar{2}} \right\rangle \quad \text{for} \quad \mathbf{t} \in \text{classes } (i), (ii), \text{ and } (iii),
$$

and that no $(2,2)$-form with a non-zero component in $\varphi_t^{12\bar{1}\bar{2}}$ can be $\tilde{\Delta}_{BC_{J_t}}$-harmonic. For $H^{2,2}_{BC}(X_t)$ with $\mathbf{t} \in$ classes *(ii)* and *(iii)*, we get a new behaviour: there are subclasses in both class *(ii)* and class *(iii)*, which can be distinguished by the dimension of $H^{2,2}_{BC}(X_t)$. Indeed, consider $(2,2)$-forms of the type $\psi = A\,\varphi_t^{13\bar{1}\bar{3}} + B\,\varphi_t^{13\bar{2}\bar{3}} + C\,\varphi_t^{23\bar{1}\bar{3}} + D\,\varphi_t^{23\bar{2}\bar{3}}$; a straightforward computation shows that such a ψ is $\tilde{\Delta}_{BC_{J_t}}$-harmonic if and only if A, B, C, $D \in \mathbb{C}$ satisfy the linear system

$$
\begin{pmatrix} \overline{\sigma_{2\bar{2}}} & -\overline{\sigma_{1\bar{2}}} & -\overline{\sigma_{2\bar{1}}} & \overline{\sigma_{1\bar{1}}} \\ \sigma_{2\bar{2}} & -\sigma_{2\bar{1}} & -\sigma_{1\bar{2}} & \sigma_{1\bar{1}} \end{pmatrix} \cdot \begin{pmatrix} A \\ B \\ C \\ D \end{pmatrix} = \begin{pmatrix} 0 \\ 0 \end{pmatrix} \, .
$$

As one can straightforwardly note, the rank of the matrix involved is 0 for $\mathbf{t} \in$ class *(i)*, while it is 1 or 2 depending on the values of the parameters within class *(ii)*, or within class *(iii)*. Therefore

$$
\dim_{\mathbb{C}} H^{2,2}_{BC}(X_t) = 7 \quad \text{for} \quad \mathbf{t} \in \text{subclasses } (ii.a) \text{ and } (iii.a)
$$

and

$$
\dim_{\mathbb{C}} H^{2,2}_{BC}(X_t) = 6 \quad \text{for} \quad \mathbf{t} \in \text{subclasses } (ii.b) \text{ and } (iii.b) \, .
$$

- **5-forms** Finally, let us compute $H_{BC}^{3,2}(X_t)$. It is straightforward to check that

$$H_{BC}^{3,2}(X_t) = \mathbb{C}\left\langle \varphi_t^{123\bar{1}\bar{2}}, \varphi_t^{123\bar{1}\bar{3}}, \varphi_t^{123\bar{2}\bar{3}} \right\rangle \quad \text{for} \quad \mathbf{t} \in \text{classes } (i), (ii), \text{ and } (iii):$$

in particular, it does not depend on $\mathbf{t} \in \Delta(\mathbf{0}, \varepsilon)$.

We summarize the results of the computations above in the following theorem, and the dimensions of the Bott-Chern cohomology and of the Aeppli cohomology of the Iwasawa manifold and of its small deformations in Table 3.5.

Theorem 3.9 ([Ang11, Theorem 5.1]). *Consider the Iwasawa manifold* $\mathbb{I}_3 := \mathbb{H}(3; \mathbb{Z}[i]) \backslash \mathbb{H}(3; \mathbb{C})$ *and the family* $\{X_t = (\mathbb{I}_3, J_t)\}_{\mathbf{t} \in \Delta(\mathbf{0},\varepsilon)}$ *of its small deformations, where* $\varepsilon > 0$ *is small enough and* $X_0 = \mathbb{I}_3$. *Then the dimensions* $h_{BC}^{p,q} := h_{BC}^{p,q}(X_t) := \dim_{\mathbb{C}} H_{BC}^{p,q}(X_t) = \dim_{\mathbb{C}} H_A^{3-p,3-q}(X_t)$ *does not depend on* $\mathbf{t} \in \Delta(\mathbf{0}, \varepsilon)$ *whenever* $p + q$ *is odd or* $(p, q) \in \{(1, 1), (3, 1), (1, 3)\}$, *and they are equal to*

$$h_{BC}^{1,0} = h_{BC}^{0,1} = 2\,,$$

$$h_{BC}^{2,0} = h_{BC}^{0,2} \in \{1, 2, 3\}\,, \qquad\qquad h_{BC}^{1,1} = 4\,,$$

$$h_{BC}^{3,0} = h_{BC}^{0,3} = 1\,, \qquad\qquad h_{BC}^{2,1} = h_{BC}^{1,2} = 6\,,$$

$$h_{BC}^{3,1} = h_{BC}^{1,3} = 2\,, \qquad\qquad h_{BC}^{2,2} \in \{6, 7, 8\}\,,$$

$$h_{BC}^{3,2} = h_{BC}^{2,3} = 3\,.$$

Remark 3.16 ([Ang11, Remark 5.2]). As a consequence of the computations above, we notice that the Bott-Chern cohomology yields a finer classification of the small deformations of $\mathbb{I}_3$ than the Dolbeault cohomology: indeed, note that $\dim_{\mathbb{C}} H_{BC}^{2,2}(X_t)$ assumes different values according to different parameters in class *(ii)*, respectively in class *(iii)*; in a sense, this says that the Bott-Chern cohomology "carries more informations" about the complex structure that the Dolbeault one. Note also that most of the dimensions of Bott-Chern cohomology groups are invariant under small deformations: this happens for example for the odd-degree Bott-Chern cohomology groups.

In order to have a complete view, we summarize in Table 3.5 the dimensions of the de Rham, Dolbeault, Bott-Chern, and Aeppli cohomologies of the Iwasawa manifold and of its small deformations, see [Ang11, Appendix].

3.3 Cohomologies of Six-Dimensional Nilmanifolds

In this section, we provide the results of the computations, obtained jointly with M.G. Franzini and F.A. Rossi in [AFR12], of the Bott-Chern cohomology for each of the complex structures on six-dimensional nilmanifolds in M. Ceballos,

A. Otal, L. Ugarte, and R. Villacampa's classification, [COUV11], see Table 1.4. In view of [Rol11a, Corollary 3.10], and noting that the unique (up to equivalence) left-invariant complex structure on the nilmanifold corresponding to $\mathfrak{h}_7$ is a rational complex structure, one is reduced to study left-invariant $\tilde{\Delta}_{BC}$-harmonic forms. In Figs. 3.4 and 3.5, we provide a graphical visualization of the non-Kählerianity of such six-dimensional nilmanifolds by means of the degrees Δ^1, Δ^2, and Δ^3, where

$$\Delta^k := \sum_{p+q=k} \left(\dim_\mathbb{C} H_{BC}^{p,q}(X) + \dim_\mathbb{C} H_A^{p,q}(X) \right) - 2 \cdot b_k \in \mathbb{N}.$$

Remark 3.17. Note that the Dolbeault, respectively Bott-Chern, cohomology being left-invariant is not an up-to-equivalence property: that is, the Bott-Chern numbers in Table 3.6 provide a complete picture *except for* nilmanifolds with Lie algebra isomorphic to $\mathfrak{h}_7$.

Theorem 3.10 ([AFR12, Theorem 2.1], [LUV12, §3]). *Consider a six-dimensional nilpotent Lie algebra endowed with a linear integrable complex structure. Equivalently, consider a six-dimensional nilmanifold with Lie algebra non-isomorphic to $\mathfrak{h}_7 = (0, 0, 0, 12, 13, 23)$ and endowed with any left-invariant complex structure, or a nilmanifold with Lie algebra isomorphic to $\mathfrak{h}_7$ and endowed with the left-invariant complex structure in M. Ceballos, A. Otal, L. Ugarte, and R. Villacampa's classification, [COUV11]. Then the dimensions of the Bott-Chern cohomology are provided in Table 3.6.*

Remark 3.18. Similar computations have been performed by A. Latorre, L. Ugarte, and R. Villacampa, [LUV12]: more precisely, they computed the dimensions of the Bott-Chern cohomology for any complex structure on a six-dimensional nilpotent Lie algebra using the classification of [COUV11], with the aim to study variations of the cohomology under deformations of the complex structure, and the behaviour of the cohomology in relation to the existence of balanced Hermitian metrics or strongly-Gauduchon metrics.

Balanced metrics (that is, Hermitian metrics whose fundamental form is co-closed, [Mic82]) on six-dimensional nilmanifolds have been studied by A. Latorre, L. Ugarte, and R. Villacampa in [LUV12]; see also [Uga07]; see also [Fra11] by M.G. Franzini for some results concerning balanced metrics on nilmanifolds and solvmanifolds. On the other hand,[1] as regards *pluriclosed* (or *strong Kähler with torsion*, shortly SKT) *metrics*, namely, Hermitian metrics whose fundamental form ω satisfies the condition $\partial\bar{\partial}\omega = 0$, [Bis89], and the

---

[1]We recall that, on six-dimensional nilmanifolds endowed with left-invariant complex structures, the existence of pluriclosed metrics and the existence of balanced metrics are complementary properties: indeed, by [FPS04, Proposition 1.4], see also [AI01, Remark 1], any Hermitian metric being both pluriclosed and balanced is in fact Kähler, and by [FPS04, Theorem 1.2] the pluriclosed property is satisfied by either all left-invariant Hermitian metrics or by none.

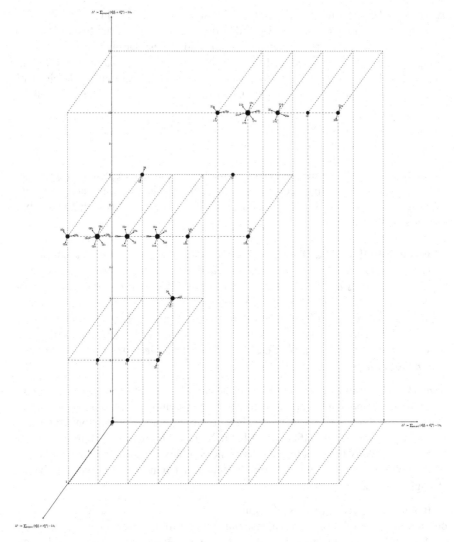

Fig. 3.4 Non-Kählerianity of six-dimensional nilmanifolds endowed with the left-invariant complex structures in M. Ceballos, A. Otal, L. Ugarte, and R. Villacampa's classification, [COUV11], from [AFR12]

possible connection between the existence of such metrics and cohomological properties, we recall here some results and examples obtained in [AFR12]. See also [Uga07].

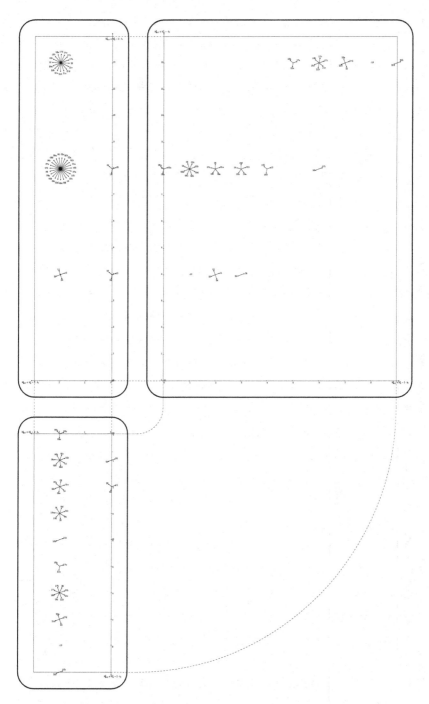

Fig. 3.5 Non-Kählerianity of six-dimensional nilmanifolds endowed with the left-invariant complex structures in M. Ceballos, A. Otal, L. Ugarte, and R. Villacampa's classification, [COUV11]

Table 3.6 Dimensions of the Bott-Chern cohomology of the six-dimensional nilpotent Lie algebras endowed with the linear integrable complex structure classified by M. Ceballos, A. Otal, L. Ugarte, and R. Villacampa in [COUV11], see [AFR12, Table 2], see also [LUV12]

Parameters:
$$\lambda \ge 0,\ c \ge 0,\ B \in \mathbb{C},\ D \in \mathbb{C}$$
$$\left(S(B,c) := c^4 - 2\big((|B|^2+1)c^2 + (|B|^2-1)^2\big)\right)$$

	$\mathfrak{h}$	J	condition	SKT [AFR12, Uga07]	Balanced [LUV12, Uga07]	$h_{BC}^{1,0}$	$h_{BC}^{0,1}$	$h_{BC}^{2,0}$	$h_{BC}^{1,1}$	$h_{BC}^{0,2}$	$h_{BC}^{3,0}$	$h_{BC}^{2,1}$	$h_{BC}^{1,2}$	$h_{BC}^{0,3}$	$h_{BC}^{3,1}$	$h_{BC}^{2,2}$	$h_{BC}^{1,3}$	$h_{BC}^{3,2}$	$h_{BC}^{2,3}$	b_1	b_2	b_3	Δ^1	Δ^2	Δ^3
00	$\mathfrak{h}_1$	J		✓	✓	3	3	3	9	3	1	9	9	1	3	9	3	3	3	6	15	20	0	0	0
01a	$\mathfrak{h}_2$	J_1^D	$D \ne i,\ \Im m D = 1$	✗		2	2	1	4	1	1	6	6	1	3	6	3	3	3	4	8	10	2	2	8
01b	$\mathfrak{h}_2$	J_1^D	$D = i$	✓		2	2	1	4	1	1	6	6	1	3	7	3	3	3	4	8	10	2	3	8
02a	$\mathfrak{h}_2$	J_2^D	$\Re e D \ne 1,\ \lvert D\rvert^2 + 2\Re e D \ne 0,\ \Im m D > 0$	✗		2	2	1	4	1	1	6	6	1	2	6	2	3	3	4	8	10	2	0	8
02b	$\mathfrak{h}_2$	J_2^D	$\Re e D = 1,\ \Im m D > 0$	✓		2	2	1	4	1	1	6	6	1	2	7	2	3	3	4	8	10	2	1	8
02c	$\mathfrak{h}_2$	J_2^D	$\Re e D \ne 1,\ \lvert D\rvert^2 + 2\Re e D = 0,\ \Im m D > 0$	✗		2	2	1	5	1	1	6	6	1	2	6	2	3	3	4	8	10	2	1	8
03	$\mathfrak{h}_3$	J_1		✗		2	2	1	4	1	1	6	6	1	3	7	3	3	3	5	9	10	0	1	8
04	$\mathfrak{h}_3$	J_2		✗		2	2	1	4	1	1	6	6	1	3	7	3	3	3	5	9	10	0	1	8
05	$\mathfrak{h}_4$	J_1		✗		2	2	1	4	1	1	6	6	1	3	6	3	3	3	4	8	10	2	2	8
06a	$\mathfrak{h}_4$	J_2^D	$D \in \mathbb{R}\setminus\{-2,0,1\}$	✗		2	2	1	4	1	1	6	6	1	2	6	2	3	3	4	8	10	2	0	8
06b	$\mathfrak{h}_4$	J_2^D	$D = 1$	✓		2	2	1	4	1	1	6	6	1	2	7	2	3	3	4	8	10	2	1	8
06c	$\mathfrak{h}_4$	J_2^D	$D = -2$	✗		2	2	1	5	1	1	6	6	1	3	6	3	3	3	4	8	10	2	2	8
07a	$\mathfrak{h}_5$	J_2^D	$D \in (0, \tfrac{1}{4})$	✗		2	2	2	6	2	1	6	6	1	3	6	3	3	3	4	8	10	2	2	8
07b	$\mathfrak{h}_5$	J_1^D	$D = 0$	✗		2	2	2	4	3	1	6	6	1	2	6	2	3	3	4	8	10	2	6	8
08	$\mathfrak{h}_5$	J_2		✗		2	2	3	4	3	1	6	6	1	2	8	2	3	3	4	8	10	2	6	8
09a	$\mathfrak{h}_5$	$J_3^{(\lambda, D)}$	$\{\lambda = 0,\ \Re e D \ne \tfrac{1}{2},\ 0 < 4(\Im m D)^2 < 1 + 4\Re e D,\ \Im m D > 0\}$ $\cup \left\{0 < \lambda^2 < \tfrac{1}{2},\ \Re e D = 0,\ 0 < \Im m D < \tfrac{\lambda^2}{2}\right\}$	✗		2	2	1	4	1	1	6	6	1	2	6	2	3	3	4	8	10	2	0	8

		Condition																				
096*	b_5	$J_3^{(\lambda,D)}$ $\cup\{\tfrac12 \le \lambda^2 <1,\ \mathfrak{Re}D=0,\ 0<\mathfrak{Im}D<\tfrac{1-\lambda^2}{2}\}$ $\cup\{\lambda^2>1,\ \mathfrak{Re}D=0,\ 0<\mathfrak{Im}D<\tfrac{\lambda^2-1}{2},\ (\mathfrak{Im}D)^2\ne\lambda^2-1\}$ $\{\lambda=0,\ \mathfrak{Re}D=\tfrac12,\ 0<4(\mathfrak{Im}D)^2<3,\ \mathfrak{Im}D>0\}$	$\vee$	2	2	1	4	1	6	6	1	2	7	2	3	3	4	8	10	2	1	8
096''		$\cup\{\lambda=0,\ \mathfrak{Re}D\notin\{0,\tfrac12\},\ 1+4\mathfrak{Re}D>0,\ \mathfrak{Im}D=0\}$	$\times$																			
09c	b_5	$J_3^{(\lambda,D)}$ $\lambda=0,\ D=\tfrac12$	$\vee$	2	2	1	4	1	6	6	1	2	8	2	3	3	4	8	10	2	2	8
09d	b_5	$\lambda^2>1,\ \mathfrak{Re}D=0,\ 0<\mathfrak{Im}D<\tfrac{\lambda^2-1}{2},\ (\mathfrak{Im}D)^2=\lambda^2-1$	$\times$	2	2	1	5	1	6	6	1	2	6	2	3	3	4	8	10	2	1	8
09e	b_5	$J_3^{(\lambda,D)}$ $\{0<\lambda^2<\tfrac12,\ D=0\}$ $\cup\{\tfrac12\le\lambda^2<1,\ D=0\}$ $\cup\{\lambda^2>1,\ D=0\}$	$\times$	2	2	2	4	1	6	6	1	2	6	2	3	3	4	8	10	2	2	8
09f	b_5	$J_3^{(\lambda,D)}$ $\lambda=0,\ D=0$	$\times$ $\vee$	2	2	2	4	1	6	6	1	2	7	2	3	3	4	8	10	2	3	8
10	b_6	J	$\times$ $\times$	2	2	2	5	1	6	6	1	2	6	2	3	3	4	9	12	2	1	4
11	b_7	J	$\times$ $\times$	1	1	2	5	1	6	6	1	2	5	2	3	3	3	8	12	2	2	4
12	b_8	J	$\vee$ $\times$	2	2	2	6	1	7	7	1	3	8	3	3	3	5	11	14	0	2	4
13	b_9	J	$\times$ $\times$	1	1	1	4	1	5	5	1	3	6	3	3	3	4	7	8	0	4	8
14	b_{10}	J	$\times$ $\times$	1	1	1	4	1	5	5	1	2	5	2	3	3	3	6	8	2	4	8
15a	b_{11}	J^B $B\in\mathbb{R}\setminus\{0,\tfrac12,1\}$	$\times$ $\times$	1	1	1	4	1	5	5	1	2	5	2	3	3	3	6	8	2	3	8
15b	b_{11}	J^B $B=\tfrac12$	$\times$ $\times$	1	1	1	4	1	5	5	1	2	6	2	3	3	3	6	8	2	3	8
16a	b_{12}	J^B $\mathfrak{Re}B\ne\tfrac12,\ \mathfrak{Im}B\ne0$	$\times$ $\times$	1	1	1	4	1	5	5	1	2	5	2	3	3	3	6	8	2	4	8
16b	b_{12}	J^B $\mathfrak{Re}B=\tfrac12,\ \mathfrak{Im}B\ne0$	$\times$ $\times$	1	1	1	4	1	5	5	1	2	6	2	3	3	3	6	8	2	4	8

(continued)

Table 3.6 (continued)

Heading for condition column:
$\lambda \geq 0$, $c \geq 0$, $B \in \mathbb{C}$, $D \in \mathbb{C}$
$\left(S(B,c) := c^4 - 2(|B|^2 + 1)c^2 + (|B|^2 - 1)^2 \right)$

	$\mathfrak{h}$	J	condition	SKT [AFR12, Uga07]	Balanced [LUV12, Uga07]	$h_{BC}^{1,0}$	$h_{BC}^{0,1}$	$h_{BC}^{2,0}$	$h_{BC}^{1,1}$	$h_{BC}^{0,2}$	$h_{BC}^{3,0}$	$h_{BC}^{2,1}$	$h_{BC}^{1,2}$	$h_{BC}^{0,3}$	$h_{BC}^{3,1}$	$h_{BC}^{2,2}$	$h_{BC}^{1,3}$	$h_{BC}^{3,2}$	$h_{BC}^{2,3}$	b_1	b_2	b_3	Δ^1	Δ^2	Δ^3
17a	$\mathfrak{h}_{13}$	$J^{(B,c)}$	$c \notin \{|B-1|, |B|\}$, $(c, |B|) \neq (0,1)$, $S(B,c) < 0$, $B \neq 1$	×	×	1	1	1	4	1	1	5	5	1	2	5	2	3	3	3	5	6	2	5	12
17b	$\mathfrak{h}_{13}$	$J^{(B,c)}$	$c = |B| > \frac{1}{2}$, $\mathfrak{Re}\,B \neq \frac{1}{2}$, $B \neq 1$	×	×	1	1	1	4	1	1	5	5	1	2	6	2	3	3	3	5	6	2	6	12
17c	$\mathfrak{h}_{13}$	$J^{(B,c)}$	$c \notin \{0, 1\}$, $c < 2$, $B = 1$	×	×	1	1	1	5	1	1	5	5	1	2	5	2	3	3	3	5	6	2	6	12
17d	$\mathfrak{h}_{13}$	$J^{(B,c)}$	$c = 1$, $B = 1$	×	×	1	1	1	5	1	1	5	5	1	2	6	2	3	3	3	5	6	2	7	12
18a	$\mathfrak{h}_{14}$	$J^{(B,c)}$	$B \neq 1$, $c \notin \{0, |B|, |B-1|\}$, $S(B,c) = 0$	×	×	1	1	1	4	1	1	5	5	1	2	5	2	3	3	3	5	6	2	5	12
18b	$\mathfrak{h}_{14}$	$J^{(B,c)}$	$c = |B| = \frac{1}{2}$, $\mathfrak{Re}\,B \neq \frac{1}{2}$	×	×	1	1	1	4	1	1	5	5	1	2	6	2	3	3	3	5	6	2	6	12
18c	$\mathfrak{h}_{14}$	$J^{(B,c)}$	$B = 1$, $c = 2$	×	×	1	1	1	5	1	1	5	5	1	3	5	3	3	3	3	5	6	2	8	12
19	$\mathfrak{h}_{15}$	J_1		×	×	1	1	1	5	1	1	5	5	1	3	5	3	3	3	3	5	6	2	7	12
20a	$\mathfrak{h}_{15}$	J_2^c	$c \notin \{0, 1\}$	×	×	1	1	2	4	2	1	5	5	1	3	5	3	3	3	3	5	6	2	9	12
20b	$\mathfrak{h}_{15}$	J_2^c	$c = 0$	×	×	1	1	2	4	2	1	5	5	1	3	5	3	3	3	3	5	6	2	9	12
21a	$\mathfrak{h}_{15}$	$J_3^{(B,c)}$	$c \notin \{0, |B-1|, |B|\}$, $B \neq 1$, $S(B,c) > 0$	×	×	1	1	1	4	1	1	5	5	1	2	5	2	3	3	3	5	6	2	5	12
21b	$\mathfrak{h}_{15}$	$J_3^{(B,c)}$	$0 < c = |B| < \frac{1}{2}$		×	1	1	1	4	1	1	5	5	1	2	6	2	3	3	3	5	6	2	6	12
21c	$\mathfrak{h}_{15}$	$J_3^{(B,c)}$	$c > 2$, $B = 1$	×	×	1	1	1	5	1	1	5	5	1	2	5	2	3	3	3	5	6	2	6	12
21d	$\mathfrak{h}_{15}$	$J_3^{(B,c)}$	$c = 0$, $B \neq 0$, $|B| \neq 1$	×	×	1	1	2	4	2	1	5	5	1	2	5	2	3	3	3	5	6	2	7	12
21e	$\mathfrak{h}_{15}$	$J_3^{(B,c)}$	$c = 0$, $B = 0$	×	×	1	1	2	4	2	1	5	5	1	2	7	2	3	3	3	5	6	2	9	12
22	$\mathfrak{h}_{16}$	J^B	$|B| = 1$, $B \neq 1$	×	✓	1	1	2	4	2	1	5	5	1	2	5	2	3	3	3	5	6	2	7	12
23	$\mathfrak{h}_{19}^-$	J_1		×	✓	1	1	1	2	1	1	3	3	1	2	4	2	2	2	3	5	6	0	2	4
24	$\mathfrak{h}_{19}^-$	J_2		×		1	1	1	2	1	1	3	3	1	2	4	2	2	2	3	5	6	0	2	4
25	$\mathfrak{h}_{26}^+$	J_1		×	×	1	1	1	2	1	1	3	3	1	2	3	2	2	2	2	4	6	2	3	4
26	$\mathfrak{h}_{26}^+$	J_2		×	×	1	1	1	2	1	1	3	3	1	2	3	2	2	2	2	4	6	2	3	4

More precisely, the following result points out which of the complex manifolds in Table 3.6 admit pluriclosed metrics.

Theorem 3.11 ([AFR12, Theorem 4.1]). *Let X be a six-dimensional nilmanifold and J be a left-invariant complex structure on X. Then (X, J) admits a left-invariant pluriclosed metric if and only if*

[00] $(X, J) = (\mathfrak{h}_1, J)$;

[01b] $(X, J) = \left(\mathfrak{h}_2, J_1^D\right)$ *with $D = $ i;*

[02b] $(X, J) = \left(\mathfrak{h}_2, J_2^D\right)$ *with $D \in \mathbb{C}$ such that $\mathfrak{Re}\, D = 1$ and $\mathfrak{Im}\, D > 0$;*

[06b] $(X, J) = \left(\mathfrak{h}_4, J_2^D\right)$ *with $D = 1$;*

[09b'] $(X, J) = \left(\mathfrak{h}_5, J_3^{(\lambda, D)}\right)$ *with $\lambda = 0$ and $D = \frac{1}{2} + $i y such that $0 < 4y^2 < 3$;*

[09c] $(X, J) = \left(\mathfrak{h}_5, J_3^{(\lambda, D)}\right)$ *with $\lambda = 0$ and $D = \frac{1}{2}$;*

[12] $(X, J) = (\mathfrak{h}_8, J)$,

where we use the notation in Tables 1.4 and 3.6, and in [COUV11].

The following example shows a curve of complex structures admitting pluriclosed metrics and with jumping Bott-Chern cohomology numbers.

Example 3.1 (A curve of complex structures on a six-dimensional nilmanifold such that each complex structure admits a pluriclosed metric and the dimensions of the Bott-Chern cohomology groups are not constant, [AFR12, Example 4.3]). Consider the nilmanifold X with associated Lie algebra

$$\mathfrak{h}_2 := \left(0^4, 12, 34\right),$$

and consider the family $\{J_t\}_{t \in \mathbb{R}}$ of left-invariant complex structures on X such that a left-invariant co-frame for the $\mathcal{C}^\infty(X)$-module of the $(1, 0)$-forms on (X, J_t) is given by $\{\psi_t^1, \psi_t^2, \psi_t^3\}$ satisfying

$$\begin{cases} d\,\psi_t^1 = 0 \\ d\,\psi_t^2 = 0 \\ d\,\psi_t^3 = t\,\psi_t^1 \wedge \psi_t^2 + \psi_t^1 \wedge \bar\psi_t^1 + t\,\psi_t^1 \wedge \bar\psi_t^2 + \left(t^2 + i\right)\psi_t^2 \wedge \bar\psi_t^2 \end{cases}.$$

One has three different cases (we use the notation in Tables 1.4 and 3.6, and [COUV11]):

t = 0: for $t = 0$, the left-invariant complex structure J_t is equivalent to the left-invariant Abelian complex structure J_1^D with $D = $ i (case 01b in Table 3.6);

t ≠ 0: for $t \neq 0$, the left-invariant complex structure J_t is equivalent to the left-invariant non-Abelian complex structure J_2^D with $D = 1 + \frac{1}{t^2}$ i (case 02b in Table 3.6).

In particular, by Theorem 3.11, (X, J_t) admits a left-invariant pluriclosed metric for every $t \in \mathbb{R}$. Notwithstanding, note that $\dim_{\mathbb{C}} H_{BC}^{3,1} (X, J_0) = \dim_{\mathbb{C}} H_{BC}^{1,3} (X, J_0) = 3$, while $\dim_{\mathbb{C}} H_{BC}^{3,1} (X, J_t) = \dim_{\mathbb{C}} H_{BC}^{1,3} (X, J_t) = 2$ for $t \neq 0$.

For the sake of completeness, we provide the following example, compare also [COUV11, Theorem 7.9]: it shows a curve of complex structures on a six-dimensional nilmanifold admitting both pluriclosed and balanced metrics.

Example 3.2 (A curve of compact complex manifolds endowed with special metrics, [AFR12, Example 4.6], see also [COUV11, Theorem 7.9]). Let X be a nilmanifold with associated Lie algebra

$$\mathfrak{h}_4 := \left(0^4, \ 12, \ 14 + 23\right) ,$$

and consider on it the family $\left\{J_2^D\right\}_{D \in \mathbb{R} \setminus \{0\}}$ of left-invariant non-Abelian complex structures whose $(1, 0)$-forms are generated by $\left\{\omega_D^1, \omega_D^2, \omega_D^3\right\}$ as a $\mathcal{C}^\infty(X; \mathbb{C})$-module and such that

$$\begin{cases} d\omega_D^1 := 0 \\ d\omega_D^2 := 0 \\ d\omega_D^3 := \omega_D^{12} + \omega_D^{1\bar{1}} + \omega_D^{1\bar{2}} + D \, \omega_D^{2\bar{2}} \end{cases} .$$

By [FPS04, Theorem 3.2] and [Uga07, Theorem 26], the Lie algebra $\mathfrak{h}_4$ admits both pluriclosed and balanced left-invariant Hermitian structures.

Let

$$\Omega_D^{r,s,t,u,v,z} := i \left(r^2 \, \omega_D^{1\bar{1}} + s^2 \, \omega_D^{2\bar{2}} + t^2 \, \omega_D^{3\bar{3}}\right) + u \, \omega_D^{1\bar{2}} - \bar{u} \, \omega_D^{2\bar{1}} + v \, \omega_D^{2\bar{3}} - \bar{v} \, \omega_D^{3\bar{2}}$$
$$+ z \, \omega_D^{1\bar{3}} - \bar{z} \, \omega_D^{3\bar{1}}$$

the generic J_2^D-Hermitian structure, where

$$r, \ s, \ t \ \in \mathbb{R} \setminus \{0\} \qquad \text{and } u, \ v, \ z \ \in \mathbb{C}$$

satisfy

$$\begin{cases} r^2 s^2 > |u|^2 \\ s^2 t^2 > |v|^2 \\ r^2 t^2 > |z|^2 \\ r^2 s^2 t^2 + 2 \mathfrak{Re} \left(i \, \bar{u} \, \bar{v} \, z\right) > t^2 \, |u|^2 + r^2 \, |v|^2 + s^2 \, |z|^2 \end{cases} ,$$

see, e.g., [UV09, pp. 3–4].

By computing

$$\partial \bar{\partial} \, \Omega_D^{r,s,t,u,v,z} \;=\; \mathrm{i}\, t^2 \, (1 - D) \; \omega_D^{12\bar{1}\bar{2}} \,,$$

we have that $\Omega_D^{r,s,t,u,v,z}$ is pluriclosed if and only if $D = 1$, compare also [FPS04, Theorem 1.2].

Computing

$$\mathrm{d}\left(\Omega_D^{r,s,t,u,v,z}\right)^2 \;=\; \frac{1}{2}\, \rho_D^{r,s,t,u,v,z} \, \omega_D^{123\bar{1}\bar{2}} \;-\; \frac{1}{2}\, \overline{\rho_D^{r,s,t,u,v,z}} \, \omega_D^{12\bar{1}\bar{2}\bar{3}}$$

where

$$\rho_D^{r,s,t,u,v,z} \;:=\; D\, t^2\, r^2 - D\, |v|^2 + t^2\, s^2 - |z|^2 + \mathrm{i}\, t^2\, u + v\, \bar{z} \,,$$

we have that $\Omega_D^{r,s,t,u,v,z}$ is balanced if and only if D, r, s, t, u, v, z are such that $\rho_D^{r,s,t,u,v,z} = 0$.

Firstly, suppose that $0 < D < \frac{1}{4}$. Let $s, t \in \mathbb{R} \setminus \{0\}$ be such that $s^2 > \left(D + s^2\right)^2$ and consider the J_2^D-Hermitian metric

$$\Omega_D^{s,t} \;:=\; \Omega_D^{1,s,t,\mathrm{i}\,(D+s^2),0,0} \;=\; \mathrm{i}\left(\omega^{1\bar{1}} + s^2\, \omega^{2\bar{2}} + t^2\, \omega^{3\bar{3}}\right) + \mathrm{i}\,\left(D + s^2\right)\cdot\left(\omega^{1\bar{2}} + \omega^{2\bar{1}}\right);$$

a straightforward computation shows that $\Omega_D^{s,t}$ is balanced, see also [COUV11, Sect. 7].

In the case $D = \frac{1}{4}$, it turns out that J_2^D admits no balanced metric, see also [COUV11, Theorem 7.9]. Indeed, for the sake of completeness, we note that, if $\tilde{\Omega}$ were a balanced metric, then one should have

$$0 = \int_X \tilde{\Omega}^2 \wedge \mathrm{d}\left(\omega_{\frac{1}{4}}^3 - \overline{\omega_{\frac{1}{4}}^3}\right) = \int_X \tilde{\Omega}^2 \wedge \left(\sqrt{2}\omega_{\frac{1}{4}}^1 + \frac{\sqrt{2}}{2}\omega_{\frac{1}{4}}^2\right) \wedge \overline{\left(\sqrt{2}\omega_{\frac{1}{4}}^1 + \frac{\sqrt{2}}{2}\omega_{\frac{1}{4}}^2\right)} > 0,$$

which yields an absurd.

Summarizing, we have a curve such that:

- for $0 < D < \frac{1}{4}$, there exists a balanced metric on $\left(X, \, J_2^D\right)$;
- for $D = \frac{1}{4}$, the complex manifold $\left(X, \, J_2^{\frac{1}{4}}\right)$ admits no balanced metric and no pluriclosed metric;
- for $D = 1$, there exists a pluriclosed metric on $\left(X, \, J_2^1\right)$.

Compare also [COUV11, Theorem 7.9], where the same example is studied in order to prove that the existence of balanced metrics and the existence of strongly-Gauduchon metrics are not closed under deformations of the complex structure.

Remark 3.19. For the study of the behaviour of the cohomology in relation to the existence of Hermitian balanced metrics or strongly-Gauduchon metrics, we refer to [LUV12] by A. Latorre, L. Ugarte, and R. Villacampa, where, for any balanced structure, they also determine L.-S. Tseng and S.-T. Yau's spaces parametrizing the deformations in type IIB supergravity, see [TY11].

Appendix: Cohomology of Solvmanifolds

Enlarging the class of nilmanifolds to solvmanifolds, it turns out that the left-invariant forms are usually not enough to recover the whole de Rham cohomology: see, for example, the non-completely-solvable solvmanifold provided in [dBT06, Sects. 3 and 4, Corollary 4.2]. The de Rham cohomology of solvmanifolds has been studied by several author, e.g., A. Hattori [Hat60], G.D. Mostow [Mos54], D. Guan [Gua07], S. Console and A.M. Fino [CF11], and by H. Kasuya [Kas13a, Kas12a, CFK13]; several results concerning the Dolbeault cohomology (and Hodge symmetry, Hodge decomposition, formality, and the Hodge and Frölicher spectral sequence on solvmanifolds) have been proven by H. Kasuya [Kas13b, Kas12a, Kas11, Kas12d, Kas12b].

In this appendix, we briefly summarize some results concerning the Dolbeault and Bott-Chern cohomologies of a class of solvmanifolds, [Kas13a, Kas13b, AK12], recalling the results on the computations of the cohomologies of the Nakamura manifold as an example.

As regards the de Rham cohomology of solvmanifolds, H. Kasuya proved the following result in [Kas13a], see also [Kas12a].

Theorem 3.12 ([Kas13a, Corollary 7.6], [Kas12a, Theorem 1.3]). *Let $\Gamma\backslash G$ be a solvmanifold. Then there exists a finite-dimensional sub-complex $A_\Gamma^\bullet$ of the de Rham complex $(\wedge^\bullet X \otimes_\mathbb{R} \mathbb{C}, d)$ such that the inclusion*

$$\left(A_\Gamma^\bullet, \, d\right) \hookrightarrow \left(\wedge^\bullet \Gamma\backslash G \, \otimes_\mathbb{R} \mathbb{C}, \, d\right)$$

is a quasi-isomorphism.

Remark 3.20 ([Kas13a, Kas12a]). The sub-complex $A_\Gamma^\bullet$ in the previous theorem is constructed as follows, see [Kas13a, Kas12a].

Consider a connected simply-connected n-dimensional solvable Lie group G admitting a discrete co-compact subgroup Γ. Denote by $\mathfrak{g}$ the solvable Lie algebra of G and by $\mathfrak{g}_\mathbb{C} := \mathfrak{g} \otimes_\mathbb{R} \mathbb{C}$ its complexification, and let $\mathfrak{n}$ be the nilradical[2] of $\mathfrak{g}$.

---

[2]Recall that the *nilradical* of a solvable Lie algebra is the maximal nilpotent ideal in $\mathfrak{g}$.

Consider the adjoint representation ad: $\mathfrak{g} \ni X \mapsto [X, \cdot] \in \mathfrak{gl}(\mathfrak{g})$ and, for every $X \in \mathfrak{g}$, take the semi-simple part $(\mathrm{ad}_X)_\mathrm{s} \in \mathfrak{gl}(\mathfrak{g})$ of the Jordan decomposition[3] of ad_X.

By considering a vector sub-space V of $\mathfrak{g}$ such that $\mathfrak{g} = V \oplus \mathfrak{n}$ as the direct sum of vector spaces and such that, for any $A, B \in V$, it holds that $(\mathrm{ad}_A)_\mathrm{s}(B) = 0$, see, e.g., [DtER03, Proposition III.1.1], one can define the representation $\mathrm{ad}_\mathrm{s} \colon \mathfrak{g} = V \oplus \mathfrak{n} \ni (A, X) \mapsto (\mathrm{ad}_\mathrm{s})_{A+X} := (\mathrm{ad}_A)_\mathrm{s} \in \mathrm{gl}(\mathfrak{g})$. Denote by $\mathrm{Ad}_\mathrm{s} \colon G \to \mathrm{Aut}(\mathfrak{g}_\mathbb{C})$ the unique representation which lifts $\mathrm{ad}_\mathrm{s} \colon \mathfrak{g} \to \mathfrak{gl}(\mathfrak{g}_\mathbb{C})$, see, e.g., [War83, Theorem 3.27].

Set $\mathcal{C}_\Gamma := \{\beta \circ \mathrm{Ad}_\mathrm{s} \in \mathrm{Hom}(G; \mathbb{C}^*) \; : \; \beta \in \mathrm{Hom}(T; \mathbb{C}^*) \text{ s.t. } (\beta \circ \mathrm{Ad}_\mathrm{s}) \lfloor_\Gamma = 1\}$, where T is the Zariski-closure of $\mathrm{Ad}_\mathrm{s}(G)$ in $\mathrm{Aut}(\mathfrak{g}_\mathbb{C})$. Hence, define $A_\Gamma^\bullet$ as the space consisting of $\mathrm{Ad}_\mathrm{s}(G)$-invariant elements of $\bigoplus_{\alpha \in \mathcal{C}_\Gamma} \alpha \cdot \wedge^\bullet \mathfrak{g}_\mathbb{C}^*$, namely,

$$A_\Gamma^\bullet := \left\{ \varphi \in \bigoplus_{\alpha \in \mathcal{C}_\Gamma} \alpha \cdot \wedge^\bullet \mathfrak{g}_\mathbb{C}^* : (\mathrm{Ad}_\mathrm{s})_g(\varphi) = \varphi \text{ for every } g \in G \right\} \subseteq \wedge^\bullet \Gamma \backslash G \otimes_\mathbb{R} \mathbb{C} .$$

By considering a basis $\{X_1, \ldots, X_n\}$ of $\mathfrak{g}_\mathbb{C}$ with respect to which

$$\mathrm{Ad}_\mathrm{s} = \mathrm{diag}(\alpha_1, \ldots, \alpha_n) \colon G \to \mathrm{Aut}(\mathfrak{g}_\mathbb{C})$$

for some characters $\alpha_1 \in \mathrm{Hom}(G; \mathbb{C}^*), \ldots, \alpha_n \in \mathrm{Hom}(G; \mathbb{C}^*)$, and by considering its dual basis $\{x_1, \ldots, x_n\}$ of $\mathfrak{g}_\mathbb{C}^*$, then one gets, for $p \in \mathbb{Z}$,

$$A_\Gamma^p = \mathbb{C} \left\langle \alpha_{i_1} \cdots \alpha_{i_p} \, x_{i_1} \wedge \cdots \wedge x_{i_p} \, \middle| \, 1 \le i_1 < \cdots < i_p \le n \text{ such that } (\alpha_{i_1} \cdots \alpha_{i_p}) \lfloor_\Gamma = 1 \right\rangle .$$

As regards the Dolbeault cohomology of solvmanifolds, one can focus on two special classes of solvmanifolds: namely, holomorphically parallelizable solvmanifolds, [Wan54, Nak75], and solvmanifolds of splitting type, [Kas13b].

Definition 3.1 ([Kas13b, Assumption 1.1]). A solvmanifold $X = \Gamma \backslash G$ endowed with a G-left-invariant complex structure J is said to be *of splitting type* if G is a semi-direct product $\mathbb{C}^n \ltimes_\phi N$ satisfying the following assumptions:

1. N is a connected simply-connected $2m$-dimensional nilpotent Lie group endowed with an N-left-invariant complex structure J_N;
2. for any $t \in \mathbb{C}^n$, one has that $\phi(t) \in \mathrm{GL}(N)$ is a holomorphic automorphism of N with respect to J_N;
3. ϕ induces a semi-simple action on the Lie algebra of N.

---

[3]Recall that, given a vector space V, any linear map $A \in \mathfrak{gl}(V)$ admits a unique *Jordan decomposition* $A = A_\mathrm{s} + A_\mathrm{n}$, where $A_\mathrm{s} \in \mathfrak{gl}(V)$ is *semi-simple* (that is, each A_s-invariant sub-space of V admits an A_s-invariant complementary sub-space in V), and $A_\mathrm{n} \in \mathfrak{gl}(V)$ is *nilpotent* (that is, there exists $N \in \mathbb{N}$ such that $A_\mathrm{n}^N = 0$), and A_s and A_n commute, see, e.g., [DtER03, II.1.10], [Kir08, Theorem 5.59].

For such classes of solvmanifolds, H. Kasuya proved the following result in [Kas13b], see also [Kas12a].

Theorem 3.13 ([Kas13b, Corollary 4.2], [Kas12a, Corollary 6.2]). *Let $\Gamma \backslash G$ be a solvmanifold endowed with a G-left-invariant complex structure. Suppose that*

- *either $\Gamma \backslash G$ is of splitting type, with the nilpotent factor N such that the inclusion $\left(\wedge^{\bullet,\bullet} \mathfrak{n}^*, \bar{\partial} \right) \hookrightarrow \left(\wedge^{\bullet,\bullet} N, \bar{\partial} \right)$ is a quasi-isomorphism, where $\mathfrak{n}$ is the Lie algebra of N,*
- *or $\Gamma \backslash G$ is holomorphically parallelizable.*

Then there exists a finite-dimensional sub-complex $B_\Gamma^{\bullet,\bullet}$ of the Dolbeault complex $\left(\wedge^{\bullet,\bullet} X, \partial, \bar{\partial} \right)$ such that the inclusion

$$\left(B_\Gamma^{\bullet,\bullet}, \bar{\partial} \right) \hookrightarrow \left(\wedge^{\bullet,\bullet} \Gamma \backslash G, \bar{\partial} \right)$$

is a quasi-isomorphism.

Remark 3.21 ([Kas13b]). In the case of solvmanifolds of splitting type, the sub-complex $B_\Gamma^{\bullet,\bullet}$ in the previous theorem is constructed as follows, see [Kas13b].

Consider the standard basis $\{X_1, \ldots, X_n\}$ of $\mathbb{C}^n$, and consider a basis $\{Y_1, \ldots, Y_m\}$ of the i-eigen-space $\mathfrak{n}^{1,0}$ of $J_N \in \mathrm{Aut}(\mathfrak{n})$ such that the induced action ϕ on $\mathfrak{n}^{1,0}$ is represented by $\phi = \mathrm{diag}(\alpha_1, \ldots, \alpha_m)$ for $\alpha_1 \in \mathrm{Hom}(\mathbb{C}^n; \mathbb{C}^*), \ldots,$ $\alpha_m \in \mathrm{Hom}(\mathbb{C}^n; \mathbb{C}^*)$ characters of $\mathbb{C}^n$. Let $\{x_1, \ldots, x_n, \alpha_1^{-1} y_1, \ldots, \alpha_m^{-1} y_m\}$ be the basis of $\wedge^{1,0} \mathfrak{g}_\mathbb{C}^*$ which is dual to $\{X_1, \ldots, X_n, \alpha_1 Y_1, \ldots, \alpha_m Y_m\}$.

By [Kas13b, Lemma 2.2], for any $j \in \{1, \ldots, m\}$, there exist unique unitary characters β_j and γ_j of $\mathbb{C}^n$ such that $\alpha_j \beta_j^{-1}$ and $\bar{\alpha}_j \gamma_j^{-1}$ are holomorphic. (By shortening, e.g., $\alpha_I := \alpha_{i_1} \cdots \cdots \alpha_{i_k}$ for a multi-index $I = (i_1, \ldots, i_k)$) define $B_\Gamma^{\bullet,\bullet} \subset \wedge^{\bullet,\bullet} \Gamma \backslash G$, for $(p, q) \in \mathbb{Z}^2$, as

$$B_\Gamma^{p,q} := \mathbb{C} \langle x_I \wedge \left(\alpha_J^{-1} \beta_J \right) y_J \wedge \bar{x}_K \wedge \left(\bar{\alpha}_L^{-1} \gamma_L \right) \bar{y}_L \mid |I| + |J| = p$$

$$\text{and } |K| + |L| = q \text{ such that } (\beta_J \gamma_L) \lfloor_\Gamma = 1 \rangle .$$

Remark 3.22 ([Kas11]). In [Kas11], H. Kasuya studied conditions assuring that a splitting type solvmanifold satisfies Hodge symmetry and Hodge decomposition. More precisely, let $G = \mathbb{C}^n \ltimes_\phi \mathbb{C}^m$ be such that $X := \Gamma \backslash G$ is a solvmanifold of splitting type. Consider a set $\{z_1, \ldots, z_n\}$ of coordinates on $\mathbb{C}^n$ and a set $\{w_1, \ldots, w_m\}$ of coordinates on $\mathbb{C}^m$. With the same notations as in Remark 3.21, let $\alpha_1 \in \mathrm{Hom}(\mathbb{C}^n; \mathbb{C}^*), \ldots, \alpha_m \in \mathrm{Hom}(\mathbb{C}^n; \mathbb{C}^*)$ be characters such that $\phi = \mathrm{diag}(\alpha_1, \ldots, \alpha_m)$, and let $\beta_j \in \mathrm{Hom}(\mathbb{C}^n; \mathbb{C}^*)$ and $\gamma_j \in \mathrm{Hom}(\mathbb{C}^n; \mathbb{C}^*)$ be the unique unitary characters such that $\alpha_j \beta_j^{-1}$ and $\bar{\alpha}_j \gamma_j^{-1}$ are holomorphic, for $j \in \{1, \ldots, m\}$, [Kas13b, Lemma 2.2]. Consider the G-left-invariant Hermitian metric $g := \sum_{h=1}^n \mathrm{d} z_h \odot \mathrm{d} \bar{z}_h + \sum_{k=1}^m \alpha_k^{-1} \bar{\alpha}_k^{-1} \mathrm{d} w_k \odot \mathrm{d} \bar{w}_k$. H. Kasuya proved in [Kas11, Theorem 1.4] that (see also [Has06, Kas11, Remark 2]):

- if the characters $\alpha_1, \ldots, \alpha_m$ are real-valued (namely, G is completely-solvable), then $\ker \overline{\square}_g = \ker \overline{\square}_g$, that is, X is a non-Kähler manifold satisfying the Hodge symmetry;
- if, for any multi-indices $K, L \subseteq \{1, \ldots, m\}$, it holds that $(\beta_K \gamma_L)\lfloor_\Gamma = 1$ if and only if $\alpha_K \bar{\alpha}_L = 1$, then $\ker \Delta_g = \ker \overline{\square}_g$, that is, X satisfies the Hodge decomposition.

In particular, by [DGMS75, 5.21], it follows that, if both the characters $\alpha_1, \ldots,$ α_m are real-valued and, for any multi-indices $K, L \subseteq \{1, \ldots, m\}$, it holds that $(\beta_K \gamma_L)\lfloor_\Gamma = 1$ if and only if $\alpha_K \bar{\alpha}_L = 1$, then X is a non-Kähler manifold satisfying the $\partial\bar{\partial}$-Lemma, [Kas12b, Corollary 1.5].

Remark 3.23 ([Kas12a]) In the case of holomorphically parallelizable solvmanifolds, the sub-complex $B_\Gamma^{\bullet,\bullet}$ in the previous theorem is constructed as follows, see [Kas12a].

Let G be a connected simply-connected complex solvable Lie group of complex dimension n and admitting a lattice Γ, and denote the Lie algebra naturally associated to G by $\mathfrak{g}$. Let N be the nilradical of G.

There exists a connected simply-connected complex nilpotent subgroup $C \subseteq G$ such that $G = C \cdot N$, see, e.g., [Dek00, Proposition 3.3]. The diagonalizable representation Ad_s constructed in Remark 3.20 can be identified with the map $G = C \cdot N \ni c \cdot n \mapsto (\mathrm{Ad}_c)_s \in \mathrm{Aut}(\mathfrak{g})$, where $(\mathrm{Ad}_c)_s$ is the semi-simple part of the Jordan decomposition of Ad_c, for $c \in C$, see [Kas12a, Remark 4].

Consider a basis $\{X_1, \ldots, X_n\}$ of the Lie algebra $\mathfrak{g}^{1,0}$ of the G-left-invariant holomorphic vector fields on G with respect to which $(\mathrm{Ad}_c)_s = \mathrm{diag}\,(\alpha_1(c), \ldots, \alpha_n(c))$, for $c \in C$, for some characters $\alpha_1 \in \mathrm{Hom}(C; \mathbb{C}^*), \ldots, \alpha_n \in \mathrm{Hom}(C; \mathbb{C}^*)$ of C. Let $\{x_1, \ldots, x_n\}$ be the basis of $(\mathfrak{g}^{1,0})^*$ which is dual to $\{X_1, \ldots, X_n\}$.

(By regarding $\alpha_1, \ldots, \alpha_n$ as characters of G, and by shortening, e.g., $\alpha_I := \alpha_{i_1} \cdots \cdots \alpha_{i_k}$ for a multi-index $I = (i_1, \ldots, i_k)$,) define $B_\Gamma^{\bullet,\bullet} \subset \wedge^{\bullet,\bullet} \Gamma\backslash G$, for $(p, q) \in \mathbb{Z}^2$, as

$$B_\Gamma^{p,q} := \wedge^p \left(\mathfrak{g}^{1,0}\right)^* \otimes_{\mathbb{C}} \mathbb{C}\left\langle \frac{\bar{\alpha}_I}{\alpha_I}\, \bar{x}_I \; \middle| \; |I| = q \text{ such that } \left(\frac{\bar{\alpha}_I}{\alpha_I}\right)\Big|_\Gamma = 1 \right\rangle.$$

Finally, as regards Bott-Chern cohomology, in a joint work with H. Kasuya, we proved the following result in [AK12].

Theorem 3.14 ([AK12, Theorems 2.16 and 2.25]). *Let $\Gamma\backslash G$ be a solvmanifold endowed with a G-left-invariant complex structure. Suppose that*

- *either $\Gamma\backslash G$ is of splitting type, with the nilpotent factor N such that the inclusion $\left(\wedge^{\bullet,\bullet}\mathfrak{n}^*, \bar{\partial}\right) \hookrightarrow \left(\wedge^{\bullet,\bullet} N, \bar{\partial}\right)$ is a quasi-isomorphism, where $\mathfrak{n}$ is the Lie algebra of N,*
- *or $\Gamma\backslash G$ is holomorphically parallelizable.*

Then there exists a finite-dimensional sub-complex $C_\Gamma^{\bullet,\bullet}$ of the Dolbeault complex $\left(\wedge^{\bullet,\bullet} X, \partial, \bar{\partial} \right)$ such that the inclusion

$$\left(C_\Gamma^{\bullet,\bullet}, \partial, \bar{\partial} \right) \hookrightarrow \left(\wedge^{\bullet,\bullet} \Gamma \backslash G, \partial, \bar{\partial} \right)$$

induces the isomorphism

$$H_{BC}^{\bullet,\bullet} \left(\Gamma \backslash G \right) \simeq \frac{\ker \left(\partial \colon C^{\bullet,\bullet} \to C^{\bullet+1,\bullet} \right) \cap \ker \left(\bar{\partial} \colon C^{\bullet,\bullet} \to C^{\bullet,\bullet+1} \right)}{\operatorname{im} \left(\partial\bar{\partial} \colon C^{\bullet-1,\bullet-1} \to C^{\bullet,\bullet} \right)} .$$

Remark 3.24. The sub-complex $C_\Gamma^{\bullet,\bullet}$ in the previous theorem is constructed as

$$C_\Gamma^{\bullet,\bullet} := B_\Gamma^{\bullet,\bullet} + \bar{B}_\Gamma^{\bullet,\bullet},$$

see [AK12].

Remark 3.25. We refer to [AK13a] for further results extending the class of solvmanifolds whose Dolbeault and Bott-Chern cohomologies are computable by means of a finite-dimensional sub-complex of the double complex of forms. More precisely, [AK13a, Theorems 1.1 and 1.2] provide conditions under which suitable deformations of the above finite-dimensional sub-complex still allow to compute the cohomologies of small deformations of a solvmanifold. As an application, we computed the Dolbeault and Bott-Chern cohomologies of the holomorphically parallelizable Nakamura manifold and of some of its deformations in [AK13a, Sect. 4]. These examples show that the property of degeneration of the Hodge and Frölicher spectral sequence at the first level (as shown also by M.G. Eastwood and M.A. Singer in [ES93, Theorem 5.4]) and the property of satisfying the $\partial\bar{\partial}$-Lemma are not closed under deformations of the complex structure, [AK13a, Corollary 6.1].

The previous results allow to compute explicitly the de Rham, Dolbeault, and Bott-Chern cohomologies of the completely-solvable Nakamura manifold and of the holomorphically parallelizable Nakamura manifold, [Kas13b, AK12].

Example 3.3 (The completely-solvable Nakamura manifold, [Kas13b, Example 1], [AK12, Example 2.17]). The completely-solvable Nakamura manifold, firstly studied by I. Nakamura in [Nak75, p. 90], is an example of a cohomologically Kähler non-Kähler solvmanifold, [dAFdLM92, FMS03, Example 3.1], [dBT06, Sect. 3].

Consider the group

$$G := \mathbb{C} \ltimes_\phi \mathbb{C}^2, \qquad \text{where} \qquad \phi \left(x + \sqrt{-1} y \right) := \begin{pmatrix} e^x & 0 \\ 0 & e^{-x} \end{pmatrix} \in \mathrm{GL} \left(\mathbb{C}^2 \right).$$

Table 3.7 A basis of harmonic representatives for the completely-solvable Nakamura manifold with respect to the metric $g := \mathrm{d}z_1 \odot \mathrm{d}\bar{z}_1 + e^{-z_1-\bar{z}_1}\,\mathrm{d}z_2 \odot \mathrm{d}\bar{z}_2 + e^{z_1+\bar{z}_1}\,\mathrm{d}z_3 \odot \mathrm{d}\bar{z}_3$

$H_{dR}^k(X;\mathbb{C})$	g-Harmonic representatives	$\dim_{\mathbb{C}} H_{dR}^k(X;\mathbb{C})$
$k=1$	$\mathrm{d}z_1, \mathrm{d}\bar{z}_1$	2
$k=2$	$\mathrm{d}z_{23}, \mathrm{d}z_{1\bar{1}}, \mathrm{d}z_{2\bar{3}}, \mathrm{d}z_{3\bar{2}}, \mathrm{d}z_{\bar{2}\bar{3}}$	5
$k=3$	$\mathrm{d}z_{123}, \mathrm{d}z_{23\bar{1}}, \mathrm{d}z_{1\bar{2}3}, \mathrm{d}z_{13\bar{2}}, \mathrm{d}z_{1\bar{2}\bar{3}}, \mathrm{d}z_{2\bar{1}\bar{3}}, \mathrm{d}z_{3\bar{1}\bar{2}}, \mathrm{d}z_{\bar{1}\bar{2}\bar{3}}$	8
$k=4$	$\mathrm{d}z_{123\bar{1}}, \mathrm{d}z_{12\bar{1}\bar{3}}, \mathrm{d}z_{23\bar{2}\bar{3}}, \mathrm{d}z_{13\bar{1}\bar{2}}, \mathrm{d}z_{1\bar{1}\bar{2}\bar{3}}$	5
$k=5$	$\mathrm{d}z_{123\bar{2}\bar{3}}, \mathrm{d}z_{23\bar{1}\bar{2}\bar{3}}$	2

For some $a \in \mathbb{R}$, the matrix $\begin{pmatrix} e^x & 0 \\ 0 & e^{-x} \end{pmatrix}$ is conjugate to an element of $\mathrm{SL}(2;\mathbb{Z})$.

Hence there exists a discrete co-compact subgroup

$$\Gamma := \left(a\,\mathbb{Z} + b\,\sqrt{-1}\,\mathbb{Z} \right) \ltimes_\phi \Gamma_{\mathbb{C}^2}$$

of G, where $\Gamma_{\mathbb{C}^2}$ is a lattice of $\mathbb{C}^2$. The *completely-solvable Nakamura manifold* is the completely-solvable solvmanifold $X := \Gamma\backslash G$.

Consider holomorphic coordinates $\{z_1, z_2, z_3\}$ on X, where $\left\{ z_1 := x + \sqrt{-1}\,y \right\}$ is the holomorphic coordinate on $\mathbb{C}$ (as a matter of notations, we shorten, for example, $e^{-z_1}\,\mathrm{d}z_{12\bar{1}} := e^{-z_1}\,\mathrm{d}z_1 \wedge \mathrm{d}z_2 \wedge \mathrm{d}\bar{z}_1$).

By A. Hattori's theorem, [Hat60, Corollary 4.2], the de Rham cohomology of $\Gamma\backslash G$ does not depend on Γ and can be computed using just G-left-invariant forms on $\Gamma\backslash G$; in Table 3.7, we list the harmonic representatives of the de Rham cohomology classes with respect to the G-left-invariant Hermitian metric $g := \mathrm{d}z_1 \odot \mathrm{d}\bar{z}_1 + e^{-z_1-\bar{z}_1}\,\mathrm{d}z_2 \odot \mathrm{d}\bar{z}_2 + e^{z_1+\bar{z}_1}\,\mathrm{d}z_3 \odot \mathrm{d}\bar{z}_3$.

As regards Dolbeault cohomology and Bott-Chern cohomology, they depend on the lattice, see [Kas13b, AK12]. In particular, following [Kas13b, Example 1], one has to distinguish three different cases:

(i) $b \in (2\,\mathbb{Z}) \cdot \pi$;
(ii) $b \in (2\,\mathbb{Z}+1) \cdot \pi$;
(iii) $b \notin \mathbb{Z} \cdot \pi$.

The Dolbeault cohomology of the completely-solvable Nakamura manifold was computed in [Kas13b, Example 5.1]. The Bott-Chern cohomology of the completely-solvable Nakamura manifold was computed in [AK12, Example 2.17], and is summarized in Table 3.8, see [AK12, Tables 4, 5, and 3].

Finally, Table 3.9 summarizes the dimensions of the de Rham, Dolbeault, and Bott-Chern cohomologies of the completely-solvable Nakamura manifold, [Kas13b, Example 1], [AK12, Example 2.17], see [AK12, Table 6].

In particular, note that the completely-solvable Nakamura manifold in case *(iii)* satisfies the $\partial\bar{\partial}$-Lemma (compare also [Kas13b, Kas11]).

Table 3.8 The Bott-Chern cohomology of the completely-solvable Nakamura manifold, [AK12, Example 2.17, Tables 4, 5, and 3]

$H_{BC}^{\bullet,\bullet}(\Gamma\backslash G)$	Case (i)	Case (ii)	Case (iii)
(0,0)	$\mathbb{C}\langle 1\rangle$	$\mathbb{C}\langle 1\rangle$	$\mathbb{C}\langle 1\rangle$
(1,0)	$\mathbb{C}\langle[dz_1]\rangle$	$\mathbb{C}\langle[dz_1]\rangle$	$\mathbb{C}\langle dz_1\rangle$
(0,1)	$\mathbb{C}\langle[dz_{\bar 1}]\rangle$	$\mathbb{C}\langle[dz_{\bar 1}]\rangle$	$\mathbb{C}\langle dz_{\bar 1}\rangle$
(2,0)	$\mathbb{C}\langle[e^{z_1}dz_{13}],[dz_{23}]\rangle$	$\mathbb{C}\langle[dz_{23}]\rangle$	$\mathbb{C}\langle dz_{23}\rangle$
(1,1)	$\mathbb{C}\langle[dz_{1\bar 1}],[e^{-z_1}dz_{1\bar 2}],[dz_{2\bar 3}],$ $[dz_{3\bar 2}],[e^{-\bar z_1}dz_{2\bar 1}],[e^{\bar z_1}dz_{3\bar 1}]\rangle$	$\mathbb{C}\langle[dz_{1\bar 1}],[dz_{2\bar 3}],[dz_{3\bar 2}]\rangle$	$\mathbb{C}\langle dz_{1\bar 1},\ dz_{2\bar 3},\ dz_{3\bar 2}\rangle$
(0,2)	$\mathbb{C}\langle[dz_{\bar 2\bar 3}],[e^{-\bar z_1}dz_{\bar 1\bar 2}],[e^{\bar z_1}dz_{\bar 1\bar 3}]\rangle$	$\mathbb{C}\langle[dz_{\bar 2\bar 3}]\rangle$	$\mathbb{C}\langle dz_{\bar 2\bar 3}\rangle$
(3,0)	$\mathbb{C}\langle[dz_{123}]\rangle$	$\mathbb{C}\langle[dz_{123}]\rangle$	$\mathbb{C}\langle dz_{123}\rangle$
(2,1)	$\mathbb{C}\langle[e^{-z_1}dz_{12\bar 2}],[dz_{12\bar 3}],[e^{z_1}dz_{13\bar 1}],[dz_{13\bar 2}],$ $[e^{2z_1}dz_{13\bar 3}],[e^{-\bar z_1}dz_{23\bar 1}],[e^{\bar z_1}dz_{13\bar 1}]\rangle$	$\mathbb{C}\langle[dz_{23\bar 1}],[e^{-2z_1}dz_{12\bar 2}],[e^{2z_1}dz_{13\bar 3}],$ $[dz_{123\bar 3}],[dz_{13\bar 2}]\rangle$	$\mathbb{C}\langle dz_{23\bar 1},\ dz_{12\bar 3},\ dz_{13\bar 2}\rangle$
(1,2)	$\mathbb{C}\langle[e^{-z_1}dz_{1\bar 1\bar 2}],[e^{-2z_1}dz_{2\bar 1\bar 2}],[dz_{3\bar 1\bar 2}],[dz_{2\bar 1\bar 3}],$ $[e^{2z_1}dz_{3\bar 1\bar 3}],[dz_{2\bar 1\bar 3}],[e^{-z_1}dz_{2\bar 1\bar 2}],[e^{z_1}dz_{1\bar 1\bar 3}]\rangle$	$\mathbb{C}\langle[dz_{1\bar 2\bar 3}],[e^{-2\bar z_1}dz_{2\bar 1\bar 2}],[e^{2\bar z_1}dz_{3\bar 1\bar 3}],$ $[dz_{2\bar 1\bar 3}],[dz_{3\bar 1\bar 2}]\rangle$	$\mathbb{C}\langle dz_{1\bar 2\bar 3},\ dz_{2\bar 1\bar 3},\ dz_{3\bar 1\bar 2}\rangle$
(0,3)	$\mathbb{C}\langle[dz_{\bar 1\bar 2\bar 3}]\rangle$	$\mathbb{C}\langle[dz_{\bar 1\bar 2\bar 3}]\rangle$	$\mathbb{C}\langle dz_{\bar 1\bar 2\bar 3}\rangle$
(3,1)	$\mathbb{C}\langle[dz_{123\bar 1}],[e^{-\bar z_1}dz_{123\bar 2}],[e^{\bar z_1}dz_{123\bar 3}]\rangle$	$\mathbb{C}\langle[dz_{123\bar 1}]\rangle$	$\mathbb{C}\langle dz_{123\bar 1}\rangle$
(2,2)	$\mathbb{C}\langle[e^{-2z_1}dz_{12\bar 1\bar 2}],[dz_{12\bar 1\bar 3}],[e^{-z_1}dz_{12\bar 2\bar 3}],[dz_{13\bar 1\bar 2}],$ $[e^{2z_1}dz_{13\bar 1\bar 3}],[e^{z_1}dz_{13\bar 2\bar 3}],[dz_{23\bar 1\bar 3}],[e^{-2z_1}dz_{12\bar 1\bar 2}],$ $[e^{-\bar z_1}dz_{23\bar 1\bar 3}],[e^{2\bar z_1}dz_{23\bar 1\bar 3}],[e^{\bar z_1}dz_{23\bar 1\bar 3}]\rangle$	$\mathbb{C}\langle[dz_{12\bar 1\bar 3}],[e^{-2z_1}dz_{12\bar 1\bar 2}],[e^{-2\bar z_1}dz_{12\bar 1\bar 2}],$ $[e^{2z_1}dz_{13\bar 1\bar 3}],[e^{2\bar z_1}dz_{13\bar 1\bar 3}],[dz_{23\bar 2\bar 3}],[dz_{13\bar 1\bar 2}]\rangle$	$\mathbb{C}\langle dz_{12\bar 1\bar 3},\ dz_{23\bar 2\bar 3},\ dz_{13\bar 1\bar 2}\rangle$
(1,3)	$\mathbb{C}\langle[dz_{1\bar 1\bar 2\bar 3}],[e^{-\bar z_1}dz_{2\bar 1\bar 2\bar 3}],[e^{\bar z_1}dz_{3\bar 1\bar 2\bar 3}]\rangle$	$\mathbb{C}\langle[dz_{1\bar 1\bar 2\bar 3}]\rangle$	$\mathbb{C}\langle dz_{1\bar 1\bar 2\bar 3}\rangle$
(3,2)	$\mathbb{C}\langle[e^{-z_1}dz_{123\bar 1\bar 2}],[e^{z_1}dz_{123\bar 1\bar 3}],[dz_{123\bar 2\bar 3}],$ $[e^{-\bar z_1}dz_{123\bar 1\bar 2}],[e^{\bar z_1}dz_{123\bar 1\bar 3}]\rangle$	$\mathbb{C}\langle[dz_{123\bar 2\bar 3}]\rangle$	$\mathbb{C}\langle dz_{123\bar 2\bar 3}\rangle$
(2,3)	$\mathbb{C}\langle[e^{-z_1}dz_{12\bar 1\bar 2\bar 3}],[e^{z_1}dz_{13\bar 1\bar 2\bar 3}],[dz_{23\bar 1\bar 2\bar 3}],$ $[e^{-\bar z_1}dz_{12\bar 1\bar 2\bar 3}],[e^{\bar z_1}dz_{13\bar 1\bar 2\bar 3}]\rangle$	$\mathbb{C}\langle[dz_{23\bar 1\bar 2\bar 3}]\rangle$	$\mathbb{C}\langle dz_{23\bar 1\bar 2\bar 3}\rangle$
(3,3)	$\mathbb{C}\langle[dz_{123\bar 1\bar 2\bar 3}]\rangle$	$\mathbb{C}\langle[dz_{123\bar 1\bar 2\bar 3}]\rangle$	$\mathbb{C}\langle dz_{123\bar 1\bar 2\bar 3}\rangle$

Table 3.9 The dimensions of the de Rham, Dolbeault, and Bott-Chern cohomologies of the completely-solvable Nakamura manifold, [Kas13b, Example 1], [AK12, Example 2.17], see [AK12, Table 6]

$H_\sharp^{\bullet,\bullet}(X)$	dR	Case (i) $\bar\partial$	BC	Case (ii) $\bar\partial$	BC	Case (iii) $\bar\partial$	BC
$(0,0)$	1	1	1	1	1	1	1
$(1,0)$	2	3	1	1	1	1	1
$(0,1)$		3	1	1	1	1	1
$(2,0)$	5	3	3	1	1	1	1
$(1,1)$		9	7	5	3	3	3
$(0,2)$		3	3	1	1	1	1
$(3,0)$	8	1	1	1	1	1	1
$(2,1)$		9	9	5	5	3	3
$(1,2)$		9	9	5	5	3	3
$(0,3)$		1	1	1	1	1	1
$(3,1)$	5	3	3	1	1	1	1
$(2,2)$		9	11	5	7	3	3
$(1,3)$		3	3	1	1	1	1
$(3,2)$	2	3	5	1	1	1	1
$(2,3)$		3	5	1	1	1	1
$(3,3)$	1	1	1	1	1	1	1

Table 3.10 The de Rham cohomology for the holomorphically parallelizable Nakamura manifold, [Kas12a, Sect. 7]

$H_{dR}^k(X;\mathbb{C})$	Case (a)	Case (b)
$k=1$	$\mathbb{C}\langle dz_1,\, dz_{\bar 1}\rangle$	$\mathbb{C}\langle dz_1,\, dz_{\bar 1}\rangle$
$k=2$	$\mathbb{C}\langle dz_{1\bar 1},\, dz_{23},\, dz_{2\bar 3},\, dz_{3\bar 2},\, dz_{\bar2\bar3}\rangle$	$\mathbb{C}\langle dz_{1\bar 1},\, dz_{23},\, dz_{\bar2\bar3}\rangle$
$k=3$	$\mathbb{C}\langle dz_{123},\, dz_{12\bar3},\, dz_{13\bar2},\, dz_{3\bar1\bar2},\, dz_{2\bar1\bar3},\, dz_{\bar1\bar2\bar3},$ $dz_{\bar123},\, dz_{1\bar2\bar3}\rangle$	$\mathbb{C}\langle dz_{123},\, dz_{\bar1\bar2\bar3},\, dz_{\bar123},\, dz_{1\bar2\bar3}\rangle$
$k=4$	$\mathbb{C}\langle dz_{123\bar1},\, dz_{13\bar1\bar2},\, dz_{23\bar2\bar3},\, dz_{12\bar1\bar3},\, dz_{1\bar1\bar2\bar3}\rangle$	$\mathbb{C}\langle dz_{123\bar1},\, dz_{23\bar2\bar3},\, dz_{1\bar1\bar2\bar3}\rangle$
$k=5$	$\mathbb{C}\langle dz_{23\bar1\bar2\bar3},\, dz_{123\bar2\bar3}\rangle$	$\mathbb{C}\langle dz_{23\bar1\bar2\bar3},\, dz_{123\bar2\bar3}\rangle$

Example 3.4 (The holomorphically parallelizable Nakamura manifold, [Kas12a, Sect. 7], [Kas13b, Example 1], [AK12, Example 2.26]). The holomorphically parallelizable Nakamura manifold was firstly considered by I. Nakamura in [Nak75, p. 90], where it is provided as an example of a three-dimensional holomorphically parallelizable solvmanifold in the third class of Nakamura classification, [Nak75, Sect. 2].

Consider the group

$$G = \mathbb{C}\ltimes_\phi \mathbb{C}^2 \qquad \text{where} \qquad \phi(z) = \begin{pmatrix} e^z & 0 \\ 0 & e^{-z} \end{pmatrix} \cdots$$

Table 3.11 The Bott-Chern cohomology of the holomorphically parallelizable Nakamura manifold, [AK12, Example 2.26, Tables 8 and 11]

$H^{\bullet,\bullet}_{BC}(X)$	Case (a)	Case (b)
(1,0)	$\mathbb{C}\langle[dz_1]\rangle$	$\mathbb{C}\langle[dz_{\bar1}]\rangle$
(0,1)	$\mathbb{C}\langle[dz_{\bar1}]\rangle$	$\mathbb{C}\langle[dz_{\bar1}]\rangle$
(2,0)	$\mathbb{C}\langle[e^{-z_1}\,dz_{12}],\ [e^{z_1}\,dz_{13}],\ [dz_{23}]\rangle$	$\mathbb{C}\langle[e^{-z_1}\,dz_{12}],\ [e^{z_1}\,dz_{13}],\ [dz_{23}]\rangle$
(1,1)	$\mathbb{C}\langle[dz_{1\bar1}],\ [e^{-z_1}\,dz_{1\bar2}],\ [e^{z_1}\,dz_{1\bar3}],\ [dz_{2\bar3}],\ [dz_{3\bar2}],\ [e^{-z_1}\,dz_{2\bar1}],\ [e^{\bar z_1}\,dz_{3\bar1}]\rangle$	$\mathbb{C}\langle[dz_{1\bar1}]\rangle$
(0,2)	$\mathbb{C}\langle[dz_{\bar2\bar3}],\ [e^{-\bar z_1}\,dz_{1\bar2}],\ [e^{\bar z_1}\,dz_{1\bar3}]\rangle$	$\mathbb{C}\langle[e^{-\bar z_1}\,dz_{1\bar2}],\ [e^{\bar z_1}\,dz_{1\bar3}],\ [dz_{2\bar3}]\rangle$
(3,0)	$\mathbb{C}\langle[dz_{123}]\rangle$	$\mathbb{C}\langle[dz_{123}]\rangle$
(2,1)	$\mathbb{C}\langle[e^{-z_1}\,dz_{12\bar1}],\ [e^{-2z_1}\,dz_{12\bar2}],\ [dz_{12\bar3}],\ [e^{z_1}\,dz_{13\bar1}],\ [dz_{13\bar2}],\ [e^{2z_1}\,dz_{13\bar3}],$ $[dz_{23\bar1}],\ [e^{-\bar z_1}\,dz_{12\bar1}],\ [e^{\bar z_1}\,dz_{13\bar1}]\rangle$	$\mathbb{C}\langle[e^{-z_1}\,dz_{12\bar1}],\ [e^{z_1}\,dz_{13\bar1}],\ [dz_{23\bar1}]\rangle$
(1,2)	$\mathbb{C}\langle[e^{-z_1}\,dz_{1\bar2\bar1}],\ [e^{-2z_1}\,dz_{1\bar2\bar2}],\ [dz_{1\bar2\bar3}],\ [e^{\bar z_1}\,dz_{1\bar3\bar2}],\ [e^{\bar z_1}\,dz_{1\bar1\bar3}],\ [dz_{2\bar1\bar3}],\ [e^{2\bar z_1}\,dz_{3\bar1\bar3}],$ $[dz_{1\bar2\bar3}],\ [e^{-\bar z_1}\,dz_{1\bar1\bar2}],\ [e^{z_1}\,dz_{1\bar1\bar3}]\rangle$	$\mathbb{C}\langle[e^{-\bar z_1}\,dz_{1\bar1\bar2}],\ [e^{\bar z_1}\,dz_{1\bar1\bar3}],\ [dz_{1\bar2\bar3}],\rangle$
(0,3)	$\mathbb{C}\langle[dz_{\bar1\bar2\bar3}]\rangle$	$\mathbb{C}\langle[dz_{\bar1\bar2\bar3}]\rangle$
(3,1)	$\mathbb{C}\langle[dz_{123\bar1}],\ [e^{-z_1}\,dz_{123\bar2}],\ [e^{z_1}\,dz_{123\bar3}]\rangle$	$\mathbb{C}\langle[dz_{123\bar1}]\rangle$
(2,2)	$\mathbb{C}\langle[e^{-2z_1}\,dz_{12\bar1\bar2}],\ [dz_{12\bar1\bar3}],\ [e^{-z_1}\,dz_{12\bar2\bar3}],\ [dz_{13\bar1\bar2}],\ [e^{2z_1}\,dz_{13\bar1\bar3}],\ [e^{z_1}\,dz_{13\bar2\bar3}],$ $[dz_{23\bar1\bar2}],\ [e^{-\bar z_1}\,dz_{23\bar1\bar3}],\ [e^{2\bar z_1}\,dz_{23\bar1\bar3}],\ [e^{\bar z_1}\,dz_{23\bar1\bar3}]\rangle$	$\mathbb{C}\langle[e^{-z_1}\,dz_{12\bar2\bar3}],\ [e^{z_1}\,dz_{13\bar2\bar3}],\ [dz_{23\bar2\bar3}],\ [e^{-\bar z_1}\,dz_{23\bar1\bar2}],\ [e^{\bar z_1}\,dz_{23\bar1\bar3}]\rangle$
(1,3)	$\mathbb{C}\langle[dz_{1\bar1\bar2\bar3}],\ [e^{-\bar z_1}\,dz_{2\bar1\bar2\bar3}],\ [e^{\bar z_1}\,dz_{3\bar1\bar2\bar3}]\rangle$	$\mathbb{C}\langle[dz_{1\bar1\bar2\bar3}]\rangle$
(3,2)	$\mathbb{C}\langle[e^{-z_1}\,dz_{123\bar1\bar2}],\ [e^{z_1}\,dz_{123\bar1\bar3}],\ [dz_{123\bar2\bar3}],\ [e^{-\bar z_1}\,dz_{123\bar1\bar2}],\ [e^{\bar z_1}\,dz_{123\bar1\bar3}]\rangle$	$\mathbb{C}\langle[e^{-\bar z_1}\,dz_{123\bar1\bar2}],\ [e^{\bar z_1}\,dz_{123\bar1\bar3}],\ [dz_{123\bar2\bar3}]\rangle$
(2,3)	$\mathbb{C}\langle[e^{-z_1}\,dz_{12\bar1\bar2\bar3}],\ [e^{z_1}\,dz_{13\bar1\bar2\bar3}],\ [dz_{23\bar1\bar2\bar3}]\rangle$	$\mathbb{C}\langle[e^{-z_1}\,dz_{12\bar1\bar2\bar3}],\ [e^{z_1}\,dz_{13\bar1\bar2\bar3}],\ [dz_{23\bar1\bar2\bar3}]\rangle$

Table 3.12 The dimensions of the de Rham, Dolbeault, and Bott-Chern cohomologies of the holomorphically parallelizable Nakamura manifold, [Kas12a, Sect. 7], [Kas13b, Example 2], [AK12, Example 2.26], see [AK12, Table 13]

$\dim_{\mathbb{C}} H_{\sharp}^{\bullet,\bullet}(X)$	Case (a)			Case (b)		
	dR	$\bar{\partial}$	BC	dR	$\bar{\partial}$	BC
(0,0)	1	1	1	1	1	1
(1,0)	2	3	1	2	3	1
(0,1)		3	1		1	1
(2,0)	5	3	3	3	3	3
(1,1)		9	7		3	1
(0,2)		3	3		1	3
(3,0)	8	1	1	4	1	1
(2,1)		9	9		3	3
(1,2)		9	9		3	3
(0,3)		1	1		1	1
(3,1)	5	3	3	3	1	1
(2,2)		9	11		3	5
(1,3)		3	3		3	1
(3,2)	2	3	5	2	1	3
(2,3)		3	5		3	3
(3,3)	1	1	1	1	1	1

There exist $a + \sqrt{-1}\, b \in \mathbb{C}$ and $c + \sqrt{-1}\, d \in \mathbb{C}$ such that $\mathbb{Z}(a + \sqrt{-1}\, b) + \mathbb{Z}(c + \sqrt{-1}\, d)$ is a lattice in $\mathbb{C}$ and $\phi(a + \sqrt{-1}\, b)$ and $\phi(c + \sqrt{-1}\, d)$ are conjugate to elements of $\mathrm{SL}(4; \mathbb{Z})$, where we regard $\mathrm{SL}(2; \mathbb{C}) \subset \mathrm{SL}(4; \mathbb{R})$, see [Has10]. Hence there exists a lattice

$$\Gamma := \left(\mathbb{Z}\left(a + \sqrt{-1}\, b\right) + \mathbb{Z}\left(c + \sqrt{-1}\, d\right) \right) \ltimes_{\phi} \Gamma_{\mathbb{C}^2}$$

of G, where $\Gamma_{\mathbb{C}^2}$ is a lattice of $\mathbb{C}^2$. The *holomorphically parallelizable Nakamura manifold* is the holomorphically parallelizable solvmanifold $X := \Gamma \backslash G$, [Nak75, Sect. 2].

Consider holomorphic coordinates $\{z_1, z_2, z_3\}$ on X, where $\{z_1\}$ is the holomorphic coordinate on $\mathbb{C}$ (as a matter of notations, we shorten, for example, $e^{-2z_1}\, dz_{12\bar{2}} := e^{-2z_1}\, dz_1 \wedge dz_2 \wedge d\bar{z}_2$).

According to [Kas12a], the de Rham, Dolbeault, and Bott-Chern cohomologies depend on the lattice, see [Kas12a, Theorems 1.3 and 1.4], [AK12, Theorem 2.25]. In particular, following [Kas12a, § 7.2], one has to distinguish two different cases[4]:

(a) $b \in \mathbb{Z} \cdot \pi$ and $d \in \mathbb{Z} \cdot \pi$;
(b) either $b \notin \mathbb{Z} \cdot \pi$ or $d \notin \mathbb{Z} \cdot \pi$.

---

[4] Note that the case $b \in (2\mathbb{Z}) \cdot \pi$ and $d \in (2\mathbb{Z}) \cdot \pi$ can be identified with the case (i) of the completely-solvable Nakamura manifold, see [Yam05, Sect. 3].

The de Rham cohomology and the Dolbeault cohomology of the holomorphically parallelizable Nakamura manifold were computed in [Kas12a, Sect. 7] and [Kas13b, Example 2] (compare also [Nak75, pp. 96/100]); in Table 3.10 we recall the de Rham cohomology for the holomorphically parallelizable Nakamura manifold as computed in [Kas12a, Sect. 7]. The Bott-Chern cohomology of the completely-solvable Nakamura manifold was computed in [AK12, Example 2.26], and is summarized in Table 3.11, see [AK12, Tables 8 and 11].

Finally, Table 3.12 summarizes the dimensions of the de Rham, Dolbeault, and Bott-Chern cohomologies of the holomorphically parallelizable Nakamura manifold, [Kas12a, §], [Kas13b, Example 2], [AK12, Example 2.26], see [AK12, Table 13].

Chapter 4
Cohomology of Almost-Complex Manifolds

Abstract Let X be a differentiable manifold endowed with an almost-complex structure J. Note that if J is not integrable, then the Dolbeault cohomology is not defined. In this chapter, we are concerned with studying some subgroups of the de Rham cohomology related to the almost-complex structure: these subgroups have been introduced by T.-J. Li and W. Zhang in (Comm. Anal. Geom. 17(4):651–683, 2009), in order to study the relation between the compatible and the tamed symplectic cones on a compact almost-complex manifold, with the aim to throw light on a question by S.K. Donaldson, (Two-forms on four-manifolds and elliptic equations, Inspired by S. S. Chern, Nankai Tracts Math., vol. 11, World Sci. Publ., Hackensack, NJ, 2006, pp. 153–172, Question 2) (see Sect. 4.4.2), and it would be interesting to consider them as a sort of counterpart of the Dolbeault cohomology groups in the non-integrable (or at least in the non-Kähler) case, see Drăghici et al. (Int. Math. Res. Not. IMRN 1:1–17, 2010, Lemma 2.15, Theorem 2.16). In particular, we are interested in studying when they let a splitting of the de Rham cohomology, and their relations with cones of metric structures.

4.1 Subgroups of the de Rham (Co)Homology of an Almost-Complex Manifold

Let X be a $2n$-dimensional (differentiable) manifold endowed with an almost-complex structure J. In this section, we set the notation concerning $\mathcal{C}^\infty$-pure-and-full and pure-and-full almost-complex structures, as introduced in [LZ09]; then we briefly review some results to motivate the study of these topics, see Remark 4.3, and we study the relations between $\mathcal{C}^\infty$-pure-and-fullness and pure-and-fullness.

D. Angella, *Cohomological Aspects in Complex Non-Kähler Geometry*,
Lecture Notes in Mathematics 2095, DOI 10.1007/978-3-319-02441-7_4,
© Springer International Publishing Switzerland 2014

4.1.1 C^∞-Pure-and-Full and Pure-and-Full Almost-Complex Structures

Let $S \subseteq \mathbb{N} \times \mathbb{N}$ and define

$$H_J^S(X; \mathbb{R}) := \left\{ [\alpha] \in H_{dR}^\bullet(X; \mathbb{R}) \; : \; \alpha \in \left(\bigoplus_{(p,q) \in S} \wedge^{p,q} X \right) \cap \wedge^\bullet X \right\} ;$$

note that a real differential form α with a component of type (p, q) has also a component of type (q, p), and hence we are interested in studying the sets S such that whenever $(p, q) \in S$, also $(q, p) \in S$. As a matter of notation, we will usually list the elements of S instead of writing S itself.

Note that, for every $k \in \mathbb{N}$, one has

$$\sum_{\substack{p+q=k \\ p \le q}} H_J^{(p,q),(q,p)}(X; \mathbb{R}) \subseteq H_{dR}^k(X; \mathbb{R}) ,$$

but, in general, the sum is neither direct nor the equality holds: several examples of these facts will be provided in the sequel.

The subgroups $H_J^{(2,0),(0,2)}(X; \mathbb{R})$ and $H_J^{(1,1)}(X; \mathbb{R})$ of $H_{dR}^2(X; \mathbb{R})$ are of special interest for their interpretation as the J-anti-invariant, respectively, J-invariant part of the second de Rham cohomology group. Indeed, note that the endomorphism $J\lfloor_{\wedge^2 X} \in \mathrm{End}(\wedge^\bullet X)$ naturally extending $J \in \mathrm{End}(TX)$ (that is, $J\alpha := \alpha(J\cdot, J\cdot\cdot)$ for every $\alpha \in \wedge^2 X$) satisfies $(J\lfloor_{\wedge^2 X})^2 = \mathrm{id}_{\wedge^2 X}$; hence, one has the splitting

$$\wedge^2 X \; = \; \wedge_J^+ X \oplus \wedge_J^- X ,$$

where, for $\pm \in \{+, -\}$,

$$\wedge_J^\pm X \; := \; \left\{ \alpha \in \wedge^2 X \; : \; J\alpha = \pm\alpha \right\} .$$

Since $H_{dR}^2(X; \mathbb{R})$ contains, in particular, the classes represented by the symplectic forms, and $H_J^{(1,1)}(X; \mathbb{R})$ contains, in particular, the classes represented by the $(1, 1)$-forms associated to the Hermitian metrics on X, in [LZ09], T.-J. Li and W. Zhang were interested in studying the J-*invariant subgroup* of $H_{dR}^2(X; \mathbb{R})$, namely,

$$H_J^+(X) \; := \; H_J^{(1,1)}(X; \mathbb{R}) \; = \; \left\{ [\alpha] \in H_{dR}^2(X; \mathbb{R}) \; : \; J\alpha = \alpha \right\} ,$$

and the J-*anti-invariant subgroup* of $H_{dR}^2(X; \mathbb{R})$, namely,

$$H_J^-(X) \; := \; H_J^{(2,0),(0,2)}(X; \mathbb{R}) \; = \; \left\{ [\alpha] \in H_{dR}^2(X; \mathbb{R}) \; : \; J\alpha = -\alpha \right\} .$$

Note that, as in the general case, one has that

$$H_J^+(X) + H_J^-(X) \subseteq H_{dR}^2(X;\mathbb{R})$$

but, in general, the sum is neither direct nor equal to $H_{dR}^2(X;\mathbb{R})$. The following definition, by T.-J. Li and W. Zhang, singles out the almost-complex structures whose subgroups $H_J^+(X)$ and $H_J^-(X)$ provide a decomposition of $H_{dR}^2(X;\mathbb{R})$.

Definition 4.1 ([LZ09, Definitions 2.2, 2.3, Lemma 2.2]). An almost-complex structure J on a manifold X is said to be

- $\mathcal{C}^\infty$-*pure* if $H_J^-(X) \cap H_J^+(X) = \{0\}$;
- $\mathcal{C}^\infty$-*full* if $H_J^-(X) + H_J^+(X) = H_{dR}^2(X;\mathbb{R})$;
- $\mathcal{C}^\infty$-*pure-and-full* if it is both $\mathcal{C}^\infty$-pure and $\mathcal{C}^\infty$-full, i.e., if the following cohomology decomposition holds:

$$H_{dR}^2(X;\mathbb{R}) \;=\; H_J^-(X) \oplus H_J^+(X) \,.$$

We will also use the following definition, which is a natural generalization of the notion of $\mathcal{C}^\infty$-pure-and-fullness to higher degree cohomology groups.

Definition 4.2. Let X be a manifold endowed with an almost-complex structure J, and fix $k \in \mathbb{N}$. Consider $H_{dR}^k(X;\mathbb{R}) \supseteq \sum_{\substack{p+q=k \\ p \leq q}} H_J^{(p,q),(q,p)}(X;\mathbb{R})$:

- if

$$\bigoplus_{\substack{p+q=k \\ p \leq q}} H_J^{(p,q),(q,p)}(X;\mathbb{R}) \;\subseteq\; H_{dR}^k(X;\mathbb{R})$$

(namely, the sum is direct), then J is called $\mathcal{C}^\infty$-*pure at the kth stage*;
- if

$$H_{dR}^k(X;\mathbb{R}) \;=\; \sum_{\substack{p+q=k \\ p \leq q}} H_J^{(p,q),(q,p)}(X;\mathbb{R}) \,,$$

then J is called $\mathcal{C}^\infty$-*full at the kth stage*;
- if J is both $\mathcal{C}^\infty$-pure at the kth stage and $\mathcal{C}^\infty$-full at the kth stage, that is,

$$H_{dR}^k(X;\mathbb{R}) \;=\; \bigoplus_{\substack{p+q=k \\ p \leq q}} H_J^{(p,q),(q,p)}(X;\mathbb{R}) \,;$$

then J is called $\mathcal{C}^\infty$-*pure-and-full at the kth stage*.

Analogous definitions can be given for the de Rham cohomology with complex coefficients. More precisely, let $S \subseteq \mathbb{N} \times \mathbb{N}$ and define

$$H_J^S(X;\mathbb{C}) := \left\{ [\alpha] \in H_{dR}^\bullet(X;\mathbb{C}) \; : \; \alpha \in \bigoplus_{(p,q)\in S} \wedge^{p,q} X \right\}$$

(as previously, we will usually list the elements of S instead of writing S itself); with such notation, one has in particular that $H_J^S(X;\mathbb{R}) = H_J^S(X;\mathbb{C}) \cap H_{dR}^\bullet(X;\mathbb{R})$.

Remark 4.1. Note that, when X is a compact manifold endowed with an integrable almost-complex structure J, then, for any $(p,q) \in \mathbb{N} \times \mathbb{N}$,

$$H_J^{(p,q)}(X;\mathbb{C}) = \mathrm{im}\left(H_{BC}^{p,q}(X) \to H_{dR}^{p+q}(X;\mathbb{C}) \right),$$

where the map $H_{BC}^{p,q}(X) \to H_{dR}^{p+q}(X;\mathbb{C})$ is the one induced by the identity (note that $\ker\partial \cap \ker\bar\partial \subseteq \ker d$ and $\mathrm{im}\,\partial\bar\partial \subseteq \mathrm{im}\,d$). Indeed, any d-closed (p,q)-form is both ∂-closed and $\bar\partial$-closed.

Note that, for every $k \in \mathbb{N}$, one has

$$\sum_{p+q=k} H_J^{(p,q)}(X;\mathbb{C}) \subseteq H_{dR}^k(X;\mathbb{C}),$$

but, in general, the sum is neither direct nor the equality holds. We can then give the following definition.

Definition 4.3. Let X be a manifold endowed with an almost-complex structure J, and fix $k \in \mathbb{N}$. Consider $H_{dR}^k(X;\mathbb{C}) \supseteq \sum_{p+q=k} H_J^{(p,q)}(X;\mathbb{C})$:

- if

$$\bigoplus_{p+q=k} H_J^{(p,q)}(X;\mathbb{C}) \subseteq H_{dR}^k(X;\mathbb{C})$$

  (namely, the sum is direct), then J is called *complex-C^∞-pure at the kth stage*;
- if

$$H_{dR}^k(X;\mathbb{C}) = \sum_{p+q=k} H_J^{(p,q)}(X;\mathbb{C}),$$

  then J is called *complex-C^∞-full at the kth stage*;
- if J is both complex-C^∞-pure at the kth stage and complex-C^∞-full at the kth stage, that is,

$$H_{dR}^k(X;\mathbb{C}) = \bigoplus_{p+q=k} H_J^{(p,q)}(X;\mathbb{C});$$

  then J is called *complex-C^∞-pure-and-full at the kth stage*.

Remark 4.2. In general, being complex-$\mathcal{C}^\infty$-full at the 2nd stage is a stronger condition that being $\mathcal{C}^\infty$-full. Furthermore, if J is integrable, then being complex-$\mathcal{C}^\infty$-pure-and-full at the 2nd stage is stronger than being $\mathcal{C}^\infty$-pure-and-full. More precisely, for any (possibly non-integrable) almost-complex structure J, it holds, [DLZ10, Lemma 2.11],

$$\begin{cases} H_J^+(X) = H_J^{(1,1)}(X;\mathbb{C}) \cap H_{dR}^2(X;\mathbb{R}) \\ H_J^{(1,1)}(X;\mathbb{C}) = H_J^+(X) \otimes_\mathbb{R} \mathbb{C} \end{cases} ,$$

and

$$H_J^{(2,0)}(X;\mathbb{C}) + H_J^{(0,2)}(X;\mathbb{C}) \subseteq H_J^-(X) \otimes_\mathbb{R} \mathbb{C} ,$$

and, if J is integrable, it holds

$$\begin{cases} H_J^-(X) = \left(H_J^{(2,0)}(X;\mathbb{C}) + H_J^{(0,2)}(X;\mathbb{C}) \right) \cap H_{dR}^2(X;\mathbb{R}) \\ H_J^{(2,0)}(X;\mathbb{C}) + H_J^{(0,2)}(X;\mathbb{C}) = H_J^-(X) \otimes_\mathbb{R} \mathbb{C} \end{cases} ;$$

indeed, $d \wedge^{2,0} X \subseteq \wedge^{3,0} X \oplus \wedge^{2,1} X$ and $d \wedge^{0,2} X \subseteq \wedge^{1,2} X \oplus \wedge^{0,3} X$. (Compare also [DLZ10, Lemma 2.12] for further results in the case of four-dimensional manifolds.)

Note also that, if J is $\mathcal{C}^\infty$-pure, then

$$H_J^{(1,1)}(X;\mathbb{R}) \cap \left(H_J^{(2,0)}(X;\mathbb{R}) + H_J^{(0,2)}(X;\mathbb{R}) \right) = \{0\} .$$

The construction of the subgroups $H_J^S(X;\mathbb{R}) \subseteq H_{dR}^\bullet(X;\mathbb{R})$ and the notion of $\mathcal{C}^\infty$-pure-and-full almost-complex structures can be repeated using the complex of currents $\left(\mathcal{D}_\bullet X := \mathcal{D}^{2n-\bullet} X, d \right)$ instead of the complex of differential forms $(\wedge^\bullet X, d)$ and the de Rham homology $H_\bullet^{dR}(X;\mathbb{R})$ instead of the de Rham cohomology $H_{dR}^\bullet(X;\mathbb{R})$. (We refer to Sect. 1.6 for notations and references concerning currents and de Rham homology.)

As in the smooth case, accordingly to T.-J. Li and W. Zhang, [LZ09], given $S \subseteq \mathbb{N} \times \mathbb{N}$, let

$$H_S^J(X;\mathbb{R}) := \left\{ [\alpha] \in H_\bullet^{dR}(X;\mathbb{C}) : \alpha \in \left(\bigoplus_{(p,q) \in S} \mathcal{D}_{p,q} X \right) \cap \mathcal{D}_\bullet X \right\} .$$

In particular, the almost-complex structures on X for which $H_{(2,0),(0,2)}^J(X;\mathbb{R})$ and $H_{(1,1)}^J(X;\mathbb{R})$ provide a decomposition of $H_2^{dR}(X;\mathbb{R})$ are emphasized by the following definition by T.-J. Li and W. Zhang.

Definition 4.4 ([LZ09, Definition 2.15, Lemma 2.16]). An almost-complex structure J on a manifold X is said to be:

- *pure* if

$$H^J_{(2,0),(0,2)}(X;\mathbb{R}) \cap H^J_{(1,1)}(X;\mathbb{R}) = \{0\} ;$$

- *full* if

$$H^J_{(2,0),(0,2)}(X;\mathbb{R}) + H^J_{(1,1)}(X;\mathbb{R}) = H^{dR}_2(X;\mathbb{R}) ;$$

- *pure-and-full* if it is both pure and full, i.e., if the following decomposition holds:

$$H^J_{(2,0),(0,2)}(X;\mathbb{R}) \oplus H^J_{(1,1)}(X;\mathbb{R}) = H^{dR}_2(X;\mathbb{R}) .$$

The following are natural generalizations of the notion of pure-and-fullness.

Definition 4.5. Let X be a manifold endowed with an almost-complex structure J, and fix $k \in \mathbb{N}$. Consider $H^k_{dR}(X;\mathbb{R}) \supseteq \sum_{\substack{p+q=k\\p\leq q}} H^J_{(p,q),(q,p)}(X;\mathbb{R})$:

- if

$$\bigoplus_{\substack{p+q=k\\p\leq q}} H^J_{(p,q),(q,p)}(X;\mathbb{R}) \subseteq H^{dR}_k(X;\mathbb{R})$$

(namely, the sum is direct), then J is called *pure at the kth stage*;
- if

$$H^{dR}_k(X;\mathbb{R}) = \sum_{\substack{p+q=k\\p\leq q}} H^J_{(p,q),(q,p)}(X;\mathbb{R}) ,$$

then J is called *full at the kth stage*;
- if J is both pure at the kth stage and full at the kth stage, that is,

$$H^{dR}_k(X;\mathbb{R}) = \bigoplus_{\substack{p+q=k\\p\leq q}} H^J_{(p,q),(q,p)}(X;\mathbb{R}) ;$$

then J is called *pure-and-full at the kth stage*.

As regards de Rham homology with complex coefficients, given $S \subseteq \mathbb{N} \times \mathbb{N}$, let

$$H^J_S(X;\mathbb{C}) := \left\{ [\alpha] \in H^{dR}_\bullet(X;\mathbb{C}) \ : \ \alpha \in \bigoplus_{(p,q)\in S} \mathcal{D}_{p,q} X \right\} ,$$

so that $H^J_S(X;\mathbb{R}) = H^J_S(X;\mathbb{C}) \cap H^{dR}_\bullet(X;\mathbb{R})$.

Definition 4.6. Let X be a manifold endowed with an almost-complex structure J, and fix $k \in \mathbb{N}$. Consider $H_k^{dR}(X;\mathbb{C}) \supseteq \sum_{p+q=k} H_{(p,q)}^J(X;\mathbb{C})$:

- if

$$\bigoplus_{p+q=k} H_{(p,q)}^J(X;\mathbb{C}) \subseteq H_k^{dR}(X;\mathbb{C})$$

(namely, the sum is direct), then J is called *complex-pure at the kth stage*;
- if

$$H_k^{dR}(X;\mathbb{C}) = \sum_{p+q=k} H_{(p,q)}^J(X;\mathbb{C}),$$

then J is called *complex-full at the kth stage*;
- if J is both complex-pure at the kth stage and complex-full at the kth stage, that is,

$$H_k^{dR}(X;\mathbb{C}) = \bigoplus_{p+q=k} H_{(p,q)}^J(X;\mathbb{C});$$

then J is called *complex-pure-and-full at the kth stage*.

Remark 4.3. The study of the subgroups $H_J^{(p,q),(q,p)}(X;\mathbb{R})$ and the notion of $\mathcal{C}^\infty$-pure-and-full almost-complex structure have been introduced by T.-J. Li and W. Zhang in [LZ09], in order to study the relations between the compatible and the tamed symplectic cones on a compact almost-complex manifold, and inspired by a question by S.K. Donaldson, [Don06, Question 2]: whether, on a compact four-dimensional manifold endowed with an almost-complex structure J tamed by a symplectic form, there exists also a symplectic form compatible with J, see Sect. 4.4.2. In [DLZ10], T. Drăghici, T.-J. Li, and W. Zhang investigated the four-dimensional case, proving, in particular, that every almost-complex structure on a compact four-dimensional manifold is $\mathcal{C}^\infty$-pure-and-full; they also obtained further results for four-dimensional almost-complex manifolds in [DLZ11], where they studied the dimensions of the subgroups $H_J^+(X)$ and $H_J^-(X)$. In [FT10], A.M. Fino and A. Tomassini studied the $\mathcal{C}^\infty$-pure-and-fullness in connection with other properties on almost-complex manifolds: in particular, by studying almost-complex solvmanifolds, they provided the first explicit example of a non-$\mathcal{C}^\infty$-pure-and-full almost-complex structure. Jointly with A. Tomassini, we studied in [AT11] the behaviour of $\mathcal{C}^\infty$-pure-and-fullness under small deformations of the complex structure or along curves of almost-complex structures, proving in particular its instability. In [AT12a] we continued the study of the cohomological properties related to the existence of an almost-complex structure, focusing, in particular, on the study of the cone of semi-Kähler structures on a compact semi-Kähler manifold. In [ATZ12], jointly with A. Tomassini and W. Zhang, we further studied cohomological properties of almost-Kähler manifolds, especially in relation with W. Zhang's Lefschetz-type property; in particular, an example of a non-C^∞-full almost-Kähler structure on a compact manifold is provided. In [DZ12], T. Drăghici and W. Zhang reformulated the S.K. Donaldson "tamed to compatible" question in terms of spaces of exact forms, proving, in particular, that an almost-complex structure J on a compact four-dimensional manifold admits a compatible symplectic form if and only if it admits tamed symplectic forms with any arbitrarily given J-anti-invariant component. For further results on the study of J-anti-invariant forms and J-anti-invariant de Rham cohomology classes on a (possibly non-compact) manifold endowed with an almost-complex structure J, see [HMT11] by R.K. Hind, C. Medori, and A. Tomassini, where a result concerning analytic continuation for J-anti-invariant forms is proven. In [LT12], T.-J. Li and A. Tomassini studied the analogue of the above problems for linear (possibly non-integrable) complex structures on four-dimensional unimodular Lie algebras; in particular, they proved that an analogue of the decomposition in [DLZ10, Theorem 2.3] holds for every four-dimensional unimodular Lie algebra endowed with a linear (possibly non-integrable) complex structure; furthermore, they considered the linear counterpart of Donaldson's "tamed to compatible" question, and of the tamed and compatible symplectic cones, studying, in particular, a sufficient condition on a four-dimensional Lie algebra g (which holds, for example, for four-dimensional unimodular Lie algebras) in order that a linear (possibly non-integrable) complex structure admits a taming linear symplectic form if and only if it admits a compatible linear symplectic form. The case of non-unimodular Lie algebras was studied in [DL13] by T. Drăghici and H. Leon. The paper [DLZ12] by T. Drăghici, T.-J. Li, and W. Zhang furnishes a survey on the known results concerning the subgroups $H_J^+(X)$ and $H_J^-(X)$, especially in dimension 4, and their application to S.K. Donaldson's "tamed to compatible" question.

Remark 4.4. The notion of C^∞-pure-and-fullness can be restated also in the symplectic context. By adapting T.-J. Li and W. Zhang theory, [LZ09], to the symplectic case, one can define subgroups of the de Rham cohomology connected with the symplectic structure ω on X as, for $r, s \in \mathbb{N}$,

$$H_\omega^{(r,s)}(X;\mathbb{R}) := \left\{ \left[L^r \beta^{(s)} \right] \in H_{dR}^{2r+s}(X;\mathbb{R}) \; : \; \beta^{(s)} \in \mathrm{P}\wedge^s X \right\} \subseteq H_{dR}^{2r+s}(X;\mathbb{R}) ,$$

and ask when the Lefschetz decomposition on the space of forms moves to cohomology, that is, when, in

$$\sum_{2r+s=k} H_\omega^{(r,s)}(X;\mathbb{R}) \subseteq H_{dR}^k(X;\mathbb{R}) ,$$

the inclusion is actually an equality and the sum is actually a direct sum. We note the relations between the above subgroups and the primitive cohomologies introduced by L.-S. Tseng and S.-T. Yau in [TY12a] (see Sect. 1.2.2.2). As regards L.-S. Tseng and S.-T. Yau's primitive $(d+d^\Lambda)$-cohomology $PH_{d+d^\Lambda}^\bullet(X;\mathbb{R})$, note that, for every $r, s \in \mathbb{N}$,

$$\text{im}\left(L^r PH^s_{d+d^A}(X;\mathbb{R}) \to H^\bullet_{dR}(X;\mathbb{R})\right) = L^r H^{(0,s)}_\omega(X;\mathbb{R}) \subseteq H^{(r,s)}_\omega(X;\mathbb{R}) .$$

In [TY12a, Sect. 4.1], L.-S. Tseng and S.-T. Yau have introduced also the primitive cohomology groups

$$PH^s_d(X;\mathbb{R}) := \frac{\ker d \cap \ker d^A \cap P\wedge^s X}{\text{im } d\lfloor_{P\wedge^{s-1} X \cap \ker d^A}},$$

where $s \in \mathbb{N}$, proving that the homology on co-isotropic chains is naturally dual to $PH^{2n-\bullet}_d(X;\mathbb{R})$, see [TY12a, pp. 40–41]; in [Lin13, Proposition A.5], Y. Lin proved that, if the Hard Lefschetz Condition holds on X, then

$$H^{(0,\bullet)}_\omega(X;\mathbb{R}) = PH^\bullet_d(X;\mathbb{R}) .$$

For results on $\mathcal{C}^\infty$-pure-and-fullness for symplectic structures, see [AT12b]. In particular, we note that a symplectic counterpart of [DLZ10, Theorem 2.3] holds: more precisely, for every compact symplectic manifold (non-necessarily of dimension 4), the decomposition $H^2_{dR}(X;\mathbb{R}) = H^{(1,0)}_\omega(X;\mathbb{R}) \oplus H^{(0,2)}_\omega(X;\mathbb{R})$ holds, [AT12b, Theorem 2.6].

Remark 4.5. The notion of $\mathcal{C}^\infty$-pure-and-fullness can be restated also in the **D**-complex context, where it assumes further significance due to the fact that the **D**-complex counterpart of Dolbeault cohomology[1] is, in general, non-finite

---

[1]Recall that, given $K \in \text{End}(TX)$ such that $K^2 = \text{id}_{TX}$, one can define, by duality, an endomorphism $K \in \text{End}(T^*X)$ such that $K^2 = \text{id}_{T^*X}$, and hence one gets a natural decomposition $T^*X = (T^+X)^* \oplus (T^-X)^*$ into eigen-bundles. Extending $K \in \text{End}(T^*X)$ to $K \in \text{End}(\wedge^\bullet X)$, one gets a decomposition of the bundle of differential ℓ-forms, for $\ell \in \mathbb{N}$: more precisely,

$$\wedge^\ell X = \bigoplus_{p+q=\ell} \wedge^{p,q}_{+-} X \qquad \text{where} \qquad \wedge^{p,q}_{+-} X := \wedge^p(T^+X)^* \otimes \wedge^q(T^-X)^* ;$$

If the almost-**D**-complex structure K is integrable, then the exterior differential splits as

$$d = \partial_+ + \partial_-$$

where

$$\partial_+ := \pi_{\wedge^{p+1,q}_{+-} X} \circ d \colon \wedge^{p,q}_{+-} X \to \wedge^{p+1,q}_{+-} X \qquad \text{and} \qquad \partial_- := \pi_{\wedge^{p,q+1}_{+-} X} \circ d \colon \wedge^{p,q}_{+-} X \to \wedge^{p,q+1}_{+-} X$$

(where $\pi_{\wedge^{r,s}_{+-} X} \colon \wedge^{\bullet,\bullet}_{+-} X \to \wedge^{r,s}_{+-} X$ denotes the natural projection onto $\wedge^{r,s}_{+-} X$, for every r, $s \in \mathbb{N}$). In particular, the condition $d^2 = 0$ is rewritten as

dimensional,[2] [Ros13, Proposition 3.1.11]. In fact, **D**-complex geometry is, in a sense, the "hyperbolic analogue" of complex geometry: in particular, there is no Hodge theory available in the non-elliptic context of **D**-complex geometry.

For results on $\mathcal{C}^\infty$-pure-and-fullness for (almost-)**D**-complex structures, see [AR12]; see also [Ros13]. In particular, we note that an analogue of [DLZ10, Theorem 2.3] holds for left-invariant **D**-complex structures on four-dimensional nil-manifolds, [AR12, Theorem 3.17], while counterexamples exist removing either the assumption on the dimension, [AR12, Examples 3.1 and 3.2], or on the nilpotency, [AR12, Example 4.1], or on the integrability, [AR12, Example 3.4], see [AR12, Remark 3.18]. Furthermore, as a difference between the complex and the **D**-complex cases (compare [DLZ10, Lemma 2.15, Theorem 2.16], or [LZ09, Proposition 2.1]), we note that admitting a **D**-Kähler structure does not yield decomposition in cohomology with respect to the **D**-complex structure, [AR12, Examples 3.1, 3.2, Proposition 3.3]. Finally, in [AR12], explicit examples of small deformations of the **D**-complex structure on nilmanifolds and solvmanifolds have been studied,[3] proving in particular that being **D**-Kähler is not a stable property under small deformations of the **D**-complex structure, [AR12, Example 4.1, Theorem 4.2] (compare with [KS60, Theorem 15] in the complex case), and studying the behaviour of $\mathcal{C}^\infty$-pure-and-fullness under small deformations of the **D**-complex structure, [AR12, Example 4.1, Proposition 4.3] (see Sect. 4.3.1.1 for the almost-complex case), and the semi-continuity properties of the dimensions of the K-(anti-)invariant subgroups of cohomology, [AR12, Examples 4.1, 4.4, 4.5, Proposition 4.6] (see Sect. 4.3.2 for the almost-complex case).

$$\begin{cases} \partial_+^2 = 0 \\ \partial_+\partial_- + \partial_-\partial_+ = 0 \ , \\ \partial_-^2 = 0 \end{cases}$$

and hence one can define a **D**-complex counterpart of the Dolbeault cohomology by considering the cohomology of the differential complex $\left(\wedge_{+-}^{\bullet, q} X, \partial_+\right)$ for every $q \in \mathbb{N}$, that is,

$$H_{\partial_+}^{\bullet, \bullet}(X; \mathbb{R}) := \frac{\ker \partial_+}{\operatorname{im} \partial_+} .$$

---

[2]For example, [AR12, p. 533], consider two compact manifolds X^+ and X^- having the same dimension and consider the natural **D**-complex structure on $X^+ \times X^-$, i.e., the **D**-complex structure given by the decomposition $T\left(X^+ \times X^-\right) = TX^+ \oplus TX^-$, where $K\lfloor_{TX^\pm} = \pm \operatorname{id}_{T(X^+ \times X^-)}$, for $\pm \in \{+, -\}$ (recall that every **D**-complex manifold is locally of this form, see, e.g., [CMMS04, Proposition 2]); one has that the vector space $H_{\partial_+}^{0,0}\left(X^+ \times X^-\right) \simeq \mathcal{C}^\infty(X^-)$ does not have finite dimension.

[3]See [MT11, Ros12a] for more results concerning deformations of (almost-)**D**-complex structures.

4.1.2 Relations Between $\mathcal{C}^\infty$-Pure-and-Fullness and Pure-and-Fullness

The following result[4] summarizes the relations between $\mathcal{C}^\infty$-pure-and-fullness and pure-and-fullness, and between complex-$\mathcal{C}^\infty$-pure-and-fullness and complex-pure-and-fullness.

Theorem 4.1 ([AT11, Theorem 2.1], see [LZ09, Proposition 2.5]). *Let J be an almost-complex structure on a compact $2n$-dimensional manifold X. The following relations between (complex-)$\mathcal{C}^\infty$-pure-and-full and (complex-)pure-and-full notions hold: for any $k \in \mathbb{N}$,*

$$
\begin{array}{ccc}
\mathcal{C}^\infty\text{-full at the } k\text{th stage} & \Longrightarrow & \text{pure at the } k\text{th stage} \\
\Big\Downarrow & & \Big\Downarrow \\
\text{full at the } (2n-k)\text{ th stage} & \Longrightarrow & \mathcal{C}^\infty\text{-pure at the } (2n-k)\text{ th stage ,}
\end{array}
$$

and

$$
\begin{array}{ccc}
\text{complex-}\mathcal{C}^\infty\text{-full at the } k\text{th stage} & \Longrightarrow & \text{complex-pure at the } k\text{th stage} \\
\Big\Downarrow & & \Big\Downarrow \\
\text{complex-full at the } (2n-k)\text{ th stage} & \Longrightarrow & \text{complex-}\mathcal{C}^\infty\text{-pure at the } (2n-k)\text{ th stage .}
\end{array}
$$

Proof. The horizontal implications follow by considering the non-degenerate duality pairing

$$\langle \cdot, \cdot \rangle : H_{dR}^\bullet(X; \mathbb{R}) \times H_\bullet^{dR}(X; \mathbb{R}) \to \mathbb{R}, \quad \text{respectively} \quad \langle \cdot, \cdot \rangle : H_{dR}^\bullet(X; \mathbb{C}) \times H_\bullet^{dR}(X; \mathbb{C}) \to \mathbb{C},$$

and noting that, for any $(p, q) \in \mathbb{N}^2$,

$$\ker \left\langle H_J^{(p,q),(p,q)}(X; \mathbb{R}), \cdot \right\rangle \supseteq \sum_{\{(r,s),(s,r)\} \neq \{(p,q),(q,p)\}} H_{(r,s),(s,r)}^J(X; \mathbb{R})$$

and

$$\ker \left\langle \cdot, H_{(p,q),(q,p)}^J(X; \mathbb{R}) \right\rangle \supseteq \sum_{\{(r,s),(s,r)\} \neq \{(p,q),(q,p)\}} H_J^{(r,s),(s,r)}(X; \mathbb{R}) ,$$

---

[4]For an analogous result in the setting of almost-D-complex structures in the sense of F. R. Harvey and H. B. Lawson, see [AR12, Proposition 1.4]; for an analogous result in the setting of symplectic structures, see [AT12b, Proposition 2.4].

respectively

$$\ker\left\langle H_J^{(p,q)}(X;\mathbb{C}),\,\cdot\,\right\rangle \supseteq \sum_{(r,s)\neq(p,q)} H_{(r,s)}^J(X;\mathbb{C}) \qquad \text{and}$$

$$\ker\left\langle\,\cdot\,,\,H_{(p,q)}^J(X;\mathbb{C})\right\rangle \supseteq \sum_{(r,s)\neq(p,q)} H_J^{(r,s)}(X;\mathbb{C}) .$$

As an example, we give the details to prove that if J is C^∞-full at the kth stage then it is also pure at the kth stage, when $k=2$. Let

$$\mathfrak{c} \in H_{(2,0),(0,2)}^J(X;\mathbb{R}) \cap H_{(1,1)}^J(X;\mathbb{R}) ,$$

with $\mathfrak{c} \neq [0]$. Hence,

$$\langle\mathfrak{c},\,\cdot\,\rangle\,\lfloor_{H_J^{(2,0),(0,2)}(X;\mathbb{R})} = 0 \qquad \text{and} \qquad \langle\mathfrak{c},\,\cdot\,\rangle\,\lfloor_{H_J^{(1,1)}(X;\mathbb{R})} = 0 ;$$

since J is C^∞-full, it follows that $\langle\mathfrak{c},\,\cdot\,\rangle\,\lfloor_{H_{dR}^2(X;\mathbb{R})} = 0$, and hence $\mathfrak{c} = [0]$.

To prove the vertical implications, it is enough to note that the quasi-isomorphism $T.\colon \wedge^\bullet X \to \mathcal{D}_{2n-\bullet}X$ defined as $T_\varphi := \int_X \varphi \wedge \cdot$ (see Sect. 1.6) induces an injective map

$$H_J^{(p,q),(q,p)}(X;\mathbb{R}) \to H_{(n-p,n-q),(n-q,n-p)}^J(X;\mathbb{R}) , \qquad\text{respectively}$$

$$H_J^{(p,q)}(X;\mathbb{C}) \to H_{(n-p,n-q)}^J(X;\mathbb{C}) ,$$

for any $(p,q) \in \mathbb{N}^2$. $\square$

Remark 4.6. On a compact $2n$-dimensional manifold X endowed with an almost-complex structure J, further linkings between $H_{dR}^2(X;\mathbb{R})$ and $H_{dR}^{2n-2}(X;\mathbb{R})$ could provide further relations between C^∞-pure-and-full and pure-and-full notions: for example, A.M. Fino and A. Tomassini proved in [FT10, Theorem 3.7] that, given a J-Hermitian metric g on X, if there exists a basis of g-harmonic representatives for $H_{dR}^2(X;\mathbb{R})$ being of pure type with respect to J, then J is both C^∞-pure-and-full and pure-and-full. Furthermore, A.M. Fino and A. Tomassini proved in [FT10, Theorem 4.1] that, given a J-compatible symplectic form ω on X satisfying the Hard Lefschetz Condition (that is, the map $[\omega^k]\smile\cdot\colon H_{dR}^{n-k}(X;\mathbb{R}) \overset{\cong}{\to} H_{dR}^{n+k}(X;\mathbb{R})$ is an isomorphism for every $k \in \mathbb{N}$), if J is C^∞-pure-and-full, then J is also pure-and-full (compare Proposition 4.4 for a similar result).

Setting $2n=4$ and $k=2$ in Theorem 4.1, it follows that, on compact four-dimensional almost-complex manifolds, C^∞-fullness implies C^∞-pureness. The following result states that, for higher dimensional manifolds, C^∞-pureness and C^∞-fullness are not, in general, related properties.

Proposition 4.1 ([AT12a, Proposition 1.4]). *There exist both examples of compact manifolds endowed with almost-complex structures being C^∞-full and non-C^∞-pure, and examples of compact manifolds endowed with almost-complex structures being C^∞-pure and non-C^∞-full.*

Proof. The proof follows from the following examples, [AT12a, Examples 1.2 and 1.3].

Step 1—*Being C^∞-full does not imply being C^∞-pure.* Take a nilmanifold N_1 with associated Lie algebra

$$\mathfrak{h}_{16} := \left(0^3,\ 12,\ 14,\ 24\right).$$

Consider the left-invariant complex structure on N_1 whose space of $(1,0)$-forms is generated, as a $C^\infty\left(N_1;\mathbb{C}\right)$-module, by

$$\begin{cases} \varphi^1 := e^1 + ie^2 \\ \varphi^2 := e^3 + ie^4 \\ \varphi^3 := e^5 + ie^6 \end{cases}.$$

Writing the structure equations in terms of $\{\varphi^1,\ \varphi^2,\ \varphi^3\}$,

$$\begin{cases} 2\,d\varphi^1 = 0 \\ 2\,d\varphi^2 = \varphi^{1\bar{1}} \\ 2\,d\varphi^3 = -i\varphi^{12} + i\varphi^{1\bar{2}} \end{cases},$$

the integrability condition is easily verified.

K. Nomizu's theorem [Nom54, Theorem 1] makes the computation of the cohomology straightforward: in fact, listing the harmonic representatives with respect to the left-invariant Hermitian metric $g := \sum_j \varphi^j \odot \bar{\varphi}^j$ instead of their classes, one finds

$$H^2_{dR}(N_1;\mathbb{C}) = \mathbb{C}\left\langle\varphi^{13},\ \varphi^{\bar{1}\bar{3}}\right\rangle \oplus \mathbb{C}\left\langle\varphi^{1\bar{3}} - \varphi^{3\bar{1}}\right\rangle \oplus \mathbb{C}\left\langle\varphi^{12} + \varphi^{1\bar{2}},\ \varphi^{2\bar{1}} - \varphi^{\bar{1}\bar{2}}\right\rangle,$$

where

$$H_J^{(2,0),(0,2)}(N_1;\mathbb{C}) = \mathbb{C}\left\langle\varphi^{13},\ \varphi^{\bar{1}\bar{3}}\right\rangle \oplus \mathbb{C}\left\langle\varphi^{12} + \varphi^{1\bar{2}},\ \varphi^{2\bar{1}} - \varphi^{\bar{1}\bar{2}}\right\rangle$$

and

$$H_J^{(1,1)}(N_1;\mathbb{C}) = \mathbb{C}\left\langle\varphi^{1\bar{3}} - \varphi^{3\bar{1}}\right\rangle \oplus \mathbb{C}\left\langle\varphi^{12} + \varphi^{1\bar{2}},\ \varphi^{2\bar{1}} - \varphi^{\bar{1}\bar{2}}\right\rangle.$$

In particular, J is a C^∞-full, non-C^∞-pure complex structure.

Step 2—*Being C^∞-pure does not imply being C^∞-full.* Take a nilmanifold N_2 with associated Lie algebra

$$\mathfrak{h}_2 := \left(0^4,\, 12,\, 34\right)\,.$$

and consider on it the left-invariant complex structure given requiring that the forms

$$\begin{cases} \varphi^1 := e^1 + i\,e^2 \\ \varphi^2 := e^3 + i\,e^4 \\ \varphi^3 := e^5 + i\,e^6 \end{cases}$$

are of type $(1,0)$.

The integrability condition follows from the structure equations

$$\begin{cases} 2\,\mathrm{d}\varphi^1 = 0 \\ 2\,\mathrm{d}\varphi^2 = 0 \\ 2\,\mathrm{d}\varphi^3 = i\,\varphi^{1\bar{1}} - i\,\varphi^{2\bar{2}} \end{cases}\,.$$

K. Nomizu's theorem [Nom54, Theorem 1] gives

$$H^2_{dR}\left(N_2; \mathbb{C}\right) = \mathbb{C}\left\langle \varphi^{12},\, \varphi^{\bar{1}\bar{2}} \right\rangle \oplus \mathbb{C}\left\langle \varphi^{1\bar{2}},\, \varphi^{2\bar{1}} \right\rangle$$

$$\oplus\, \mathbb{C}\left\langle \varphi^{13} + \varphi^{1\bar{3}},\, \varphi^{3\bar{1}} - \varphi^{1\bar{3}},\, \varphi^{3\bar{2}} - \varphi^{\bar{2}\bar{3}},\, \varphi^{23} - \varphi^{2\bar{3}} \right\rangle\,,$$

where

$$H_J^{(2,0),(0,2)}\left(N_2; \mathbb{C}\right) = \mathbb{C}\left\langle \varphi^{12},\, \varphi^{\bar{1}\bar{2}} \right\rangle$$

and

$$H_J^{(1,1)}\left(N_2; \mathbb{C}\right) = \mathbb{C}\left\langle \varphi^{1\bar{2}},\, \varphi^{2\bar{1}} \right\rangle\,;$$

this can be proven arguing as follows: with respect to the left-invariant Hermitian metric $g := \sum_j \varphi^j \odot \bar{\varphi}^j$, one computes

$$\partial^* \varphi^{13} = \partial^* \varphi^{23} = \partial^* \varphi^{12} = 0\,,$$

that is, φ^{13}, φ^{12} and φ^{23} are g-orthogonal to the space $\partial \wedge^{1,0} N_2$; in the same way, one computes

$$\partial^* \varphi^{1\bar{2}} = \bar{\partial}^* \varphi^{1\bar{2}} = \partial^* \varphi^{1\bar{3}} = \bar{\partial}^* \varphi^{1\bar{3}} = 0$$

(compare also Proposition 4.2). In particular, J is a $\mathcal{C}^\infty$-pure, non-$\mathcal{C}^\infty$-full complex structure.   □

4.2 $\mathcal{C}^\infty$-Pure-and-Fullness for Special Manifolds

In this section, we study the property of being $\mathcal{C}^\infty$-pure-and-full on special classes of (almost-)complex manifolds. After recalling some motivating results by T. Drăghici, T.-J. Li, and W. Zhang, we study $\mathcal{C}^\infty$-pure-and-fullness for left-invariant complex-structures on solvmanifolds, providing some examples in dimension 4 or higher; furthermore, we consider almost-complex manifolds endowed with special metric structures, namely, *semi-Kähler*, and *almost-Kähler* structures.

4.2.1 Special Classes of $\mathcal{C}^\infty$-Pure-and-Full (Almost-)Complex Manifolds

In this section, we recall some results by T. Drăghici, T.-J. Li, and W. Zhang, providing classes of $\mathcal{C}^\infty$-pure-and-full and pure-and-full (almost-)complex manifolds. They could be considered as motivations to study $\mathcal{C}^\infty$-pure-and-fullness: in fact, [DLZ10, Lemma 2.15, Theorem 2.16] suggests that the subgroups $H_J^{(\bullet,\bullet)}(X;\mathbb{C})$ can be viewed as a generalization of the Dolbeault cohomology groups for non-Kähler, and non-integrable, almost-complex manifolds X. On the other hand, [DLZ10, Theorem 2.3] states that, on a compact four-dimensional almost-complex manifold X, the subgroups $H_J^+(X)$ and $H_J^-(X)$ induce always a decomposition of $H_{dR}^2(X;\mathbb{R})$: this could be intended as a generalization of the Hodge decomposition theorem for compact four-dimensional almost-complex manifolds.

According to the following result, the groups $H_J^{(\bullet,\bullet)}(X;\mathbb{C})$ can be considered as the counterpart of the Dolbeault cohomology groups in the non-Kähler and non-integrable cases.

Theorem 4.2 ([DLZ10, Lemma 2.15, Theorem 2.16]). *Let X be a compact complex manifold. If the Hodge and Frölicher spectral sequence degenerates at the first step and the natural filtration associated with the structure of double complex of $\left(\wedge^{\bullet,\bullet}X,\,\partial,\,\bar\partial\right)$ induces a Hodge decomposition of weight k on $H_{dR}^k(X;\mathbb{C})$ for some $k \in \mathbb{N}$, then X is complex-$\mathcal{C}^\infty$-pure-and-full at the kth stage, and*

$$H_J^{(p,q)}(X;\mathbb{C}) \simeq H_{\bar\partial}^{p,q}(X)$$

for every $(p,q) \in \mathbb{N}^2$ such that $p + q = k$.

Recall that, for compact complex surfaces, the Hodge and Frölicher spectral sequence degenerates at the first step, [Kod64], and the natural filtration associated with the structure of double complex of $\left(\wedge^{\bullet,\bullet} X, \partial, \bar{\partial}\right)$ induces a Hodge decomposition of weight 2 on $H^2_{dR}(X;\mathbb{C})$, see, e.g., [BHPVdV04, Theorem IV.2.8, Proposition IV.2.9]. (We recall that, in [Rol08], for every $n \in \mathbb{N} \setminus \{0,1\}$, a nilmanifold of real dimension $2n$ with left-invariant complex structure such that the Hodge and Frölicher spectral sequence does not degenerate at the nth step was constructed, showing that the Hodge and Frölicher spectral sequence can be arbitrarily non-degenerate.) Note also that, for compact complex manifolds satisfying the $\partial\bar{\partial}$-Lemma, the assumptions of Theorem 4.2 are satisfied, for any $k \in \mathbb{N}$, by [DGMS75, 5.21], and recall that compact complex manifolds admitting a Kähler metric satisfy the $\partial\bar{\partial}$-Lemma, [DGMS75, Lemma 5.11]. Therefore, as a corollary of [DLZ10, Lemma 2.15, Theorem 2.16], and of Remark 4.2 and Theorem 4.1, one gets the following result.

Corollary 4.1 ([LZ09, Proposition 2.1], [DLZ10, Theorem 2.16, Proposition 2.17]). *One has that:*

 (i) *every compact complex surface is complex-$\mathcal{C}^\infty$-pure-and-full at the 2nd stage, and hence, in particular, $\mathcal{C}^\infty$-pure-and-full and pure-and-full;*
 (ii) *every compact complex manifold satisfying the $\partial\bar{\partial}$-Lemma is complex-$\mathcal{C}^\infty$- pure-and-full at every stage, and hence complex-pure-and-full at every stage;*
 (iii) *every compact complex manifold admitting a Kähler structure is complex-$\mathcal{C}^\infty$- pure-and-full at every stage, and hence complex-pure-and-full at every stage.*

Actually, T. Drăghici, T.-J. Li, and W. Zhang proved in [DLZ10] the following result, which one can consider as a sort of Hodge decomposition theorem in the non-Kähler case.

Theorem 4.3 ([DLZ10, Theorem 2.3]). *Every almost-complex structure on a compact four-dimensional manifold is $\mathcal{C}^\infty$-pure-and-full and pure-and-full.*

Proof. The proof of the previous theorem rests on the very special properties of four-dimensional manifolds. For the sake of completeness, we recall here the argument by T. Drăghici, T.-J. Li, and W. Zhang in [DLZ10]. Firstly, note that, by Theorem 4.1, it suffices to prove that an almost-complex structure J on a compact four-dimensional manifold is $\mathcal{C}^\infty$-full. Suppose that J is not $\mathcal{C}^\infty$-full. Fix a Hermitian metric g on X, and denote its associated $(1,1)$-form by ω. Recall that the Hodge-$*$-operator $*_g\lfloor_{\wedge^2 X}: \wedge^2 X \to \wedge^2 X$ satisfies $\left(*_g\lfloor_{\wedge^2 X}\right)^2 = \mathrm{id}_{\wedge^2 X}$, hence it induces a splitting

$$\wedge^2 X = \wedge^+_g X \oplus \wedge^-_g X \,,$$

where $\wedge^\pm_g X := \left\{\varphi \in \wedge^2 X : *_g\varphi = \pm\varphi\right\}$, for $\pm \in \{+,-\}$. Setting $P\wedge^\bullet X := \ker \Lambda = \ker L^{2-\bullet+1}\lfloor_{\wedge^\bullet X}$ the space of primitive forms, where Λ is the adjoint operator of the Lefschetz operator $L := \omega \wedge \cdot : \wedge^\bullet X \to \wedge^{\bullet+2} X$ with respect to the pairing induced by ω (see Sect. 1.2), one has

$$\wedge_g^+ X = L\left(C^\infty(X;\mathbb{R})\right) \oplus \left((\wedge^{2,0} X \oplus \wedge^{0,2} X) \cap \wedge^2 X\right) \qquad \text{and}$$

$$\wedge_g^- X = P\wedge^2 X \cap \wedge^{1,1} X \; ;$$

indeed, recall that, on a compact $2n$-dimensional manifold X endowed with an almost-complex structure J and a Hermitian metric g with associated $(1,1)$-form ω, one has, for every $j \in \mathbb{N}$, for every $k \in \mathbb{N}$, the Weil identity, [Wei58, Théorème 2],

$$*_g L^j \big\lfloor_{P\wedge^k X} = (-1)^{\frac{k(k+1)}{2}} \frac{j!}{(n-k-j)!} L^{n-k-j} J \,,$$

see, e.g., [Huy05, Proposition 1.2.31]. Since the Laplacian operator Δ and the Hodge-$*$-operator $*_g$ commute, the splitting $\wedge^2 X = \wedge_g^+ X \oplus \wedge_g^- X$ induces a decomposition in cohomology,

$$H^2_{dR}(X;\mathbb{R}) = H_g^+(X) \oplus H_g^-(X) \,,$$

where $H_g^\pm(X) := \left\{ [\varphi] \in H^2_{dR}(X;\mathbb{R}) \,:\, \varphi \in \wedge_g^\pm X \right\}$ for $\pm \in \{+,-\}$. Consider the non-degenerate pairing

$$\langle \cdot, \cdot \rangle : H^2_{dR}(X;\mathbb{R}) \times H^2_{dR}(X;\mathbb{R}) \to \mathbb{R} \,, \qquad \langle \varphi, \psi \rangle := \int_X \varphi \wedge \psi \,,$$

and take $\mathfrak{a} \in \left(H_J^+(X) + H_J^-(X)\right)^\perp \subseteq H^2_{dR}(X;\mathbb{R})$. Since $\wedge_g^- X \subseteq \wedge^{1,1} X$, one can reduce to consider $\mathfrak{a} \in H_g^+(X)$; let $\alpha \in \wedge_g^+ X$ be such that $\mathfrak{a} = [\alpha]$. According to the decomposition $\wedge_g^+ X = L\left(C^\infty(X;\mathbb{R})\right) \oplus \left((\wedge^{2,0} X \oplus \wedge^{0,2} X) \cap \wedge^2 X\right)$, let $f\omega$ be the component of α in $L\left(C^\infty(X;\mathbb{R})\right)$. Consider the Hodge decomposition

$$f\omega = h_{f\omega} + \mathrm{d}\vartheta + \mathrm{d}^* \eta$$

of $f\omega \in \wedge^2 X$, where $h_{f\omega} \in \ker\Delta \cap \wedge^2 X$, $\vartheta \in \wedge^1 X$, and $\eta \in \wedge^3 X$. Since $f\omega \in \wedge_g^+ X$ and by the uniqueness of the Hodge decomposition, one has

$$h_{f\omega} + 2\,\mathrm{d}\vartheta = f\omega + 2\pi_{\wedge_g^- X}(\mathrm{d}\vartheta) \in \wedge^{1,1} X \cap \wedge^2 X$$

(where $\pi_{\wedge_g^\pm X} : \wedge^2 X \to \wedge_g^\pm X$ denotes the natural projection onto $\wedge_g^\pm X$, for $\pm \in \{+,-\}$). Therefore, noting also that $H_g^+(X)$ is orthogonal to $H_g^-(X)$ with respect to $\langle \cdot, \cdot \rangle$, one has

$$0 \; = \; \left\langle \mathfrak{a}, \left[h_{f \omega} + 2 \, \mathrm{d} \vartheta \right] \right\rangle \; = \; \left\langle \mathfrak{a}, \left[f \, \omega + 2 \pi_{\wedge_J^- X} \left(\mathrm{d} \vartheta \right) \right] \right\rangle \; = \; \int_X f^2 \, \omega^2 \, ,$$

from which it follows that $f = 0$, and hence $\mathfrak{a} = 0$.   □

Remark 4.7. The result in [DLZ10, Theorem 2.3] does not hold anymore true in dimension greater than or equal to 6, or without the compactness assumption: the first example of a non-$\mathcal{C}^\infty$-pure almost-complex structure was provided by A.M. Fino and A. Tomassini in [FT10, Example 3.3] using a six-dimensional nilmanifold (for other examples, even in the integrable case, see Propositions 4.1, 4.9, 4.10, Example 4.9, Theorem 4.8), while non-$\mathcal{C}^\infty$-pure-and-full almost-complex structures on non-compact four-dimensional manifolds arise from [DLZ11, Theorem 3.24] by T. Drăghici, T.-J. Li, and W. Zhang.

4.2.2 $\mathcal{C}^\infty$-Pure-and-Full Solvmanifolds

Let $X = \Gamma \backslash G$ be a solvmanifold, and denote the Lie algebra naturally associated to G by $\mathfrak{g}$, and its complexification by $\mathfrak{g}_{\mathbb{C}} := \mathfrak{g} \otimes_{\mathbb{R}} \mathbb{C}$. (We refer to Sect. 1.7 for notations and results concerning solvmanifolds.)

We recall that if X is a nilmanifold or, more in general, a completely-solvable solvmanifold, the inclusion of the sub-complex given by the G-left-invariant differential forms, which is isomorphic to the complex $\wedge^\bullet \mathfrak{g}^*$ of linear forms on the dual of the Lie algebra $\mathfrak{g}$ associated to G, into the de Rham complex of X turns out to be a quasi-isomorphism, in view of K. Nomizu's theorem [Nom54, Theorem 1], respectively A. Hattori's theorem [Hat60, Corollary 4.2].

Let J be a G-left-invariant almost-complex structure on X. In this case, one can study the problem of cohomological decomposition both on X and on $\mathfrak{g}$: in this section, we investigate the relations between the cohomological decompositions at the level of the solvmanifold and at the level of the associated Lie algebra, Proposition 4.2, Corollary 4.2.

Firstly, we set some notations. Consider $H_{dR}^\bullet (\mathfrak{g}; \mathbb{R}) := H^\bullet (\wedge^\bullet \mathfrak{g}^*, \mathrm{d})$. Being J a G-left-invariant almost-complex structure, it induces a bi-graded splitting also on the vector space $\wedge^\bullet \mathfrak{g}_{\mathbb{C}}^*$. For every $S \subset \mathbb{N} \times \mathbb{N}$, and for $\mathbb{K} \in \{\mathbb{R}, \mathbb{C}\}$, set

$$H_J^S (\mathfrak{g}; \mathbb{K}) \; := \; \left\{ [\alpha] \in H_{dR}^\bullet (\mathfrak{g}; \mathbb{K}) \; : \; \alpha \in \bigoplus_{(p,q) \in S} \wedge^{p,q} \mathfrak{g}_{\mathbb{C}}^* \cap \left(\wedge^\bullet \mathfrak{g}^* \otimes_{\mathbb{R}} \mathbb{K} \right) \right\} \, ,$$

see [LT12, Definition 0.3] by T.-J. Li and A. Tomassini.

The following are the natural linear counterparts of the corresponding definitions for manifolds.

Definition 4.7. Let $X = \Gamma \backslash G$ be a solvmanifold, and denote the Lie algebra naturally associated to G by $\mathfrak{g}$. Fixed $k \in \mathbb{N}$, a G-left-invariant almost-complex structure J on X is called

- *linear-C^∞-pure at the kth stage* if

$$\bigoplus_{\substack{p+q=k \\ p \leq q}} H_J^{(p,q),(q,p)}(\mathfrak{g}; \mathbb{R}) \subseteq H_{dR}^k(\mathfrak{g}; \mathbb{R}) \ ,$$

  namely, if the sum is direct;
- *linear-C^∞-full at the kth stage* if

$$H_{dR}^k(\mathfrak{g}; \mathbb{R}) = \sum_{\substack{p+q=k \\ p \leq q}} H_J^{(p,q),(q,p)}(\mathfrak{g}; \mathbb{R}) \ ,$$

- *linear-C^∞-pure-and-full at the kth stage* if J is both linear-C^∞-pure at the kth stage and linear-C^∞-full at the kth stage, that is, if the cohomological decomposition

$$H_{dR}^k(\mathfrak{g}; \mathbb{R}) = \bigoplus_{\substack{p+q=k \\ p \leq q}} H_J^{(p,q),(q,p)}(\mathfrak{g}; \mathbb{R})$$

holds.

Furthermore, J is called

- *linear-complex-C^∞-pure at the kth stage* if

$$\bigoplus_{p+q=k} H_J^{(p,q)}(\mathfrak{g}; \mathbb{C}) \subseteq H_{dR}^k(\mathfrak{g}; \mathbb{C}) \ ,$$

  namely, if the sum is direct;
- *linear-complex-C^∞-full at the kth stage* if

$$H_{dR}^k(\mathfrak{g}; \mathbb{C}) = \sum_{p+q=k} H_J^{(p,q)}(\mathfrak{g}; \mathbb{C}) \ ,$$

- *linear-complex-C^∞-pure-and-full at the kth stage* if J is both linear-complex-C^∞-pure at the kth stage and linear-complex-C^∞-full at the kth stage, that is, if the cohomological decomposition

$$H_{dR}^k(\mathfrak{g}; \mathbb{C}) = \bigoplus_{p+q=k} H_J^{(p,q)}(\mathfrak{g}; \mathbb{C})$$

holds.

(In any case, when $k = 2$, the specification "at the 2nd stage" will be understood.)

It is natural to ask what relations link the subgroups $H_J^{(\bullet,\bullet)}(X; \mathbb{R})$ and the subgroups $H_J^{(\bullet,\bullet)}(\mathfrak{g}; \mathbb{R})$, and whether a G-left-invariant linear-C^∞-pure-and-full almost-complex structure on $X = \Gamma \backslash G$ is also C^∞-pure-and-full.

The following lemma is the F.A. Belgun symmetrization trick, [Bel00, Theorem 7], in the almost-complex setting.[5]

Lemma 4.1 ([Bel00, Theorem 7]). *Let $X = \Gamma \backslash G$ be a solvmanifold, and denote the Lie algebra naturally associated to G by $\mathfrak{g}$. Let J be a G-left-invariant almost-complex structure on X. Let η be the G-bi-invariant volume form on G given by J. Milnor's Lemma, [Mil76, Lemma 6.2], and such that $\int_X \eta = 1$. Up to identifying G-left-invariant forms on X and linear forms over $\mathfrak{g}^*$ through left-translations, consider the Belgun symmetrization map*

$$\mu \colon \wedge^\bullet X \to \wedge^\bullet \mathfrak{g}^* , \qquad \mu(\alpha) := \int_X \alpha \lfloor_m \eta(m) .$$

Then one has that

$$\mu\lfloor_{\wedge^\bullet \mathfrak{g}^*} = \mathrm{id}\lfloor_{\wedge^\bullet \mathfrak{g}^*} ,$$

and that

$$\mathrm{d}\left(\mu(\cdot)\right) = \mu\left(\mathrm{d}\,\cdot\right) \qquad and \qquad J\left(\mu(\cdot)\right) = \mu\left(J\,\cdot\right) .$$

Using the previous lemma, we can prove the following Nomizu-type result[6], which relates the subgroups $H_J^{(r,s)}(X; \mathbb{R})$ with their left-invariant part $H_J^{(r,s)}(\mathfrak{g}; \mathbb{R})$.

Proposition 4.2 ([ATZ12, Theorem 5.4]). *Let $X = \Gamma \backslash G$ be a solvmanifold endowed with a G-left-invariant almost-complex structure J, and denote the Lie algebra naturally associated to G by $\mathfrak{g}$. For any $S \subset \mathbb{N} \times \mathbb{N}$, and for $\mathbb{K} \in \{\mathbb{R}, \mathbb{C}\}$, the map*

$$j \colon H_J^S(\mathfrak{g}; \mathbb{K}) \to H_J^S(X; \mathbb{K})$$

induced by left-translations is injective, and, if $H_{dR}^\bullet(\mathfrak{g}; \mathbb{K}) \simeq H_{dR}^\bullet(X; \mathbb{K})$ (for instance, if X is a completely-solvable solvmanifold), then $j \colon H_J^S(\mathfrak{g}; \mathbb{K}) \to H_J^S(X; \mathbb{K})$ is in fact an isomorphism.

Proof. Since J is G-left-invariant, left-translations induce the map $j \colon H_J^S(\mathfrak{g}; \mathbb{K}) \to H_J^S(X; \mathbb{K})$. Consider the Belgun symmetrization map $\mu \colon \wedge^\bullet X \otimes \mathbb{K} \to \wedge^\bullet \mathfrak{g}^* \otimes_\mathbb{R} \mathbb{K}$, [Bel00, Theorem 7]: since μ commutes with d by [Bel00, Theorem 7], it induces the

---

[5]For an analogous result in the setting of almost-**D**-complex structures in the sense of F.R. Harvey and H.B. Lawson, see [AR12, Lemma 2.3]; for an analogous result in the setting of symplectic structures, see [AT12b, Lemma 3.2].

[6]For analogous results in the context of almost-**D**-complex structures in the sense of F. R. Harvey and H. B. Lawson, see [AR12, Proposition 2.4]; for analogous results in the context of symplectic structures, see [AT12b, Proposition 3.3]; compare also with [FT10, Theorem 3.4], by A. M. Fino and A. Tomassini, for almost-complex structures.

map $\mu\colon H^\bullet_{dR}(X;\mathbb{K}) \to H^\bullet_{dR}(\mathfrak{g};\mathbb{K})$, and, since μ commutes with J, it preserves the bi-graduation; therefore it induces the map $\mu\colon H^S_J(X;\mathbb{K}) \to H^S_J(\mathfrak{g};\mathbb{K})$. Moreover, since μ is the identity on the space of G-left-invariant forms by [Bel00, Theorem 7], we get the commutative diagram

$$H^S_J(\mathfrak{g};\mathbb{K}) \xrightarrow{\ j\ } H^S_J(X;\mathbb{K}) \xrightarrow{\ \mu\ } H^S_J(\mathfrak{g};\mathbb{K})$$
$$\text{id}$$

hence $j\colon H^S_J(\mathfrak{g};\mathbb{K}) \to H^S_J(X;\mathbb{K})$ is injective, and $\mu\colon H^S_J(X;\mathbb{K}) \to H^S_J(\mathfrak{g};\mathbb{K})$ is surjective.

Furthermore, when $H^\bullet_{dR}(\mathfrak{g};\mathbb{K}) \sim H^\bullet_{dR}(X;\mathbb{K})$ (for instance, when X is a completely-solvable solvmanifold, by A. Hattori's theorem [Hat60, Theorem 4.2]), since

$$\mu\big\lfloor_{\wedge^\bullet \mathfrak{g}^* \otimes_\mathbb{R} \mathbb{K}} = \text{id}\big\lfloor_{\wedge^\bullet \mathfrak{g}^* \otimes_\mathbb{R} \mathbb{K}}$$

by [Bel00, Theorem 7], we get that $\mu\colon H^\bullet_{dR}(X;\mathbb{K}) \to H^\bullet_{dR}(\mathfrak{g};\mathbb{K})$ is the identity map, and hence $\mu\colon H^S_J(X;\mathbb{K}) \to H^S_J(\mathfrak{g};\mathbb{K})$ is also injective, and hence an isomorphism.

$\square$

As a straightforward consequence, we get the following result.

Corollary 4.2. *Let $X = \Gamma \backslash G$ be a solvmanifold endowed with a G-left-invariant almost-complex structure J, and denote the Lie algebra naturally associated to G by $\mathfrak{g}$. Suppose that $H^\bullet_{dR}(\mathfrak{g};\mathbb{R}) \simeq H^\bullet_{dR}(X;\mathbb{R})$ (for instance, suppose that X is a completely-solvable solvmanifold). For every $k \in \mathbb{N}$, the almost-complex structure J is linear-C^∞-pure (respectively, linear-C^∞-full, linear-C^∞-pure-and-full, linear-complex-C^∞-pure, linear-complex-C^∞-full, linear-complex-C^∞-pure-and-full) at the kth stage if and only if it is C^∞-pure (respectively, C^∞-full, C^∞-pure-and-full, complex-C^∞-pure, complex-C^∞-full, complex-C^∞-pure-and-full) at the kth stage.*

As an example, we provide here an explicit C^∞-pure-and-full almost-complex structure on a six-dimensional solvmanifold.

Example 4.1 ([AT11, Example 2.1]). A C^∞-pure-and-full and pure-and-full almost-complex structure on a compact six-dimensional completely-solvable solvmanifold.
Let G be the six-dimensional simply-connected completely-solvable Lie group defined by

$$G := \left\{ \begin{pmatrix} e^{x^1} & 0 & x^2\,e^{x^1} & 0 & 0 & x^3 \\ 0 & e^{-x^1} & 0 & x^2\,e^{-x^1} & 0 & x^4 \\ 0 & 0 & e^{x^1} & 0 & 0 & x^5 \\ 0 & 0 & 0 & e^{-x^1} & 0 & x^6 \\ 0 & 0 & 0 & 0 & 1 & x^1 \\ 0 & 0 & 0 & 0 & 0 & 1 \end{pmatrix} \in \mathrm{GL}(6;\mathbb{R}) : x^1, \ldots, x^6 \in \mathbb{R} \right\}.$$

According to [FdLS96, Sect. 3], there exists a discrete co-compact subgroup $\Gamma \subset G$: therefore $X := \Gamma \backslash G$ is a six-dimensional completely-solvable solvmanifold.

The G-left-invariant 1-forms on G defined as

$$e^1 := dx^1, \qquad\qquad\qquad e^2 := dx^2,$$
$$e^3 := \exp(-x^1) \cdot (dx^3 - x^2 dx^5), \qquad e^4 := \exp(x^1) \cdot (dx^4 - x^2 dx^6),$$
$$e^5 := \exp(-x^1) \cdot dx^5; \qquad\qquad e^6 := \exp(x^1) \cdot dx^6$$

give rise to G-left-invariant 1-forms on X. With respect to the co-frame $\{e^1, \dots, e^6\}$, the structure equations are given by

$$\begin{cases} de^1 = 0 \\ de^2 = 0 \\ de^3 = -e^1 \wedge e^3 - e^2 \wedge e^5 \\ de^4 = e^1 \wedge e^4 - e^2 \wedge e^6 \\ de^5 = -e^1 \wedge e^5 \\ de^6 = e^1 \wedge e^6 \end{cases}.$$

Since G is completely-solvable, by A. Hattori's theorem [Hat60, Corollary 4.2], it is straightforward to compute

$$H^2(X; \mathbb{R}) = \mathbb{R}\langle e^1 \wedge e^2, e^5 \wedge e^6, e^3 \wedge e^6 + e^4 \wedge e^5 \rangle.$$

Therefore, setting

$$\begin{cases} \varphi^1 := e^1 + i e^2 \\ \varphi^2 := e^3 + i e^4 \\ \varphi^3 := e^5 + i e^6 \end{cases},$$

we have that the almost-complex structure J whose $\mathcal{C}^\infty (X; \mathbb{C})$-module of complex $(1,0)$-forms is generated by $\{\varphi^1, \varphi^2, \varphi^3\}$ is $\mathcal{C}^\infty$-full: indeed,

$$H_J^{(1,1)} (X; \mathbb{R}) = \mathbb{R}\langle -\tfrac{1}{2i} \varphi^1 \wedge \bar{\varphi}^1, -\tfrac{1}{2i} \varphi^3 \wedge \bar{\varphi}^3 \rangle,$$
$$H_J^{(2,0),(0,2)} (X; \mathbb{R}) = \mathbb{R}\langle \tfrac{1}{2i} (\varphi^2 \wedge \varphi^3 - \bar{\varphi}^2 \wedge \bar{\varphi}^3) \rangle.$$

Since

$$d_{\wedge^1 \mathfrak{g}_{\mathbb{C}}^*} = \mathbb{C}\langle \varphi^{13} - \varphi^{1\bar{3}}, \varphi^{3\bar{1}} + \varphi^{1\bar{3}}, \varphi^{13} + \varphi^{1\bar{3}}, \varphi^{3\bar{1}} - \varphi^{1\bar{3}}, \varphi^{12} - \varphi^{2\bar{1}}, \varphi^{1\bar{2}} + \varphi^{\bar{1}\bar{2}} \rangle,$$

then J is linear-$\mathcal{C}^\infty$-pure-and-full. Since X is a completely-solvable solvmanifold, one gets that J is also $\mathcal{C}^\infty$-pure by Corollary 4.2. (Note that the $\mathcal{C}^\infty$-pureness of J can be proven also by using a different argument: according to [FT10, Theorem 3.7], since the above basis of harmonic representatives with respect to the G-left-invariant Hermitian metric $\sum_{j=1}^3 \varphi^j \odot \bar{\varphi}^j$ consists of pure type forms with respect to the almost-complex structure, J is both $\mathcal{C}^\infty$-pure-and-full and pure-and-full.)

Remark 4.8. Further results concerning linear (possibly non-integrable) complex structures on four-dimensional unimodular Lie algebra and their cohomological properties have been obtained by T.-J. Li and A. Tomassini in [LT12]. (See [DL13] for the case of non-unimodular Lie algebras.) In particular, they proved an analogous of [DLZ10, Theorem 2.3], namely, that for every four-dimensional unimodular Lie algebra $\mathfrak{g}$ endowed with a linear (possibly non-integrable) complex structure J, one has the cohomological decomposition $H_{dR}^2(\mathfrak{g}; \mathbb{R}) = H_J^{(2,0),(0,2)}(\mathfrak{g}; \mathbb{R}) \oplus H_J^{(1,1)}(\mathfrak{g}; \mathbb{R})$, [LT12, Theorem 3.3]. Furthermore, they studied the linear counterpart of S. K. Donaldson's question [Don06, Question 2] (see Sect. 4.4.2.1), proving that, on a four-dimensional Lie algebra $\mathfrak{g}$ satisfying the condition $B \wedge B = 0$, where $B \subseteq \wedge^2 \mathfrak{g}$ denotes the space of boundary 2-vectors, a linear (possibly non-integrable) complex structure admits a taming linear symplectic form if and only if it admits a compatible linear symplectic form, [LT12, Theorem 2.5]; note that four-dimensional unimodular Lie algebras satisfy the assumption $B \wedge B = 0$. Finally, given a linear (possibly non-integrable) complex structure on a four-dimensional Lie algebra, they studied the convex cones composed of the classes of J-taming, respectively J-compatible, linear symplectic forms, comparing them by means of $H_J^{(2,0),(0,2)}(\mathfrak{g}; \mathbb{R})$, [LT12, Theorem 3.10]: this result is the linear counterpart of [LZ09, Theorem 1.1].

4.2.3 Complex-$\mathcal{C}^\infty$-Pure-and-Fullness for Four-Dimensional Manifolds

By [DLZ10, Lemma 2.15, Theorem 2.16], or [LZ09, Proposition 2.1], every compact complex surface is complex-$\mathcal{C}^\infty$-pure-and-full at the 2nd stage; on the other hand, a compact complex surface is complex-$\mathcal{C}^\infty$-pure-and-full at the 1st stage if and only if its first Betti number b_1 is even, that is, if and only if it admits a Kähler structure, see [Kod64, Miy74, Siu83], or [Lam99, Corollaire 5.7], or [Buc99, Theorem 11].

One may wonder about the relations between being complex-$\mathcal{C}^\infty$-pure-and-full and being integrable for an almost-complex structure on a compact four-dimensional manifold; this is the matter of the following result.

Proposition 4.3 ([AT12a, Proposition 1.7]). *There exist*

- *non-complex-$\mathcal{C}^\infty$-pure-and-full at the 1st stage non-integrable almost-complex structures, and*

- *complex-C^∞-pure-and-full at the 1st stage non-integrable almost-complex structures*

on compact four-dimensional manifolds with b_1 even.

Proof. The proof follows from the following examples, [AT12a, Examples 1.5 and 1.6].

Step 1—*There exists a non-complex-C^∞-pure-and-full at the 1st stage non-integrable almost-complex structure on a four-dimensional manifold.* Consider the standard Kähler structure (J_0, ω_0) on the four-dimensional torus $\mathbb{T}^4$ with coordinates $\{x^j\}_{j \in \{1,\dots,4\}}$, that is,

$$J_0 := \begin{pmatrix} -1 & & & \\ & -1 & & \\ 1 & & & \\ & 1 & & \end{pmatrix} \in \mathrm{End}\,(\mathbb{T}^4) \qquad \text{and}$$

$$\omega_0 := \mathrm{d}x^1 \wedge \mathrm{d}x^3 + \mathrm{d}x^2 \wedge \mathrm{d}x^4 \in \wedge^2 \mathbb{T}^4 \,,$$

and, for $\varepsilon > 0$ small enough, let $\{J_t\}_{t \in (-\varepsilon, \varepsilon)}$ be the curve of almost-complex structures defined by

$$J_t := J_{t,\ell} := (\mathrm{id} - t\,L)\,J_0\,(\mathrm{id} - t\,L)^{-1} = \begin{pmatrix} -\frac{1-t\,\ell}{1+t\,\ell} & & & \\ & -1 & & \\ \frac{1+t\,\ell}{1-t\,\ell} & & & \\ & 1 & & \end{pmatrix} \in \mathrm{End}\,(\mathbb{T}^4)\,,$$

where

$$L = \begin{pmatrix} \ell & & & \\ & 0 & & \\ & & -\ell & \\ & & & 0 \end{pmatrix} \in \mathrm{End}\,(\mathbb{T}^4)$$

and $\ell = \ell(x_2) \in C^\infty(\mathbb{R}^4; \mathbb{R})$ is a $\mathbb{Z}^4$-periodic non-constant function.

For $t \in (-\varepsilon, \varepsilon) \setminus \{0\}$, a straightforward computation yields

$$H^{(1,0)}_{J_t}\left(\mathbb{T}^2_\mathbb{C}; \mathbb{C}\right) = \mathbb{C}\langle \mathrm{d}x^2 + \mathrm{i}\,\mathrm{d}x^4\rangle \,, \qquad H^{(0,1)}_{J_t}\left(\mathbb{T}^2_\mathbb{C}; \mathbb{C}\right) = \mathbb{C}\langle \mathrm{d}x^2 - \mathrm{i}\,\mathrm{d}x^4\rangle$$

therefore

$$\dim_\mathbb{C} H^{(1,0)}_{J_t}\left(\mathbb{T}^2_\mathbb{C}; \mathbb{C}\right) + \dim_\mathbb{C} H^{(0,1)}_{J_t}\left(\mathbb{T}^2_\mathbb{C}; \mathbb{C}\right) = 2 < 4 = b_1\left(\mathbb{T}^2_\mathbb{C}\right) \,,$$

that is, J_t is not complex-C^∞-pure-and-full at the 1st stage.

Step 2—*There exists a complex-C^∞-pure-and-full at the 1st stage non-integrable almost-complex structure on a four-dimensional manifold.* Consider a compact four-dimensional nilmanifold $X = \Gamma\backslash G$, quotient of the simply-connected nilpotent Lie group G whose associated Lie algebra is

$$\mathfrak{g} := \left(0^2,\ 14,\ 12\right) ;$$

let J be the G-left-invariant almost-complex structure defined by

$$Je^1 := -e^2 , \qquad Je^3 := -e^4 ;$$

note that J is not integrable, since $\mathrm{Nij}(e_1, e_3) \neq 0$, where $\{e_i\}_{i\in\{1,2,3,4\}}$ is the dual basis of $\{e^i\}_{i\in\{1,2,3,4\}}$. In fact, X has no integrable almost-complex structure: indeed, since $b_1(X) = 2$ is even, if there were a complex structure on X, then X should carry a Kähler metric; this is not possible for compact non-tori nilmanifolds, by [Has89, Theorem 1, Corollary], or [BG88, Theorem A].

By K. Nomizu's theorem [Nom54, Theorem 1], one computes

$$H^1_{dR}(X;\mathbb{C}) = \mathbb{C}\langle\varphi^1,\ \bar\varphi^1\rangle \qquad \text{and} \qquad H^2_{dR}(X;\mathbb{C}) = \mathbb{C}\langle\varphi^{12} + \varphi^{\bar1\bar2},\ \varphi^{1\bar2} - \varphi^{2\bar1}\rangle ;$$

in particular, it follows that J is complex-C^∞-pure-and-full at the 1st stage. Note that J is not complex-C^∞-pure-and-full at the 2nd stage but just C^∞-pure-and-full: indeed, using Proposition 4.2, one can prove that the class $\left[\varphi^{12} + \varphi^{\bar1\bar2}\right]$ admits no pure type representative with respect to J. Moreover, observe that the G-left-invariant almost-complex structure

$$J'e^1 := -e^3 , \qquad J'e^2 := -e^4 ,$$

is complex-C^∞-pure-and-full at the 2nd stage and non-complex-C^∞-pure-and-full at the 1st stage (obviously, in this case, $h^-_{J'} = 0$, according to [DLZ10, Corollary 2.14]).   □

Remark 4.9. T. Drăghici, T.-J. Li, and W. Zhang proved in [DLZ10, Corollary 2.14] that an almost-complex structure on a compact four-dimensional manifold X is complex-C^∞-pure-and-full at the 2nd stage if and only if J is integrable or $\dim_\mathbb{R} H^-_J(X) = 0$.

4.2.4 Almost-Complex Manifolds with Large Anti-invariant Cohomology

Given an almost-complex structure J on a compact manifold X, it is natural to ask how large the cohomology subgroup $H^-_J(X)$ can be.

In [DLZ11, Theorem 1.1], T. Drăghici, T.-J. Li, and W. Zhang, starting with a compact complex surface X endowed with the complex structure J, proved that the dimension $h_{\tilde{J}}^- := \dim_{\mathbb{R}} H_{\tilde{J}}^-(X)$ of the $\tilde{J}$-anti-invariant subgroup $H_{\tilde{J}}^-(X)$ of $H_{dR}^2(X;\mathbb{R})$ associated to any *metric related* almost-complex structures $\tilde{J}$ on X (that is, the almost-complex structures $\tilde{J}$ on X inducing the same orientation as J and with a common compatible metric with J), such that $\tilde{J} \neq \pm J$, satisfies $h_{\tilde{J}}^- \in \{0, 1, 2\}$, and they provided a description of such almost-complex structures $\tilde{J}$ having $h_{\tilde{J}}^- \in \{1, 2\}$.

In this direction, T. Drăghici, T.-J. Li, and W. Zhang proposed the following conjecture.

Conjecture 4.1 ([DLZ11, Conjecture 2.5]). On a compact four-dimensional manifold endowed with an almost-complex structure J, if $\dim_{\mathbb{R}} H_J^-(X) \geq 3$, then J is integrable.

In [DLZ11], it was conjectured that $h_J^- = 0$ for a generic almost-complex structure J on a compact four-dimensional manifold, [DLZ11, Conjecture 2.4].

In [ATZ12, Sect. 5], a 1-parameter family $\{J_t\}_{t\in(-\varepsilon,\varepsilon)}$ of (non-integrable) almost-complex structures on the six-dimensional torus $\mathbb{T}^6$, where $\varepsilon > 0$ is small enough, having $\dim_{\mathbb{R}} H_{J_t}^-(\mathbb{T}^6)$ greater than 3 has been provided, see also [AT11, Sect. 4].

Example 4.2 ([ATZ12, Sect. 5]). A family of almost-complex structures on the six-dimensional torus with anti-invariant cohomology of dimension larger than 3. Consider the six-dimensional torus $\mathbb{T}^6$, with coordinates $\{x^j\}_{j\in\{1,\ldots,6\}}$. For $\varepsilon > 0$ small enough, choose a function $\alpha : (-\varepsilon, \varepsilon) \times \mathbb{T}^6 \to \mathbb{R}$ such that $\alpha_t := \alpha(t, \cdot) \in \mathcal{C}^\infty(\mathbb{T}^6)$ depends just on x^3 for any $t \in (-\varepsilon, \varepsilon)$, namely $\alpha_t = \alpha_t(x^3)$, and that $\alpha_0(x^3) = 1$. Define the almost-complex structure J_t in such a way that

$$
\begin{cases}
\varphi_t^1 := dx^1 + i\,\alpha_t\,dx^4 \\
\varphi_t^2 := dx^2 + i\,dx^5 \\
\varphi_t^3 := dx^3 + i\,dx^6
\end{cases}
$$

provides a co-frame for the $\mathcal{C}^\infty(\mathbb{T}^6;\mathbb{C})$-module of $(1,0)$-forms on $\mathbb{T}^6$ with respect to J_t. In terms of this co-frame, the structure equations are

$$
\begin{cases}
d\varphi_t^1 = i\,d\alpha_t \wedge dx^4 \\
d\varphi_t^2 = 0 \\
d\varphi_t^3 = 0
\end{cases}.
$$

Straightforward computations give that the J_t-anti-invariant real closed 2-forms are of the type

$$\psi = \frac{C}{\alpha_t}\left(\mathrm{d}x^{13} - \alpha_t\,\mathrm{d}x^{46}\right) + D\left(\mathrm{d}x^{16} - \alpha_t\,\mathrm{d}x^{34}\right) + E\left(\mathrm{d}x^{23} - \mathrm{d}x^{56}\right)$$

$$+F\left(\mathrm{d}x^{26} - \mathrm{d}x^{35}\right),$$

where C, D, E, $F \in \mathbb{R}$ (we shorten $\mathrm{d}x^{jk} := \mathrm{d}x^j \wedge \mathrm{d}x^k$). Moreover, the forms $\mathrm{d}x^{23} - \mathrm{d}x^{56}$ and $\mathrm{d}x^{26} - \mathrm{d}x^{35}$ are clearly harmonic with respect to the standard Riemannian metric $\sum_{j=1}^6 \mathrm{d}x^j \otimes \mathrm{d}x^j$, while the classes of $\mathrm{d}x^{16} - \alpha_t\,\mathrm{d}x^{34}$ and $\mathrm{d}x^{13} - \alpha_t\,\mathrm{d}x^{46}$ are non-zero, being their harmonic parts non-zero. Therefore, we get that

$$h^-_{J_t} = 4 \qquad \text{for small } t \neq 0,$$

while $h^-_{J_0} = 6$.

The natural generalization of [DLZ10, Conjecture 2.5] to higher dimensional manifolds yields the following question.

Question 4.1 ([ATZ12, Question 5.2]). Are there compact $2n$-dimensional manifolds X endowed with non-integrable almost-complex structures J with $\dim_{\mathbb{R}} H^-_J(X) > n\,(n-1)$?

Remark 4.10. Note that, when $X = \Gamma \backslash G$ is a $2n$-dimensional completely-solvable solvmanifold endowed with a G-left-invariant almost-complex structure J, then, by Proposition 4.2, it follows that

$$\dim_{\mathbb{R}} H^-_J(X) \leq n\,(n-1) \qquad \text{and} \qquad \dim_{\mathbb{R}} H^+_J(X) \leq n^2.$$

4.2.5   Semi-Kähler Manifolds

As already recalled, A.M. Fino and A. Tomassini's [FT10, Theorem 4.1] proves that, given an almost-Kähler structure on a compact manifold, if the almost-complex structure is $\mathcal{C}^\infty$-pure-and-full and the symplectic structure satisfies the Hard Lefschetz Condition, then the almost-complex structure is pure-and-full too; moreover, by [FT10, Proposition 3.2], see also [DLZ10, Proposition 2.8], the almost-complex structure of every almost-Kähler structure on a compact manifold is $\mathcal{C}^\infty$-pure.

To study the cohomology of balanced manifolds X and the duality between $H^{(\bullet,\bullet)}_J(X;\mathbb{C})$ and $H^J_{(\bullet,\bullet)}(X;\mathbb{C})$, we get the following result, which can be considered as the semi-Kähler counterpart of [FT10, Theorem 4.1].

Proposition 4.4 ([AT12a, Proposition 3.1]). *Let X be a compact $2n$-dimensional manifold endowed with an almost-complex structure J and a semi-Kähler form ω. Suppose that $\left[\omega^{n-1}\right] \smile \cdot\colon H^1_{dR}(X;\mathbb{R}) \to H^{2n-1}_{dR}(X;\mathbb{R})$ is an isomorphism. If J is* *complex-C^∞-pure-and-full at the 1st stage, then it is also complex-pure-and-full at the 1st stage, and*

$$H_J^{(1,0)}(X;\mathbb{C}) \simeq H_{(0,1)}^J(X;\mathbb{C}) \,.$$

Proof. Firstly, note that J is complex-pure at the 1st stage. Indeed, if

$$\mathfrak{a} \in H_{(1,0)}^J(X;\mathbb{C}) \cap H_{(0,1)}^J(X;\mathbb{C}) \,,$$

then

$$\mathfrak{a}\big\lfloor_{H_J^{(1,0)}(X;\mathbb{C})} = 0 = \mathfrak{a}\big\lfloor_{H_J^{(0,1)}(X;\mathbb{C})} \,.$$

Therefore, by the assumption

$$H_{dR}^1(X;\mathbb{C}) = H_J^{(1,0)}(X;\mathbb{C}) \oplus H_J^{(0,1)}(X;\mathbb{C}) \,,$$

we get that

$$\mathfrak{a} = 0 \,.$$

Now, note that, since

$$\left[\omega^{n-1}\right] \smile H_J^{(1,0)}(X;\mathbb{C}) \subseteq H_J^{(n,n-1)}(X;\mathbb{C}) \qquad \text{and}$$

$$\left[\omega^{n-1}\right] \smile H_J^{(0,1)}(X;\mathbb{C}) \subseteq H_J^{(n-1,n)}(X;\mathbb{C}) \,,$$

the isomorphism

$$H_{dR}^1(X;\mathbb{C}) \xrightarrow{\left[\omega^{n-1}\right]\smile\cdot} H_{dR}^{2n-1}(X;\mathbb{C}) \xrightarrow{T.} H_{dR}^1(X;\mathbb{C})$$

yields the injective maps

$$H_J^{(1,0)}(X;\mathbb{C}) \hookrightarrow H_{(0,1)}^J(X;\mathbb{C}) \qquad \text{and} \qquad H_J^{(0,1)}(X;\mathbb{C}) \hookrightarrow H_{(1,0)}^J(X;\mathbb{C}) \,.$$

Since, by hypothesis, J is complex-C^∞-pure-and-full at the 1st stage, namely, $H_{dR}^1(X;\mathbb{C}) = H_J^{(1,0)}(X;\mathbb{C}) \oplus H_J^{(0,1)}(X;\mathbb{C})$, we get the proof.   □

We provide here some explicit examples, [AT12a, Examples 3.2 and 3.3], checking the validity of the hypothesis of $\left[\omega^{n-1}\right] \smile \cdot : H_{dR}^1(X;\mathbb{R}) \to H_{dR}^{2n-1}(X;\mathbb{R})$ being an isomorphism in Proposition 4.4.

Example 4.3 ([AT12a, Example 3.2]). A balanced structure on the Iwasawa manifold.

On the Iwasawa manifold $\mathbb{I}_3$ (see Sect. 3.2.1), consider the balanced structure

$$\omega := \frac{i}{2}\left(\varphi^1 \wedge \bar{\varphi}^1 + \varphi^2 \wedge \bar{\varphi}^2 + \varphi^3 \wedge \bar{\varphi}^3\right) .$$

Since

$$H^1_{dR}(\mathbb{I}_3;\mathbb{C}) = \mathbb{C}\left\langle \varphi^1,\ \varphi^2,\ \bar{\varphi}^1,\ \bar{\varphi}^2\right\rangle \qquad \text{and}$$

$$H^5_{dR}(\mathbb{I}_3;\mathbb{C}) = \mathbb{C}\left\langle \varphi^{123\bar{1}3},\ \varphi^{123\bar{2}3},\ \varphi^{13\bar{1}23},\ \varphi^{23\bar{1}23}\right\rangle ,$$

it is straightforward to check that

$$\left[\omega^2\right] \smile \cdot : H^1_{dR}(\mathbb{I}_3;\mathbb{C}) \to H^5_{dR}(\mathbb{I}_3;\mathbb{C})$$

is an isomorphism. Therefore, by Proposition 4.4, $\mathbb{I}_3$ is complex-C^∞-pure-and-full at the 1st stage and complex-pure-and-full at the 1st stage (the same result follows also arguing as in [FT10, Theorem 3.7], the above harmonic representatives of $H^1_{dR}(\mathbb{I}_3;\mathbb{C})$, with respect to the Hermitian metric $\sum_{j=1}^3 \varphi^j \odot \bar{\varphi}^j$, being of pure type with respect to the complex structure).

Example 4.4 ([AT12a, Example 3.3]). A six-dimensional manifold endowed with a semi-Kähler structure not inducing an isomorphism in cohomology. Consider the six-dimensional nilmanifold

$$X = \Gamma\backslash G := \left(0^4,\ 12,\ 13\right) .$$

In [FT10, Example 3.3], the almost-complex structure

$$J'e^1 := -e^2 , \qquad J'e^3 := -e^4 , \qquad J'e^5 := -e^6$$

is provided as a first example of non-C^∞-pure almost-complex structure. Note that J' is not even C^∞-full: indeed, the cohomology class $\left[e^{15} + e^{16}\right]$ admits neither J'-invariant nor J'-anti-invariant G-left-invariant representatives, and hence, by Proposition 4.2, it admits neither J'-invariant nor J'-anti-invariant representatives.

Consider now the almost-complex structure

$$Je^1 := -e^5 , \qquad Je^2 := -e^3 , \qquad Je^4 := -e^6$$

and the non-degenerate J-invariant 2-form

$$\omega := e^{15} + e^{23} + e^{46} .$$

A straightforward computation shows that

$$d\omega = -e^{134} \neq 0 \quad \text{and} \quad d\omega^2 = d\left(e^{1235} - e^{1456} + e^{2346}\right) = 0 .$$

By K. Nomizu's theorem [Nom54, Theorem 1], it is straightforward to compute

$$H^1_{dR}(X;\mathbb{R}) \;=\; \mathbb{R}\langle e^1,\, e^2,\, e^3,\, e^4 \rangle\,.$$

Since

$$\omega^2\, e^1 \;=\; e^{12346} \;=\; \mathrm{d}\, e^{3456}\,,$$

we get that $\left[\omega^2\right] \smile \cdot : H^1_{dR}(X;\mathbb{R}) \to H^5_{dR}(X;\mathbb{R})$ is not injective.

We give two explicit examples of $2n$-dimensional complex manifolds endowed with a balanced structure, with $2n = 10$, respectively $2n = 6$, such that the $(n-1)$th power of the associated $(1,1)$-form induces an isomorphism in cohomology, and admitting small balanced deformations, [AT12a, Examples 3.4 and 3.5].

Example 4.5 ([AT12a, Example 3.4]). A curve of balanced structures on $\eta\beta_5$ inducing an isomorphism in cohomology.

We recall the construction of the ten-dimensional nilmanifold $\eta\beta_5$, introduced and studied in [AB90] by L. Alessandrini and G. Bassanelli to prove that being p-Kähler is not a stable property under small deformations of the complex structure; more in general, in [AB91], the manifold $\eta\beta_{2n+1}$, for any $n \in \mathbb{N}\setminus\{0\}$, has been provided as a generalization of the Iwasawa manifold $\mathbb{I}_3$, and the existence of p-Kähler structures on $\eta\beta_{2n+1}$ has been investigated. (For definitions and results concerning p-Kähler structures, see [AB91], or, e.g., [Sil96, Ale11].)

For $n \in \mathbb{N}\setminus\{0\}$, consider the complex Lie group

$$G_{2n+1} \;:=\; \left\{ A \in \mathrm{GL}(n+2;\mathbb{C}) \;:\; A = \left(\begin{array}{c|ccc|c} 1 & x^1 & \cdots & x^n & z \\ \hline 0 & 1 & & & y^1 \\ \vdots & & \ddots & & \vdots \\ 0 & & & 1 & y^n \\ \hline 0 & 0 & \cdots & 0 & 1 \end{array}\right) \right\}\;;$$

equivalently, one can identify G_{2n+1} with $(\mathbb{C}^{2n+1}, *)$, where the group structure $*$ is defined as

$$\left(x^1, \ldots, x^n, y^1, \ldots, y^n, z\right) * \left(u^1, \ldots, u^n, v^1, \ldots, v^n, w\right)$$

$$:= \left(x^1 + u^1, \ldots, x^n + u^n, y^1 + v^1, \ldots, y^n + v^n, z + w + x^1 \cdot v^1 + \cdots + x^n \cdot v^n\right).$$

Since the subgroup

$$\Gamma_{2n+1} \;:=\; G_{2n+1} \cap \mathrm{GL}\left(n+2;\mathbb{Z}[\mathrm{i}]\right) \;\subset\; G_{2n+1}$$

is a discrete co-compact subgroup of the nilpotent Lie group G_{2n+1}, one gets a compact complex manifold, of complex dimension $2n + 1$,

$$\eta\beta_{2n+1} := \Gamma_{2n+1} \backslash G_{2n+1} ,$$

which is a holomorphically parallelizable nilmanifold and admits no Kähler metric, [Wan54, Corollary 2], or [BG88, Theorem A], or [Has89, Theorem 1, Corollary]; note that $\eta\beta_3 = \mathbb{I}_3$ is the Iwasawa manifold (see Sect. 3.2.1). In fact, one has that $\eta\beta_{2n+1}$ is not p-Kähler for $1 < p \le n$ and it is p-Kähler for $n + 1 \le p \le 2n + 1$, [AB91, Theorem 4.2]; furthermore, $\eta\beta_{2n+1}$ has complex submanifolds of any complex dimension less than or equal to $2n + 1$, and hence it follows that the p-Kähler forms on $\eta\beta_{2n+1}$ can never be exact, [AB91, Sect. 4.4].
Setting

$$\begin{cases} \varphi^{2j-1} := d x^j , & \text{for } j \in \{1, \ldots, n\} , \\ \varphi^{2j} := d y^j , & \text{for } j \in \{1, \ldots, n\} , \\ \varphi^{2n+1} := d z - \sum_{j=1}^n x^j \, d y^j , \end{cases}$$

one gets the global co-frame $\{\varphi^j\}_{j \in \{1,\ldots,2n+1\}}$ for the space of holomorphic 1-forms, with respect to which the structure equations are

$$\begin{cases} d\varphi^1 = \cdots = d\varphi^{2n} = 0 \\ d\varphi^{2n+1} = -\sum_{j=1}^n \varphi^{2j-1} \wedge \varphi^{2j} \end{cases} .$$

Now, take $2n + 1 = 5$. With respect to the co-frame $\{\varphi^j\}_{j \in \{1,\ldots,5\}}$ for the space of holomorphic 1-forms on $\eta\beta_5$, the structure equations are

$$\begin{cases} d\varphi^1 = d\varphi^2 = d\varphi^3 = d\varphi^4 = 0 \\ d\varphi^5 = -\varphi^{12} - \varphi^{34} \end{cases}$$

(where, as usually, we shorten, e.g., $\varphi^{12} := \varphi^1 \wedge \varphi^2$).
Consider on $\eta\beta_5$ the balanced structure

$$\omega := \frac{i}{2} \sum_{j=1}^5 \varphi^j \wedge \bar\varphi^j .$$

By K. Nomizu's theorem [Nom54, Theorem 1], it is straightforward to compute

$$H^1_{dR}(\eta\beta_5; \mathbb{C}) = \mathbb{C}\langle \varphi^1, \varphi^2, \varphi^3, \varphi^4, \bar\varphi^1, \bar\varphi^2, \bar\varphi^3, \bar\varphi^4 \rangle$$

and

$$H^9_{dR}(\eta\beta_5; \mathbb{C}) = \mathbb{C}\left\langle \varphi^{12345\overline{2}\overline{3}\overline{4}\overline{5}}, \varphi^{12345\overline{1}\overline{3}\overline{4}\overline{5}}, \varphi^{12345\overline{1}\overline{2}\overline{4}\overline{5}}, \varphi^{12345\overline{1}\overline{2}\overline{3}\overline{5}}, \right.$$

$$\left. \varphi^{2345\overline{1}\overline{2}\overline{3}\overline{4}\overline{5}}, \varphi^{1345\overline{1}\overline{2}\overline{3}\overline{4}\overline{5}}, \varphi^{1245\overline{1}\overline{2}\overline{3}\overline{4}\overline{5}}, \varphi^{1235\overline{1}\overline{2}\overline{3}\overline{4}\overline{5}} \right\rangle ;$$

therefore, $\eta\beta_5$ is complex-$\mathcal{C}^\infty$-pure-and-full at the 1st stage and

$$\left[\omega^4\right] \smile \cdot : H^1_{dR}(\eta\beta_5; \mathbb{R}) \to H^9_{dR}(\eta\beta_5; \mathbb{R})$$

is an isomorphism, and so $\eta\beta_5$ is also complex-pure-and-full at the 1st stage by Proposition 4.4 (note that, the above pure type representatives being harmonic with respect to the metric $\sum_{j=1}^5 \varphi^j \odot \bar{\varphi}^j$, the same result follows also arguing as in [FT10, Theorem 3.7]).

Now, let $\{J_t\}_{t\in\Delta(0,\varepsilon)\subset\mathbb{C}}$, where $\varepsilon > 0$ is small enough, be a family of small deformations of the complex structure such that

$$\begin{cases} \varphi^1_t := \varphi^1 + t\,\bar{\varphi}^1 \\ \varphi^2_t := \varphi^2 \\ \varphi^3_t := \varphi^3 \\ \varphi^4_t := \varphi^4 \\ \varphi^5_t := \varphi^5 \end{cases}$$

is a co-frame for the J_t-holomorphic cotangent bundle. With respect to $\left\{\varphi^j_t\right\}_{j\in\{1,\dots,5\}}$, the structure equations are written as

$$\begin{cases} d\varphi^1_t = d\varphi^2_t = d\varphi^3_t = d\varphi^4_t = 0 \\ d\varphi^5_t = -\frac{1}{1-|t|^2}\varphi^{12}_t - \varphi^{34}_t - \frac{t}{1-|t|^2}\varphi^{2\bar{1}}_t \end{cases} .$$

Setting, for $t \in \Delta(0,\varepsilon) \subset \mathbb{C}$,

$$\omega_t := \frac{\mathrm{i}}{2}\sum_{j=1}^5 \varphi^j_t \wedge \overline{\varphi}^j_t ,$$

one gets a curve of balanced structures $\{(J_t, \omega_t)\}_{t\in\Delta(0,\varepsilon)}$ on the smooth manifold underlying $\eta\beta_5$. Furthermore, for every $t \in \Delta(0,\varepsilon)$, the complex structure J_t is complex-$\mathcal{C}^\infty$-pure-and-full at the 1st stage and

$$\left[\omega^4_t\right] \smile \cdot : H^1_{dR}(\eta\beta_5; \mathbb{R}) \to H^9_{dR}(\eta\beta_5; \mathbb{R})$$

is an isomorphism. Therefore, according to Proposition 4.4, it follows that, for every $t \in \Delta(0, \varepsilon)$, the complex structure J_t is complex-pure-and-full at the 1st stage, and that $H_{J_t}^{(1,0)}(\eta\beta_5; \mathbb{C}) \simeq H_{(0,1)}^{J_t}(\eta\beta_5; \mathbb{C})$.

Example 4.6 ([AT12a, Example 3.5]). A curve of semi-Kähler structures on a six-dimensional completely-solvable solvmanifold inducing an isomorphism in cohomology.

Consider a completely-solvable solvmanifold

$$X = \Gamma\backslash G := (0, -12, 34, 0, 15, 46)$$

endowed with the almost-complex structure J_0 whose holomorphic cotangent bundle has co-frame generated by

$$\begin{cases} \varphi^1 := e^1 + i\,e^4 \\ \varphi^2 := e^2 + i\,e^5 \\ \varphi^3 := e^3 + i\,e^6 \end{cases}$$

and with the J_0-compatible symplectic form

$$\omega_0 := e^{14} + e^{25} + e^{36}$$

(see also [FT10, Sect. 6.3]). The structure equations with respect to $\{\varphi^1, \varphi^2, \varphi^3\}$ are

$$\begin{cases} d\varphi^1 = 0 \\ 2\,d\varphi^2 = -\varphi^{1\bar{2}} - \varphi^{\bar{1}\bar{2}} \\ 2i\,d\varphi^3 = -\varphi^{1\bar{3}} + \varphi^{\bar{1}3} \end{cases} ;$$

using A. Hattori's theorem [Hat60, Corollary 4.2], one computes

$$H_{dR}^1(X; \mathbb{R}) = \mathbb{R}\langle e^1, e^4 \rangle,$$

$$H_{dR}^5(X; \mathbb{R}) = \mathbb{R}\langle *_{g_0} e^1, *_{g_0} e^4 \rangle = \mathbb{R}\langle e^{23456}, e^{12356} \rangle,$$

where g_0 is the J_0-Hermitian metric induced by (J_0, ω_0).

Now, consider the curve $\{J_t\}_{t \in (-\varepsilon, \varepsilon) \subset \mathbb{R}}$ of almost-complex structures on X, where $\varepsilon > 0$ is small enough and J_t is defined requiring that the J_t-holomorphic cotangent bundle is generated by

$$\begin{cases} \varphi_t^1 := \varphi^1 \\ \varphi_t^2 := \varphi^2 + i\,t\,e^6 \\ \varphi_t^3 := \varphi^3 \end{cases} ;$$

for every $t \in (-\varepsilon, \varepsilon)$, consider also the non-degenerate J_t-compatible 2-form

$$\omega_t := e^{14} + e^{25} + e^{36} + t\, e^{26} \; ;$$

for $t \neq 0$, one has that $d\omega \neq 0$, but

$$d\omega_t^2 = d\left(\omega_0^2 - t\, e^{1246}\right) = 0 \,,$$

hence $\{(J_t, \omega_t)\}_{t \in (-\varepsilon, \varepsilon)}$ gives rise to a curve of semi-Kähler structures on X. Moreover, note that

$$\omega_t^2 \wedge e^1 = e^{12356} \,, \qquad \omega_t^2 \wedge e^4 = e^{23456} \,,$$

therefore $\left[\omega_t^2\right] \smile \cdot \colon H_{dR}^1(X; \mathbb{R}) \to H_{dR}^5(X; \mathbb{R})$ is an isomorphism, for every $t \in (-\varepsilon, \varepsilon)$.

4.2.6 Almost-Kähler Manifolds and Lefschetz-Type Property

Recall that every compact manifold X endowed with a Kähler structure (J, ω) is $\mathcal{C}^\infty$-pure-and-full, in fact, complex-$\mathcal{C}^\infty$-pure-and-full at every stage, [DLZ10, Lemma 2.15, Theorem 2.16], or [LZ09, Proposition 2.1]. A natural question is whether or not the same holds true even for almost-Kähler structures, namely, without the integrability assumption on J.

In this section, we study cohomological properties for almost-Kähler structures, in connection with a Lefschetz-type property, Theorem 4.4, and we describe some explicit examples, [ATZ12].

Let X be a compact $2n$-dimensional manifold endowed with an *almost-Kähler structure* (J, ω, g), that is, J is an almost-complex structure on X and g is a J-Hermitian metric whose associated $(1, 1)$-form $\omega := g(J \cdot, \cdot\cdot) \in \wedge^{1,1}X \cap \wedge^2 X$ is d-closed.

Firstly, we recall the following result on decomposition in cohomology for almost-Kähler manifolds, proven by T. Drăghici, T.-J. Li, and W. Zhang in [DLZ10] and, in a different way, by A.M. Fino and A. Tomassini in [FT10].

Proposition 4.5 ([DLZ10, Proposition 2.8], [FT10, Proposition 3.2]). *Let X be a compact manifold and let (J, ω, g) be an almost-Kähler structure on X. Then J is $\mathcal{C}^\infty$-pure.*

Proof. For the sake of completeness, we recall here the proof by T. Drăghici, T.-J. Li, and W. Zhang in [DLZ10].

Suppose that $[\alpha] = [\beta] \in H_J^+(X) \cap H_J^-(X)$, where $\alpha \in \wedge^{1,1}X \cap \wedge^2 X$ and $\beta \in \left(\wedge^{2,0}X \oplus \wedge^{0,2}X\right) \cap \wedge^2 X$.

Since $\beta \in P\wedge^2 X$, by the Weyl identity, see, e.g., [Huy05, Proposition 1.2.31], one gets that $*_g\beta = \frac{1}{(n-2)!} L^{n-2}\beta$, where n is the complex dimension of X. Hence

$$0 = \int_X \alpha \wedge \beta \wedge \omega^{n-2} = \int_X \beta \wedge L^{n-2}\beta = (n-2)! \int_X \beta \wedge *_g\beta \,,$$

from which it follows that $\beta = 0$, and hence J is C^∞-pure.   □

Hence, one is brought to study the C^∞-fullness of almost-Kähler structures.

Note that ω is in particular a symplectic form on X. We recall that, given a compact $2n$-dimensional manifold X endowed with a symplectic form ω, and fixed $k \in \mathbb{N}$, the Lefschetz-type operator on $(n - k)$-forms associated with ω is the operator

$$L^k := L^k_\omega \colon \wedge^{n-k} X \to \wedge^{n+k} X \,, \qquad L^k(\alpha) := \omega^k \wedge \alpha$$

(see Sect. 1.2 for notations concerning symplectic structures); since $d\omega = 0$, the map $L^k \colon \wedge^{n-k} X \to \wedge^{n+k} X$ induces a map in cohomology, namely,

$$L^k \colon H^{n-k}_{dR}(X; \mathbb{R}) \to H^{n+k}_{dR}(X; \mathbb{R}) \,, \qquad L^k(\mathfrak{a}) := [\omega^k] \smile \mathfrak{a} \,,$$

Initially motivated by studying, in [Zha13], Taubes currents, which have been introduced by C.H. Taubes in [Tau11] in order to study S.K. Donaldson's "tamed to compatible" question, [Don06, Question 2], W. Zhang considered the following Lefschetz-type property, see also [DLZ12, Sect. 2.2].

Definition 4.8. Let X be a compact $2n$-dimensional manifold endowed with an almost-complex structure J and with a J-Hermitian metric g; denote by ω the $(1, 1)$-form associated to g. One says that the *Lefschetz-type property (on 2-forms)* holds on X if

$$L^{n-2}_\omega \colon \wedge^2 X \to \wedge^{2n-2} X$$

takes g-harmonic 2-forms to g-harmonic $(2n - 2)$-forms.

Since the map $L^k \colon \wedge^{n-k} X \to \wedge^{n+k} X$ is an isomorphism for every $k \in \mathbb{N}$, [Yan96, Corollary 2.7], it follows that the Lefschetz-type property on 2-forms is stronger than the Hard Lefschetz Condition on 2-classes, namely, the property that $[\omega]^{n-2} \smile \cdot \colon H^2_{dR}(X; \mathbb{R}) \to H^{2n-2}_{dR}(X; \mathbb{R})$ is an isomorphism.

In order to study the relation between the Lefschetz-type property on 2-forms and the C^∞-fullness, we prove here the following result, which states that the Lefschetz-type property on 2-forms is satisfied provided that the almost-Kähler structure admits a basis of pure type harmonic representatives for the second de Rham cohomology group. (Recall that A.M. Fino and A. Tomassini proved in [FT10, Theorem 3.7] that an almost-Kähler manifold admitting a basis of harmonic 2-forms of pure type with respect to the almost-complex structure is C^∞-pure-and-full and

pure-and-full; they also described several examples of non-Kähler solvmanifolds
satisfying the above assumption, [FT10, Sects. 5 and 6].)

Theorem 4.4 ([ATZ12, Theorem 2.3, Remark 2.4]). *Let X be a compact manifold endowed with an almost-Kähler structure (J, ω, g).*

- *Suppose that there exists a basis of $H_{dR}^2(X; \mathbb{R})$ represented by g-harmonic
  2-forms which are of pure type with respect to J. Then the Lefschetz-type
  property on 2-forms holds on X.*
- *Suppose that the Lefschetz-type property on 2-forms holds and that J is $\mathcal{C}^\infty$-full.
  Then J is $\mathcal{C}^\infty$-pure-and-full and pure-and-full.*

Proof. The first statement was proven in [ATZ12, Theorem 2.3]. We recall that,
on a compact $2n$-dimensional symplectic manifold, using the symplectic form ω
instead of a Riemannian metric and miming the Hodge theory for Riemannian
manifolds, one can define a symplectic-$\star$-operator $\star_\omega: \wedge^\bullet X \to \wedge^{2n-\bullet} X$ such
that $\alpha \wedge \star_\omega \beta = \left(\omega^{-1}\right)^k (\alpha, \beta) \frac{\omega^n}{n!}$ for every $\alpha, \beta \in \wedge^k X$, see [Bry88, Sect. 2].
(See Sect. 1.2 for further details on symplectic structures, and see Sect. 1.2.2.1 for
definitions and results concerning the Hodge theory for symplectic manifolds.) In
particular, on a compact manifold X endowed with an almost-Kähler structure
(J, ω, g), the Hodge-$*$-operator $*_g$ and the symplectic-$\star$-operator $\star_\omega$ are related by

$$\star_\omega = *_g J \,,$$

see [Bry88, Theorem 2.4.1, Remark 2.4.4]. Therefore, for forms of pure type with
respect to J, the properties of being g-harmonic and of being ω-symplectically-
harmonic (that is, both d-closed and d^Λ-closed, where d^Λ is the symplectic
co-differential operator, which is defined, for every $k \in \mathbb{N}$, as $d^\Lambda\lfloor_{\wedge^k X} := (-1)^{k+1}$
$\star_\omega \, d \, \star_\omega$) are equivalent. The statement follows noting that

$$[d, L] = 0 \qquad \text{and} \qquad [d^\Lambda, L] = -d \,,$$

see, e.g., [Yan96, Lemma 1.2]: hence L sends ω-symplectically-harmonic 2-forms
(of pure type with respect to J) to ω-symplectically-harmonic $(2n - 2)$-forms (of
pure type with respect to J).

As regards the second statement, we have already noticed that J is $\mathcal{C}^\infty$-pure by
[DLZ10, Proposition 2.8] or [FT10, Proposition 3.2]. Moreover, since J is $\mathcal{C}^\infty$-full,
J is also pure by [LZ09, Proposition 2.5]. We recall now the argument in [FT10,
Theorem 4.1] to prove that J is also full. Firstly, note that if the Lefschetz-type
property on 2-forms holds, then $\left[\omega^{n-2}\right] \smile \cdot : H_{dR}^2(X; \mathbb{R}) \to H_{dR}^{2n-2}(X; \mathbb{R})$ is an
isomorphism. Therefore, we get that

$$H_{dR}^{2n-2}(X; \mathbb{R}) = H_J^{(n,n-2),(n-2,n)}(X; \mathbb{R}) + H_J^{(n-1,n-1)}(X; \mathbb{R}) \,;$$

indeed (following the argument in [FT10, Theorem 4.1]) since $\left[\omega^{n-2}\right] \smile \cdot : H_{dR}^2$
$(X; \mathbb{R}) \to H_{dR}^{2n-2}(X; \mathbb{R})$ is in particular surjective, we have

$$H_{dR}^{2n-2}(X;\mathbb{R}) = \left[\omega^{n-2}\right] \smile H_{dR}^2(X;\mathbb{R})$$

$$= \left[\omega^{n-2}\right] \smile \left(H_J^{(2,0),(0,2)}(X;\mathbb{R}) \oplus H_J^{(1,1)}(X;\mathbb{R})\right)$$

$$\subseteq H_J^{(n,n-2),(n-2,n)}(X;\mathbb{R}) + H_J^{(n-1,n-1)}(X;\mathbb{R}),$$

yielding the above decomposition of $H_{dR}^{2n-2}(X;\mathbb{R})$. Then, it follows that J is also full by Theorem 4.1. $\qquad\qquad\qquad\qquad\qquad\qquad\qquad\qquad\qquad\qquad\qquad\square$

We describe here some examples, from [ATZ12], of almost-Kähler manifolds, studying Lefschetz-type property and C^∞-fullness on them.

In the following example, from [ATZ12, Sect. 2.2], we give a family of C^∞-full almost-Kähler manifolds satisfying the Lefschetz-type property on 2-forms.

Example 4.7 ([ATZ12, Sect. 2.2]). A family of C^∞-full almost-Kähler manifolds satisfying the Lefschetz-type property on 2-forms.
Consider the six-dimensional Lie algebra

$$\mathfrak{h}_7 := \left(0^3,\ 23,\ 13,\ 12\right).$$

By Mal'tsev's theorem [Mal49, Theorem 7], the connected simply-connected Lie group G associated with $\mathfrak{h}_7$ admits a discrete co-compact subgroup Γ: let $N := \Gamma \backslash G$ be the nilmanifold obtained as a quotient of G by Γ. Note that N is not formal by K. Hasegawa's theorem [Has89, Theorem 1, Corollary].
Fix $\alpha > 1$ and consider the G-left-invariant symplectic form ω_α on N defined by

$$\omega_\alpha := e^{14} + \alpha \cdot e^{25} + (\alpha - 1) \cdot e^{36}.$$

Consider the left-invariant almost-complex structure J on N defined by

$$
\begin{array}{lll}
J_\alpha e_1 := e_4, & J_\alpha e_2 := \alpha\, e_5, & J_\alpha e_3 := (\alpha - 1)\, e_6, \\[4pt]
J_\alpha e_4 := -e_1, & J_\alpha e_5 := -\tfrac{1}{\alpha}\, e_2, & J_\alpha e_6 := -\tfrac{1}{\alpha-1}\, e_3,
\end{array}
$$

where $\{e_1, \ldots, e_6\}$ denotes the global dual frame of the G-left-invariant co-frame $\{e^1, \ldots, e^6\}$ associated to the structure equations.
Finally, define the G-left-invariant symmetric tensor

$$g_\alpha(\cdot, \cdot\cdot) := \omega_\alpha\left(\cdot,\, J_\alpha \cdot\cdot\right).$$

It is straightforward to check that $\{(J_\alpha, \omega_\alpha, g_\alpha)\}_{\alpha>1}$ is a family of G-left-invariant almost-Kähler structures on N; moreover, setting

$$
\begin{array}{lll}
E_\alpha^1 := e^1, & E_\alpha^2 := \alpha\, e^2, & E_\alpha^3 := (\alpha - 1)\, e^3, \\[4pt]
E_\alpha^4 := e^4, & E_\alpha^5 := e^5, & E_\alpha^6 := e^6,
\end{array}
$$

we get the G-left-invariant g_α-orthonormal co-frame $\{E_\alpha^1, \ldots, E_\alpha^6\}$ on N. The structure equations with respect to the co-frame $\{E_\alpha^1, \ldots, E_\alpha^6\}$ read as follows:

$$\begin{cases} d\,E_\alpha^1 = 0 \\ d\,E_\alpha^2 = 0 \\ d\,E_\alpha^3 = 0 \\ d\,E_\alpha^4 = \frac{1}{\alpha(\alpha-1)}\,E_\alpha^{23} \\ d\,E_\alpha^5 = \frac{1}{\alpha-1}\,E_\alpha^{13} \\ d\,E_\alpha^6 = \frac{1}{\alpha}\,E_\alpha^{12} \end{cases}.$$

Then

$$\varphi_\alpha^1 := E_\alpha^1 + i\,E_\alpha^4, \qquad \varphi_\alpha^2 := E_\alpha^2 + i\,E_\alpha^5, \qquad \varphi_\alpha^3 := E_\alpha^3 + i\,E_\alpha^6$$

are $(1,0)$-forms for the almost-complex structure J_α, and

$$\omega_\alpha = E_\alpha^1 \wedge E_\alpha^4 + E_\alpha^2 \wedge E_\alpha^5 + E_\alpha^3 \wedge E_\alpha^6.$$

By K. Nomizu's theorem [Nom54, Theorem 1], the de Rham cohomology of N is straightforwardly computed:

$$H_{dR}^2(N;\mathbb{R})$$

$$= \mathbb{R}\left\langle E_\alpha^{15}, E_\alpha^{16}, E_\alpha^{24}, E_\alpha^{26}, E_\alpha^{34}, E_\alpha^{35}, E_\alpha^{14} + \frac{1}{\alpha}\,E_\alpha^{25}, \frac{1}{\alpha}\,E_\alpha^{25} + \frac{1}{\alpha-1}\,E_\alpha^{36} \right\rangle$$

$$= \mathbb{R}\left\langle i\,\alpha\,\varphi_\alpha^{1\bar{1}} + i\,\varphi_\alpha^{2\bar{2}}, i\,(\alpha-1)\,\varphi_\alpha^{2\bar{2}} + i\,\alpha\,\varphi_\alpha^{3\bar{3}}, \mathfrak{Im}\,\varphi_\alpha^{1\bar{2}}, \mathfrak{Im}\,\varphi_\alpha^{13}, \mathfrak{Im}\,\varphi_\alpha^{3\bar{2}} \right\rangle$$

$$\oplus \left\langle \mathfrak{Im}\,\varphi_\alpha^{12}, \mathfrak{Im}\,\varphi_\alpha^{13}, \mathfrak{Im}\,\varphi_\alpha^{23} \right\rangle.$$

Note that the g_α-harmonic representatives of the above basis of $H_{dR}^2(N;\mathbb{R})$ are of pure type with respect to J_α: hence, the almost-complex structure J_α is C^∞-pure-and-full and pure-and-full by [FT10, Theorem 3.7]; furthermore, by Theorem 4.4, the Lefschetz-type property on 2-forms holds on N endowed with the almost-Kähler structure $(J_\alpha, \omega_\alpha, g_\alpha)$, where $\alpha > 1$. Moreover, we get

$$h_{J_\alpha}^+(N) = 5, \qquad h_{J_\alpha}^-(N) = 3.$$

On the other hand, one can explicitly note that

$$L_{\omega_\alpha} E_\alpha^{15} = E_\alpha^{1536} = *_{g_\alpha} E_\alpha^{24} \,,$$
$$L_{\omega_\alpha} E_\alpha^{16} = E_\alpha^{1625} = *_{g_\alpha} E_\alpha^{34} \,,$$
$$L_{\omega_\alpha} E_\alpha^{24} = E_\alpha^{2436} = *_{g_\alpha} E_\alpha^{15} \,,$$
$$L_{\omega_\alpha} E_\alpha^{26} = E_\alpha^{2614} = *_{g_\alpha} E_\alpha^{35} \,,$$
$$L_{\omega_\alpha} E_\alpha^{34} = E_\alpha^{3425} = *_{g_\alpha} E_\alpha^{16} \,,$$
$$L_{\omega_\alpha} E_\alpha^{35} = E_\alpha^{3514} = *_{g_\alpha} E_\alpha^{26} \,,$$

and

$$d *_{g_\alpha} L_{\omega_\alpha} \left(E_\alpha^{14} + \frac{1}{\alpha} E_\alpha^{25} \right) = d \left(-\frac{\alpha+1}{\alpha} E_\alpha^{36} - E_\alpha^{25} - \frac{1}{\alpha} E_\alpha^{14} \right) = 0 \,,$$

and

$$d *_{g_\alpha} L_{\omega_\alpha} \left(e^{25} + e^{36} \right) = 0 \,;$$

this proves explicitly that the Lefschetz-type property on 2-forms holds on N endowed with the almost-Kähler structure $(J_\alpha, \omega_\alpha, g_\alpha)$, where $\alpha > 1$.

Note that, while $\omega_\alpha \wedge \cdot : \wedge^2 N \rightarrow \wedge^4 N$ induces an isomorphism $[\omega_\alpha] \smile \cdot : H_{dR}^2$ $(N; \mathbb{R}) \overset{\simeq}{\rightarrow} H_{dR}^4(N; \mathbb{R})$ in cohomology, the map $[\omega_\alpha]^2 \smile \cdot : H_{dR}^1(N; \mathbb{R}) \rightarrow H_{dR}^5$ $(N; \mathbb{R})$ is not an isomorphism, according to [BG88, Theorem A].

We show explicitly that the nilmanifold N is not formal, without using K. Hasegawa's theorem [Has89, Theorem 1, Corollary]. By [DGMS75, Corollary 1], every Massey product on a formal manifold is zero. Since

$$\left[E_\alpha^1 \right] \smile \left[E_\alpha^3 \right] = (\alpha - 1) \left[d E_\alpha^5 \right] = 0 \quad \text{and}$$
$$\left[E_\alpha^3 \right] \smile \left[E_\alpha^2 \right] = -\alpha \, (\alpha - 1) \left[d E_\alpha^4 \right] = 0 \,,$$

the triple Massey product

$$\langle [E_\alpha^1], [E_\alpha^3], [E_\alpha^2] \rangle = -(\alpha - 1) \left[E_\alpha^{25} + \alpha \, E_\alpha^{14} \right]$$

is not zero, and hence N is not formal.

Summarizing, we state the following result.

Proposition 4.6 ([ATZ12, Proposition 2.5]). *There exists a non-formal six-dimensional nilmanifold endowed with an 1-parameter family $\{(J_\alpha, \omega_\alpha, g_\alpha)\}_{\alpha>1}$ of left-invariant almost-Kähler structures, such that J_α is C^∞-pure-and-full and pure-and-full, and for which the Lefschetz-type property on 2-forms holds.*

In the following example, we give a $\mathcal{C}^\infty$-pure-and-full almost-Kähler structure on the completely-solvable Nakamura manifold.

Example 4.8 ([ATZ12, Sect. 3]). A $\mathcal{C}^\infty$-pure-and-full almost-Kähler structure on the completely-solvable Nakamura manifold.

Firstly, we recall the construction of the *completely-solvable Nakamura manifold*: it is a completely-solvable solvmanifold diffeomorphic to the *Nakamura manifold* studied by I. Nakamura in [Nak75, p. 90], and it is an example of a cohomologically-Kähler non-Kähler manifold, [dAFdLM92], [FMS03, Example 3.1], [dBT06, Sect. 3].

Take $A \in \mathrm{SL}(2; \mathbb{Z})$ with two different real positive eigenvalues e^λ and $e^{-\lambda}$ with $\lambda > 0$, and fix $P \in \mathrm{GL}(2; \mathbb{R})$ such that $PAP^{-1} = \mathrm{diag}\left(e^\lambda, e^{-\lambda}\right)$. For example, take

$$A := \begin{pmatrix} 2 & 1 \\ 1 & 1 \end{pmatrix}, \quad \text{and} \quad P := \begin{pmatrix} \frac{1-\sqrt{5}}{2} & 1 \\ 1 & \frac{\sqrt{5}-1}{2} \end{pmatrix},$$

and consequently $\lambda = \log \frac{3+\sqrt{5}}{2}$.

Let $M^6 := M^6(\lambda)$ be the six-dimensional completely-solvable solvmanifold

$$M^6 := \mathbb{S}^1_{x^2} \times \frac{\mathbb{R}_{x^1} \times \mathbb{T}^2_{\mathbb{C}, \left(x^3, x^4, x^5, x^6\right)}}{\langle T_1 \rangle}$$

where $\mathbb{T}^2_{\mathbb{C}}$ is the two-dimensional complex torus

$$\mathbb{T}^2_{\mathbb{C}} := \frac{\mathbb{C}^2}{P\mathbb{Z}[i]^2}$$

and T_1 acts on $\mathbb{R} \times \mathbb{T}^2_{\mathbb{C}}$ as

$$T_1\left(x^1, x^3, x^4, x^5, x^6\right) := \left(x^1 + \lambda, \; e^{-\lambda} x^3, \; e^\lambda x^4, \; e^{-\lambda} x^5, \; e^\lambda x^6\right).$$

Using coordinates x^2 on $\mathbb{S}^1$, x^1 on $\mathbb{R}$ and $\left(x^3, x^4, x^5, x^6\right)$ on $\mathbb{T}^2_{\mathbb{C}}$, we set

$$\begin{aligned}
e^1 &:= \mathrm{d}x^1, & e^2 &:= \mathrm{d}x^2, \\
e^3 &:= e^{x^1} \mathrm{d}x^3, & e^4 &:= e^{-x^1} \mathrm{d}x^4, \\
e^5 &:= e^{x^1} \mathrm{d}x^5, & e^6 &:= e^{-x^1} \mathrm{d}x^6.
\end{aligned}$$

as a basis for $\mathfrak{g}^*$, where $\mathfrak{g}$ denotes the Lie algebra naturally associated to M^6; therefore, with respect to $\{e^i\}_{i \in \{1,\dots,6\}}$, the structure equations are the following:

$$\begin{cases} \mathrm{d}\,e^1 = 0 \\ \mathrm{d}\,e^2 = 0 \\ \mathrm{d}\,e^3 = e^1 \wedge e^3 \\ \mathrm{d}\,e^4 = -e^1 \wedge e^4 \\ \mathrm{d}\,e^5 = e^1 \wedge e^5 \\ \mathrm{d}\,e^6 = -e^1 \wedge e^6 \end{cases}.$$

Let J be the almost-complex structure on M^6 defined requiring that a co-frame for the space of complex $(1,0)$-forms is given by

$$\begin{cases} \varphi^1 := \frac{1}{2}\left(e^1 + i\,e^2\right) \\ \varphi^2 := e^3 + i\,e^5 \\ \varphi^3 := e^4 + i\,e^6 \end{cases}.$$

It is straightforward to check that J is integrable.

Being M^6 a compact quotient of a completely-solvable Lie group, one computes the de Rham cohomology of M^6 by A. Hattori's theorem [Hat60, Corollary 4.2]:

$$H^0_{dR}\left(M^6; \mathbb{C}\right) = \mathbb{C}\langle 1\rangle\,,$$

$$H^1_{dR}\left(M^6; \mathbb{C}\right) = \mathbb{C}\langle \varphi^1, \bar{\varphi}^1\rangle\,,$$

$$H^2_{dR}\left(M^6; \mathbb{C}\right) = \mathbb{C}\langle \varphi^{1\bar{1}}, \varphi^{2\bar{3}}, \varphi^{3\bar{2}}, \varphi^{23}, \varphi^{\bar{2}\bar{3}}\rangle\,,$$

$$H^3_{dR}\left(M^6; \mathbb{C}\right) = \mathbb{C}\langle \varphi^{12\bar{3}}, \varphi^{13\bar{2}}, \varphi^{123}, \varphi^{1\bar{2}\bar{3}}, \varphi^{2\bar{1}\bar{3}}, \varphi^{3\bar{1}\bar{2}}, \varphi^{23\bar{1}}, \varphi^{\bar{1}\bar{2}\bar{3}}\rangle$$

(as usually, for the sake of clearness, we write, for example, $\varphi^{A\bar{B}}$ in place of $\varphi^A \wedge \bar{\varphi}^B$, and we list the harmonic representatives with respect to the metric $g := \sum_{j=1}^3 \varphi^j \odot \bar{\varphi}^j$ instead of their classes). Therefore, [FMS03, Proposition 3.2]: *(i)* M^6 is *geometrically formal*, that is, the product of g-harmonic forms is still g-harmonic, and therefore it is formal; *(ii)* furthermore,

$$\omega := e^{12} + e^{34} + e^{56}$$

is a symplectic form on M^6 satisfying the Hard Lefschetz Condition.

Note also that $\tilde{\omega} := \frac{i}{2}\left(\varphi^{1\bar{1}} + \varphi^{2\bar{2}} + \varphi^{3\bar{3}}\right)$ is not closed but $\mathrm{d}\,\tilde{\omega}^2 = 0$, from which it follows that the manifold M^6 admits a balanced metric.

Since M^6 is a compact quotient of a completely-solvable Lie group, by K. Hasegawa's theorem [Has06, Main Theorem], the manifold M^6, endowed with any integrable almost-complex structure (e.g., the J defined above), admits no Kähler structure, and it is not in class $\mathcal{C}$ of Fujiki, see also [FMS03, Theorem 3.3].

Therefore, we consider the (non-integrable) almost-complex structure J' defined by

$$J' e^1 := -e^2, \qquad J' e^3 := -e^4, \qquad J' e^5 := -e^6.$$

Set

$$\begin{cases} \psi^1 := \frac{1}{2}\left(e^1 + i e^2\right) \\ \psi^2 := e^3 + i e^4 \\ \psi^3 := e^5 + i e^6 \end{cases}$$

as a co-frame for the space of $(1,0)$-forms on M^6 with respect to J'; the structure equations with respect to this co-frame are

$$\begin{cases} d\psi^1 = 0 \\ d\psi^2 = \psi^{12} + \psi^{1\bar{2}} \\ d\psi^3 = \psi^{13} + \psi^{1\bar{3}} \end{cases},$$

from which it is clear that J' is not integrable.

The J'-compatible 2-form

$$\omega' := e^{12} + e^{34} + e^{56}$$

is d-closed; hence (J', ω') is an almost-Kähler structure on M^6.

Moreover, as already remarked, using A. Hattori's theorem [Hat60, Corollary 4.2], one gets

$$H^2_{dR}\left(M^6; \mathbb{R}\right) = \mathbb{R}\left\langle e^{12}, e^{34}, e^{56}, -e^{36} + e^{45}, e^{36} + e^{45}\right\rangle$$

$$= \underbrace{\mathbb{R}\left\langle i\,\psi^{1\bar{1}}, i\,\psi^{2\bar{2}}, i\,\psi^{3\bar{3}}, i\left(\psi^{2\bar{3}} + \psi^{3\bar{2}}\right)\right\rangle}_{\subseteq H^+_{J'}(M^6)} \oplus \underbrace{\mathbb{R}\left\langle i\left(\psi^{2\bar{3}} - \psi^{\bar{2}3}\right)\right\rangle}_{\subseteq H^-_{J'}(M^6)},$$

where we have listed the harmonic representatives with respect to the metric $g' := \sum_{j=1}^6 e^j \odot e^j$ instead of their classes; note that the above g'-harmonic representatives are of pure type with respect to J'. Therefore, J' is obviously C^∞-full; it is also C^∞-pure by [FT10, Proposition 3.2], or [DLZ10, Proposition 2.8]. Moreover, since any cohomology class in $H^+_{J'}(M^6)$, respectively in $H^-_{J'}(M^6)$, has a d-closed g'-harmonic representative in $\wedge^{1,1}_{J'} M^6 \cap \wedge^2 M^6$, respectively in $\left(\wedge^{2,0}_{J'} M^6 \oplus \wedge^{0,2}_{J'} M^6\right) \cap \wedge^2 M^6$, then J' is also pure-and-full, by [FT10, Theorem 3.7], and the Lefschetz-type property on 2-forms holds, by Theorem 4.4.

One can explicitly check that the Lefschetz-type operator

$$L_{\omega'}: \wedge^2 M^6 \to \wedge^4 M^6$$

takes g'-harmonic 2-forms to g'-harmonic 4-forms, since

$$L_{\omega'}\, e^{12} = e^{1234} + e^{1256} = *_{g'}\left(e^{34} + e^{56}\right)\,,$$

$$L_{\omega'}\, e^{34} = e^{1234} + e^{3456} = *_{g'}\left(e^{12} + e^{56}\right)\,,$$

$$L_{\omega'}\, e^{56} = e^{1256} + e^{3456} = *_{g'}\left(e^{12} + e^{34}\right)\,,$$

$$L_{\omega'}\, e^{36} = \qquad e^{1236} \qquad = \qquad *_{g'}\, e^{45}\,,$$

$$L_{\omega'}\, e^{45} = \qquad e^{1245} \qquad = \qquad *_{g'}\, e^{36}\,.$$

Summarizing, the content of the last example yields the following result, [ATZ12, Proposition 3.3].

Proposition 4.7. *The completely-solvable Nakamura manifold M^6 admits*

- *both a C^∞-pure-and-full and pure-and-full complex structure J, and*
- *a C^∞-pure-and-full and pure-and-full almost-Kähler structure $(J',\, \omega',\, g')$, for which the Lefschetz-type property on 2-forms holds.*

Finally, in the following example, we give a non-C^∞-full almost-Kähler structure, [ATZ12, Sect. 4]. In particular, this provides another strong difference between the (non-integrable) almost-Kähler case and the (integrable) Kähler case, all the compact Kähler manifolds being C^∞-pure-and-full by [DLZ10, Lemma 2.15, Theorem 2.16], or [LZ09, Proposition 2.1].

Example 4.9 ([ATZ12, Sect. 4]). An almost-Kähler non-C^∞-full structure for which the Lefschetz-type property on 2-forms does not hold.
Consider the Iwasawa manifold $\mathbb{I}_3 := \mathbb{H}\,(3;\mathbb{Z}\,[\mathrm{i}])\backslash\, \mathbb{H}(3;\mathbb{C})$, see Sect. 3.2.1. Recall that, given the standard complex structure induced by the one on $\mathbb{C}^3$ and setting $\{\varphi^1,\, \varphi^2,\, \varphi^3\}$ as a global co-frame for the $(1,0)$-forms on $\mathbb{I}_3$, by K. Nomizu's theorem [Nom54, Theorem 1] one gets

$$H_{dR}^2\,(\mathbb{I}_3;\mathbb{C}) = \mathbb{R}\left\langle\varphi^{13} + \varphi^{\bar1\bar3},\, \mathrm{i}\left(\varphi^{13} - \varphi^{\bar1\bar3}\right),\, \varphi^{23} + \varphi^{\bar2\bar3},\, \mathrm{i}\left(\varphi^{23} - \varphi^{\bar2\bar3}\right),\, \varphi^{1\bar2} - \varphi^{2\bar1},\right.$$

$$\left.\mathrm{i}\left(\varphi^{1\bar2} + \varphi^{2\bar1}\right),\, \mathrm{i}\varphi^{1\bar1},\, \mathrm{i}\varphi^{2\bar2}\right\rangle \otimes_{\mathbb{R}} \mathbb{C}\,,$$

where we have listed the harmonic representatives with respect to the metric $g := \sum_{h=1}^3 \varphi^h \odot \bar\varphi^h$ instead of their classes. Using the co-frame $\{e^1,\, \dots,\, e^6\}$ of the cotangent bundle defined by

$$\varphi^1 =: e^1 + \mathrm{i}\,e^2\,, \qquad \varphi^2 =: e^3 + \mathrm{i}\,e^4\,, \qquad \varphi^3 =: e^5 + \mathrm{i}\,e^6\,,$$

one computes the structure equations

$$d e^1 = d e^2 = d e^3 = d e^4 = 0, \quad d e^5 = -e^{13} + e^{24}, \quad d e^6 = -e^{14} - e^{23}.$$

Therefore

$$H^2_{dR}(\mathbb{I}_3; \mathbb{R}) = \mathbb{R}\langle e^{15} - e^{26}, \ e^{16} + e^{25}, \ e^{35} - e^{46}, \ e^{36} + e^{45}, \ e^{13} + e^{24}, \ e^{23} - e^{14}, \ e^{12}, \ e^{34}\rangle.$$

Consider the almost-complex structure J on X defined by

$$J e^1 := -e^6, \quad J e^2 := -e^5, \quad J e^3 := -e^4,$$

and set

$$\omega := e^{16} + e^{25} + e^{34}.$$

Then (J, ω, g) is an almost-Kähler structure on the Iwasawa manifold $\mathbb{I}_3$. We easily get that

$$\mathbb{R}\langle e^{16} + e^{25}, \ (e^{35} - e^{46}) + (e^{13} + e^{24}), \ (e^{36} + e^{45}) - (e^{23} - e^{14}), \ e^{34}\rangle \subseteq H^+_J(\mathbb{I}_3)$$

and

$$\mathbb{R}\langle e^{15} - e^{26}, \ (e^{35} - e^{46}) - (e^{13} + e^{24}), \ (e^{36} + e^{45}) + (e^{23} - e^{14})\rangle \subseteq H^-_J(\mathbb{I}_3).$$

We claim that the previous inclusions are actually equalities, and in particular that J is a non-$\mathcal{C}^\infty$-full almost-Kähler structure on $\mathbb{I}_3$. Indeed, we firstly note that, by [FT10, Proposition 3.2] or [DLZ10, Proposition 2.8], J is $\mathcal{C}^\infty$-pure, since it admits a symplectic structure compatible with it. Moreover, we recall that a $\mathcal{C}^\infty$-full almost-complex structure is also pure by [LZ09, Proposition 2.5], and therefore it is also $\mathcal{C}^\infty$-pure at the 4th stage, by Theorem 4.1, that is,

$$H^{(3,1),(1,3)}_J(\mathbb{I}_3; \mathbb{R}) \cap H^{(2,2)}_J(\mathbb{I}_3; \mathbb{R}) = \{0\}.$$

Therefore, our claim reduces to prove that J is not $\mathcal{C}^\infty$-pure at the 4th stage. Note that

$$0 \neq [e^{3456}] = [e^{3456} - d e^{135}] = [e^{3456} + e^{1234}]$$

$$= [e^{3456} + d e^{135}] = [e^{3456} - e^{1234}],$$

and that $e^{3456} + e^{1234} \in \left(\wedge^{3,1}_J \mathbb{I}_3 \oplus \wedge^{1,3}_J \mathbb{I}_3\right) \cap \wedge^4 \mathbb{I}_3$, while $e^{3456} - e^{1234} \in \wedge^{2,2}_J \mathbb{I}_3 \cap \wedge^4 \mathbb{I}_3$, and so $H^{(3,1),(1,3)}_J(\mathbb{I}_3; \mathbb{R}) \cap H^{(2,2)}_J(\mathbb{I}_3; \mathbb{R}) \ni [e^{3456}]$, therefore J is not $\mathcal{C}^\infty$-pure at the 4th stage, and hence it is not $\mathcal{C}^\infty$-full.

Let L_ω be the Lefschetz-type operator associated to the almost-Kähler structure (J, ω, g). Then, we have

$$L_\omega \left(e^{12} \right) = e^{1234} = d \left(e^{245} \right) ,$$

namely, L_ω does not take g-harmonic 2-forms to g-harmonic 4-forms.

The previous example proves the following result.

Proposition 4.8 ([ATZ12, Proposition 4.1]). *The differentiable manifold X underlying the Iwasawa manifold $\mathbb{I}_3 := \mathbb{H}(3; \mathbb{Z}[\mathrm{i}]) \backslash \mathbb{H}(3; \mathbb{C})$ admits an almost-Kähler structure (J, ω, g) which is C^∞-pure and non-C^∞-full, and for which the Lefschetz-type property on 2-forms does not hold.*

The argument of the proof of [DLZ10, Theorem 2.3] suggests the following question, [ATZ12, Question 3.4], compare also [DLZ12, Sect. 2], in accordance with Proposition 4.8.

Question 4.2 ([ATZ12, Question 3.4], see also [DLZ12, Sect. 2]). Let X be a compact $2n$-dimensional manifold endowed with an almost-Kähler structure (J, ω, g) such that the Lefschetz-type property on 2-forms holds. Is J C^∞-full?

4.3 C^∞-Pure-and-Fullness and Deformations of (Almost-)Complex Structures

In this section, we are interested in studying the behaviour of the cohomological decomposition of the de Rham cohomology of an (almost-)complex manifold under small deformations of the complex structure and along curves of almost-complex structures, [AT11, AT12a].

More precisely, we prove that being C^∞-pure-and-full is not a stable property under small deformations of the complex structure, Theorem 4.8, as a consequence of the study of the C^∞-pure-and-fullness for small deformations of the Iwasawa manifold, Theorem 4.8. Then we study some explicit examples of curves of almost-complex structures on compact manifolds: by using a construction introduced by J. Lee, [Lee04, Sect. 1], we construct a curve of almost-complex structures along which the property of being C^∞-pure-and-full remains satisfied, Theorem 4.9. In Sect. 4.3.2, we provide counterexamples to the upper-semi-continuity of $t \mapsto H^-_{J_t}(X)$, Proposition 4.9, and to the lower-semi-continuity of $t \mapsto H^+_{J_t}(X)$, Proposition 4.10, where $\{J_t\}_t$ is a curve of almost-complex structures on a compact manifold X of dimension greater than 4; we also study a stronger semi-continuity problem, Sect. 4.3.2.2.

4.3.1 Deformations of C^∞-Pure-and-Full Almost-Complex Structures

In this section, we consider the problem of the stability of the C^∞-pure-and-fullness under small deformations of the complex structure and along curves of almost-complex structures.

4.3.1.1 Instability of C^∞-Pure-and-Full Property

We recall that a property concerning compact complex (respectively, almost-complex) manifolds (e.g., admitting Kähler metrics, admitting balanced metrics, satisfying the $\partial\bar\partial$-Lemma, admitting compatible symplectic structures) is called *stable under small deformations of the complex* (respectively, *almost-complex*) *structure* if, for every complex-analytic family $\{X_t := (X, J_t)\}_{t \in B}$ of compact complex manifolds (respectively, for every smooth curve $\{J_t\}_{t \in B}$ of almost-complex structures on a compact differentiable manifold X), whenever the property holds for (X, J_t) for some $t \in B$, it holds also for (X, J_s) for any s in a neighbourhood of t in B.

The main result in the context of stability under small deformations of the complex structure is the following classical theorem by K. Kodaira and D.C. Spencer, [KS60], which actually holds for differentiable families of compact complex manifolds.

Theorem 4.5 ([KS60, Theorem 15]). *For a compact manifold, admitting a Kähler structure is a stable property under small deformations of the complex structure.*

Remark 4.11. Conditions under which the property of admitting a balanced metric is stable under small deformations of the complex structure have been studied by C.-C. Wu [Wu06, Sect. 5], and by J. Fu and S.-T. Yau [FY11]. More precisely, in [Wu06, Theorem 5.13], it is proven that small deformations of a compact complex manifold admitting balanced metric and satisfying the $\partial\bar\partial$-Lemma still admit balanced metrics and satisfy the $\partial\bar\partial$-Lemma. In [FY11, Theorem 6], it is shown that if a compact complex manifold of complex dimension n admits balanced metrics and its small deformations satisfy the so-called $(n - 1, n)$ *th weak* $\partial\bar\partial$-*Lemma* (that is, for any real $(n-1, n-1)$-form φ such that $\bar\partial\varphi$ is ∂-exact, there exists a $(n-2, n-1)$-form ψ such that $\bar\partial\varphi = i\,\partial\bar\partial\psi$), then also its small deformations admit balanced metrics; in particular, if X is a compact complex manifold admitting balanced metrics and with $\dim_\mathbb{C} H_{\bar\partial}^{2,0}(X) = 0$, then any sufficiently small deformation of X still admits balanced metrics, [FY11, Corollary 8].

Remark 4.12. As regards the **D**-complex counterpart of Kählerness, in a joint work with F.A. Rossi, [AR12], we have proven that admitting **D**-Kähler structures is not a stable property under small deformations of the **D**-complex structure, [AR12,

Theorem 4.2]: indeed, on a four-dimensional solvmanifold with structure equations $\left(0^2,\ 23,\ -24\right)$, there exists a curve

$$\left\{K_t\ :=\ \begin{pmatrix} -1 & 0 & 0 & 0 \\ 0 & 1 & 0 & -2t \\ 0 & 0 & 1 & 0 \\ 0 & 0 & 0 & -1 \end{pmatrix}\right\}_{t\in\mathbb{R}}$$

of left-invariant **D**-complex structures such that K_0 admits a **D**-Kähler structure and K_t, for $t \in \mathbb{R} \setminus \{0\}$, does not admit any **D**-Kähler structure, [AR12, Example 4.1].

Note that, by [DLZ11, Theorem 5.4], see also [Don06], on compact almost-complex manifolds of dimension 4, the property of admitting an almost-Kähler structure is stable under small deformations of the almost-complex structure. This result stands on the very special properties of four-dimensional manifolds, and does not hold true in higher dimension. More precisely, we provide here an explicit example, in dimension 6, showing that, relaxing the integrability condition in the previous theorem (namely, starting with an almost-Kähler structure), we lose the stability under small deformations of the almost-complex structure.

Example 4.10. A curve $\{J_t\}_t$ of almost-complex structures on a compact six-dimensional manifold such that J_0 admits an almost-Kähler structure and J_t, for $t \neq 0$, admits no almost-Kähler structure.

For $c \in \mathbb{R}$, consider the completely-solvable Lie group

$$\mathrm{Sol}(3)_{(x^1,y^1,z^1)}\ :=\ \left\{\begin{pmatrix} e^{c\,z^1} & & x^1 \\ & e^{-c\,z^1} & y^1 \\ & & 1 & z^1 \\ & & & 1 \end{pmatrix} \in \mathrm{GL}(4;\mathbb{R})\ :\ x^1, y^1, z^1 \in \mathbb{R}\right\}.$$

Choose a suitable $c \in \mathbb{R}$, for which there exists a co-compact discrete subgroup $\Gamma(c) \subset \mathrm{Sol}(3)$ such that

$$M(c)_{(x^1,y^1,z^1)}\ :=\ \Gamma(c) \backslash \mathrm{Sol}(3)_{(x^1,y^1,z^1)}$$

is a compact three-dimensional completely-solvable solvmanifold, [AGH63, Sect. 3].

The manifold

$$N^6(c)\ :=\ M(c)_{(x^1,y^1,z^1)} \times M(c)_{(x^2,y^2,z^2)}.$$

is cohomologically-Kähler, see [BG90, Example 1], is formal and has a symplectic structure satisfying the Hard Lefschetz Condition, but it admits no Kähler structure, see [FMS03, Theorem 3.5].

Consider $\left\{e^i\right\}_{i\in\{1,\ldots,6\}}$ as a $(\mathrm{Sol}(3)\times\mathrm{Sol}(3))$-left-invariant co-frame for $N^6(c)$, where

$$e^1 := e^{-cz^1}\,d x^1, \quad e^2 := e^{-cz^1}\,d y^1, \quad e^3 := d z^1,$$

$$e^4 := e^{-cz^2}\,d x^2, \quad e^5 := e^{-cz^2}\,d y^2, \quad e^6 := d z^2;$$

with respect to it, the structure equations are

$$\begin{cases} d e^1 = c\, e^1 \wedge e^3 \\ d e^2 = -c\, e^2 \wedge e^3 \\ d e^3 = 0 \\ d e^4 = c\, e^4 \wedge e^6 \\ d e^5 = -c\, e^5 \wedge e^6 \\ d e^6 = 0 \end{cases}.$$

By A. Hattori's theorem [Hat60, Corollary 4.2], it is straightforward to compute

$$H^2_{dR}\left(N^6(c); \mathbb{R}\right) = \mathbb{R}\left\langle e^1 \wedge e^2,\ e^3 \wedge e^6,\ e^4 \wedge e^5\right\rangle,$$

hence the space of $(\mathrm{Sol}(3)\times\mathrm{Sol}(3))$-left-invariant d-closed 2-forms is

$$\mathbb{R}\left\langle e^{12},\ e^{36},\ e^{45}\right\rangle \oplus \mathbb{R}\left\langle e^{13},\ e^{23},\ e^{45},\ e^{46}\right\rangle$$

(where, as usually, we shorten $e^{AB} := e^A \wedge e^B$).

Let $J_0 \in \mathrm{End}\left(TN^6(c)\right)$ be the almost-complex structure given, with respect to the frame $\{e_1,\ldots,e_6\}$ dual to $\{e^1,\ldots,e^6\}$, by

$$J_0 := \begin{pmatrix} & -1 & & & & \\ 1 & & & & & \\ & & & -1 & & \\ & & & & -1 & \\ & & & 1 & & \\ & & 1 & & & \end{pmatrix} \in \mathrm{End}\left(TN^6(c)\right).$$

It is straightforward to check that J_0 admits almost-Kähler structures: more precisely, the cone $\mathcal{K}^c_{J_0,\,\mathrm{inv}}$ of $(\mathrm{Sol}(3)\times\mathrm{Sol}(3))$-left-invariant almost-Kähler structures on $\left(N^6(c), J_0\right)$ is

$$\mathcal{K}^c_{J_0,\,\mathrm{inv}} = \left\{\alpha\, e^1 \wedge e^2 + \beta\, e^3 \wedge e^6 + \gamma\, e^4 \wedge e^5 \ :\ \alpha,\ \beta,\ \gamma > 0\right\}.$$

Take now

$$
L := \begin{pmatrix}
0 & 0 & 1 & 0 & 0 & 0 \\
0 & 0 & 0 & 0 & 0 & -1 \\
0 & 0 & 0 & 1 & 0 & 0 \\
0 & 0 & 0 & 0 & 0 & 0 \\
0 & 0 & 0 & 0 & 0 & 0 \\
0 & 0 & 0 & 0 & -1 & 0
\end{pmatrix} \in \mathrm{End}\left(TN^6(c)\right)
$$

and define, for $t \in \mathbb{R}$, the almost-complex structure

$$
J_t := \left(\mathrm{id}_{TN^6(c)} - t\, L\right) J_0 \left(\mathrm{id}_{IN^6(c)} - t\, L\right)^{-1}
$$

$$
= \begin{pmatrix}
 & 1 & & & 2t^2 & -2t \\
-1 & & -2t & -2t^2 & & \\
 & & & -2t & 1 & \\
 & & & 1 & & \\
 & & -1 & & & \\
 & & -1 & -2t & &
\end{pmatrix} \in \mathrm{End}\left(T^*N^6(c)\right) .
$$

We first prove that J_t admits no $(\mathrm{Sol}(3) \times \mathrm{Sol}(3))$-left-invariant almost-Kähler structure for $t \neq 0$. Indeed, for $t \neq 0$, the space of $(\mathrm{Sol}(3) \times \mathrm{Sol}(3))$-left-invariant d-closed J_t-invariant 2-forms is

$$
\mathbb{R}\left\langle e^{36} + 2t\, e^{46},\ e^{45} \right\rangle
$$

and

$$
\left(\beta\, e^3 \wedge e^6 + \gamma\, e^4 \wedge e^5 + 2t\, \beta\, e^4 \wedge e^6\right)^3 = 0 \quad \text{for every } \beta,\, \gamma \in \mathbb{R} ,
$$

hence

$$
\mathcal{K}^c_{J_t,\,\mathrm{inv}} = \varnothing \quad \text{for } t \neq 0 .
$$

Now, using F.A. Belgun's symmetrization trick, [Bel00, Theorem 7], we get that, if J_t admits an almost-Kähler structure ω, then it should admits a $(\mathrm{Sol}(3) \times \mathrm{Sol}(3))$-left-invariant almost-Kähler structure

$$
\mu(\omega) := \int_{N^6(c)} \omega \lfloor_m \eta(m) ,
$$

where η is a $(\mathrm{Sol}(3) \times \mathrm{Sol}(3))$-bi-invariant volume form on $N^6(c)$, whose existence is guaranteed by [Mil76, Lemma 6.2].

We resume the content of the previous example in the following result.

Theorem 4.6. *Being almost-Kähler is not a stable property along curves of almost-complex structures.*

In view of K. Kodaira and D.C. Spencer's theorem [KS60, Theorem 15], a natural question in non-Kähler geometry is what properties, weaker that the property of being Kähler, still remain stable under small deformations of the complex structure. This does not hold true, for example, for the balanced property, as proven in [AB90, Proposition 4.1] by L. Alessandrini and G. Bassanelli; on the other hand, the cohomological property of satisfying the $\partial\bar\partial$-Lemma is stable under small deformations of the complex structure, as we have seen in Corollary 2.2, see also [Voi02, Proposition 9.21], or [Wu06, Theorem 5.12], or [Tom08, Sect. B]. We show now that the cohomological property of C^∞-pure-and-fullness turns out to be non-stable under small deformations of the complex structure.

Theorem 4.7 ([AT11, Theorem 3.2]). *The properties of being C^∞-pure-and-full, or C^∞-pure, or C^∞-full, or pure-and-full, or pure, or full are not stable under small deformations of the complex structure.*

The proof of Theorem 4.7 follows studying explicitly C^∞-pure-and-fullness for small deformations of the standard complex structure on the Iwasawa manifold $\mathbb{I}_3$, [AT11, Theorem 3.1]. (We refer to Sect. 3.2.1 for notations and results concerning the Iwasawa manifold and its Kuranishi space; we recall here that $\mathbb{I}_3$ is a holomorphically parallelizable nilmanifold of complex dimension 3, and its Kuranishi space is smooth and depends on six effective parameters; the small deformations of $\mathbb{I}_3$ can be divided into three classes, *(i)–(iii)*, according to their Hodge numbers; in particular, the Hodge numbers of the small deformations in class *(i)* are equal to the Hodge numbers of $\mathbb{I}_3$.)

Theorem 4.8 ([AT11, Theorem 3.1]). *Let $\mathbb{I}_3 := \mathbb{H}(3;\mathbb{Z}[i])\backslash\mathbb{H}(3;\mathbb{C})$ be the Iwasawa manifold, endowed with the complex structure inherited by the standard complex structure on $\mathbb{C}^3$, and consider the small deformations in its Kuranishi space. Then:*

- *the natural complex structure on $\mathbb{I}_3$ is C^∞-pure-and-full at every stage and pure-and-full at every stage;*
- *the small deformations in class (i) are C^∞-pure-and-full at every stage and pure-and-full at every stage;*
- *the small deformations in classes (ii) and (iii) are neither C^∞-pure nor C^∞-full nor pure nor full.*

Proof. We follow the notation introduced in Sect. 3.2.1; in particular, we recall that the structure equations with respect to a certain co-frame $\{\varphi_t^1,\ \varphi_t^2,\ \varphi_t^3\}$ of the space of $(1,0)$-forms on $X_t = (\mathbb{I}_3,\ J_t)$, for $\mathbf{t} \in \Delta(0,\varepsilon) \subset \mathbb{C}^6$ with $\varepsilon > 0$ small enough, are the following:

$$\begin{cases} d\varphi_t^1 = 0 \\ d\varphi_t^2 = 0 \\ d\varphi_t^3 = \sigma_{12}\,\varphi_t^1 \wedge \varphi_t^2 + \sigma_{1\bar{1}}\,\varphi_t^1 \wedge \bar{\varphi}_t^1 + \sigma_{1\bar{2}}\,\varphi_t^1 \wedge \bar{\varphi}_t^2 + \sigma_{2\bar{1}}\,\varphi_t^2 \wedge \bar{\varphi}_t^1 + \sigma_{2\bar{2}}\,\varphi_t^2 \wedge \bar{\varphi}_t^2 \end{cases},$$

where σ_{12}, $\sigma_{1\bar{1}}$, $\sigma_{1\bar{2}}$, $\sigma_{2\bar{1}}$, $\sigma_{2\bar{2}} \in \mathbb{C}$ are complex numbers depending just on **t**. The asymptotic behaviour of σ_{12}, $\sigma_{1\bar{1}}$, $\sigma_{1\bar{2}}$, $\sigma_{2\bar{1}}$, and $\sigma_{2\bar{2}}$ for **t** near **0** is the following, see Sect. 3.2.1.3:

$$\begin{cases} \sigma_{12} = -1 + o\,(|\mathbf{t}|) \\ \sigma_{1\bar{1}} = t_{21} + o\,(|\mathbf{t}|) \\ \sigma_{1\bar{2}} = t_{22} + o\,(|\mathbf{t}|) \quad ; \\ \sigma_{2\bar{1}} = -t_{11} + o\,(|\mathbf{t}|) \\ \sigma_{2\bar{2}} = -t_{12} + o\,(|\mathbf{t}|) \end{cases}$$

more precisely, for **t** in class *(i)*, respectively class *(ii)*, we actually have

$$\begin{cases} \sigma_{12} = -1 \\ \sigma_{1\bar{1}} = 0 \\ \sigma_{1\bar{2}} = 0 \qquad \text{for} \qquad \mathbf{t} \in \text{class } (i)\,, \\ \sigma_{2\bar{1}} = 0 \\ \sigma_{2\bar{2}} = 0 \end{cases}$$

and

$$\begin{cases} \sigma_{12} = -1 + o\,(|\mathbf{t}|) \\ \sigma_{1\bar{1}} = t_{21}\,(1 + o\,(1)) \\ \sigma_{1\bar{2}} = t_{22}\,(1 + o\,(1)) \qquad \text{for} \qquad \mathbf{t} \in \text{class } (ii)\,. \\ \sigma_{2\bar{1}} = -t_{11}\,(1 + o\,(1)) \\ \sigma_{2\bar{2}} = -t_{12}\,(1 + o\,(1)) \end{cases}$$

By K. Nomizu's theorem [Nom54, Theorem 1], one computes straightforwardly the de Rham cohomology of $\mathbb{I}_3$ and of its small deformations; for the sake of clearness, we recall in Table 4.1 a basis of the space of the harmonic representatives of the de Rham cohomology classes with respect to the metric $g_0 := \sum_{j=1}^{3} \varphi_0^j \odot \bar{\varphi}_0^j$.

Note that the harmonic representatives in Table 4.1 of the classes in $H_{dR}^{\bullet}(\mathbb{I}_3; \mathbb{R})$ are of pure type with respect to J_0 and to J_t with **t** in class *(i)*: hence, by Theorem 4.1

Table 4.1 A basis of harmonic representatives for the de Rham cohomology of the Iwasawa manifold with respect to the metric $g_0 = \sum_{j=1}^{3} \varphi^j \odot \bar{\varphi}^j$

k	$\mathbb{K}$	g_0 -Harmonic representatives of $H_{dR}^k(\mathbb{I}_3; \mathbb{K})$
1	$\mathbb{C}$	$\varphi^1,\ \varphi^2,\ \bar{\varphi}^1,\ \bar{\varphi}^2$
	$\mathbb{R}$	$\varphi^1 + \bar{\varphi}^1,\ \mathrm{i}\left(\varphi^1 - \bar{\varphi}^1\right),\ \varphi^2 + \bar{\varphi}^2,\ \mathrm{i}\left(\varphi^2 - \bar{\varphi}^2\right)$
2	$\mathbb{C}$	$\varphi^{13},\ \varphi^{23},\ \varphi^{1\bar{1}},\ \varphi^{1\bar{2}},\ \varphi^{2\bar{1}},\ \varphi^{2\bar{2}},\ \varphi^{\bar{1}3},\ \varphi^{\bar{2}3}$
	$\mathbb{R}$	$\varphi^{13} + \varphi^{\bar{1}3},\ \mathrm{i}\left(\varphi^{13} - \varphi^{\bar{1}3}\right),\ \varphi^{23} + \varphi^{\bar{2}3},\ \mathrm{i}\left(\varphi^{23} - \varphi^{\bar{2}3}\right),\ \varphi^{1\bar{2}} - \varphi^{2\bar{1}},\ \mathrm{i}\left(\varphi^{1\bar{2}} + \varphi^{2\bar{1}}\right),$ $\mathrm{i}\varphi^{1\bar{1}},\ \mathrm{i}\varphi^{2\bar{2}}$
3	$\mathbb{C}$	$\varphi^{123},\ \varphi^{13\bar{1}},\ \varphi^{13\bar{2}},\ \varphi^{23\bar{1}},\ \varphi^{23\bar{2}},\ \varphi^{1\bar{1}3},\ \varphi^{1\bar{2}3},\ \varphi^{2\bar{1}3},\ \varphi^{2\bar{2}3},\ \varphi^{\bar{1}23}$
	$\mathbb{R}$	$\varphi^{123} + \varphi^{\bar{1}23},\ \mathrm{i}\left(\varphi^{123} - \varphi^{\bar{1}23}\right),\ \varphi^{13\bar{1}} + \varphi^{1\bar{1}3},\ \mathrm{i}\left(\varphi^{13\bar{1}} - \varphi^{1\bar{1}3}\right),\ \varphi^{13\bar{2}} + \varphi^{2\bar{1}3},$ $\mathrm{i}\left(\varphi^{13\bar{2}} - \varphi^{2\bar{1}3}\right),$ $\varphi^{23\bar{1}} + \varphi^{1\bar{2}3},\ \mathrm{i}\left(\varphi^{23\bar{1}} - \varphi^{1\bar{2}3}\right),\ \varphi^{23\bar{2}} + \varphi^{2\bar{2}3},\ \mathrm{i}\left(\varphi^{23\bar{2}} - \varphi^{2\bar{2}3}\right)$
4	$\mathbb{C}$	$\varphi^{123\bar{1}},\ \varphi^{123\bar{2}},\ \varphi^{13\bar{1}3},\ \varphi^{13\bar{2}3},\ \varphi^{23\bar{1}3},\ \varphi^{23\bar{2}3},\ \varphi^{1\bar{1}23},\ \varphi^{2\bar{1}23}$
	$\mathbb{R}$	$\varphi^{123\bar{1}} - \varphi^{1\bar{1}23},\ \mathrm{i}\left(\varphi^{123\bar{1}} + \varphi^{1\bar{1}23}\right),\ \varphi^{123\bar{2}} - \varphi^{2\bar{1}23},\ \mathrm{i}\left(\varphi^{123\bar{2}} + \varphi^{2\bar{1}23}\right),\ \varphi^{13\bar{1}3},$ $\varphi^{13\bar{2}3} + \varphi^{23\bar{1}3},\ \mathrm{i}\left(\varphi^{13\bar{2}3} - \varphi^{23\bar{1}3}\right),\ \varphi^{23\bar{2}3}$
5	$\mathbb{C}$	$\varphi^{123\bar{1}3},\ \varphi^{123\bar{2}3},\ \varphi^{13\bar{1}23},\ \varphi^{23\bar{1}23}$
	$\mathbb{R}$	$\varphi^{123\bar{1}3} + \varphi^{13\bar{1}23},\ \mathrm{i}\left(\varphi^{123\bar{1}3} - \varphi^{13\bar{1}23}\right),\ \varphi^{123\bar{2}3} + \varphi^{23\bar{1}23},\ \mathrm{i}\left(\varphi^{123\bar{2}3} - \varphi^{23\bar{1}23}\right)$

(or arguing as in [FT10, Theorem 3.7]), one gets that $\mathbb{I}_3$ and its small deformations in class *(i)* are C^∞-pure-and-full at every stage and pure-and-full at every stage.

Concerning small deformations J_t in class *(ii)* and in class *(iii)*, using the asymptotic behaviour of the structure equations, we obtain that

$$\left[\sigma_{12}\,\varphi_t^{12}\right] = \left[\sigma_{1\bar{1}}\,\varphi_t^{1\bar{1}} + \sigma_{1\bar{2}}\,\varphi_t^{1\bar{2}} + \sigma_{2\bar{1}}\,\varphi_t^{2\bar{1}} + \sigma_{2\bar{2}}\,\varphi_t^{2\bar{2}}\right] \neq 0$$

in $H_{dR}^2(\mathbb{I}_3; \mathbb{C})$. Therefore

$$H_{J_t}^{(1,1)}(\mathbb{I}_3; \mathbb{C}) \cap \left(H_{J_t}^{(2,0)}(\mathbb{I}_3; \mathbb{C}) + H_{J_t}^{(0,2)}(\mathbb{I}_3; \mathbb{C})\right) \neq \{0\}\ ,$$

and in particular J_t is not complex-C^∞-pure. It follows from Remark 4.2 that J_t cannot be C^∞-pure; from [LZ09, Proposition 2.30], or Theorem 4.1, it follows that J_t cannot be full.

To prove that small deformations in class *(ii)* and in class *(iii)* are non-pure and non-C^∞-full, fix **t** small enough and choose two positive complex numbers $A := A(\mathbf{t}) \in \mathbb{C}$ and $B := B(\mathbf{t}) \in \mathbb{C}$, depending just on **t**, such that

$$\left(A\,\sigma_{1\bar{2}} - B\,\sigma_{1\bar{1}},\ A\,\sigma_{2\bar{2}} - B\,\sigma_{2\bar{1}} \right) \neq (0, 0) \; ;$$

computing $-\mathrm{d}\left(A\,\varphi_t^{1 3 \bar{3}} + B\,\varphi_t^{2 3 \bar{3}} \right)$, note that

$$\left[\left(A\,\sigma_{2\bar{1}} - B\,\sigma_{1\bar{1}} \right) \varphi_t^{1 2 \bar{1} \bar{3}} + \left(A\,\sigma_{2\bar{2}} - B\,\sigma_{1\bar{2}} \right) \varphi_t^{1 2 \bar{2} \bar{3}} - A\,\bar{\sigma}_{12} \varphi_t^{1 3 \bar{1} \bar{2}} - B\,\bar{\sigma}_{12}\varphi_t^{2 3 \bar{1} \bar{2}} \right]$$

$$\left[\left(A\,\bar{\sigma}_{1\bar{2}} - B\,\sigma_{1\bar{1}} \right) \varphi_t^{1 2 3 \bar{1}} + \left(A\,\bar{\sigma}_{2\bar{2}} - B\,\bar{\sigma}_{21} \right) \varphi_t^{1 2 3 \bar{2}} \right] \neq 0 \; ,$$

in $H^4_{dR}(\mathbb{I}_3; \mathbb{C})$. As before, it follows that J_t is not C^∞-pure at the 4th stage, and consequently it is neither pure nor C^∞-full, by Theorem 4.1. $\qquad\square$

4.3.1.2 Curves of C^∞-Pure-and-Full Almost-Complex Structures

We study here some explicit examples of curves of almost-complex structures on compact manifolds, along which the property of being C^∞-pure-and-full remains satisfied. The aim of this section is to better understand the behaviour of C^∞-pure-and-fullness along curves of almost-complex structures.

Firstly, we recall some general results concerning curves of almost-complex structures on compact manifolds, referring, e.g., to [AL94].

Let J be an almost-complex structure on a compact $2n$-dimensional manifold X. Every curve $\{J_t\}_{t \in (-\varepsilon, \varepsilon) \subset \mathbb{R}}$ of almost-complex structures on X such that $J_0 = J$ can be written, for $\varepsilon > 0$ small enough, as

$$J_t = (\mathrm{id} - L_t)\, J\, (\mathrm{id} - L_t)^{-1} \in \mathrm{End}\,(TX) \; ,$$

where $L_t \in \mathrm{End}\,(TX)$, see, e.g., [AL94, Proposition 1.1.6]; the endomorphism L_t is uniquely determined further requiring that $L_t \in T_J^{1,0} X \otimes \left(T_J^{0,1} X \right)^*$, namely,

$$L_t\, J + J\, L_t = 0 \; ;$$

furthermore, set $L_t =: t\, L + \mathrm{o}(t)$: if J is compatible with a symplectic form ω, then the curves consisting of ω-compatible almost-complex structures J_t are exactly those ones for which $L^t = L$.

In [dBM10, Proposition 3.3], P. de Bartolomeis and F. Meylan computed

$$\frac{\mathrm{d}}{\mathrm{d}t}\bigg|_{t=0} \mathrm{Nij}_J = -4\,(\bar{\partial}_J L)(\cdot, \cdot\cdot) - \mathrm{Nij}_J\,(L\cdot, \cdot\cdot) - \mathrm{Nij}_J\,(\cdot, L\cdot\cdot) \; ,$$

where, by definition,

$$\left(\overline{\partial}_J L\right)(\cdot,\cdot\cdot) = \left(\overline{\partial}_J (L(\cdot\cdot))\right)(\cdot) - \left(\overline{\partial}_J L((\cdot))\right)(\cdot\cdot) - \frac{1}{2} L([\cdot,\cdot\cdot] - i J [\cdot,\cdot\cdot]) ,$$

getting a characterization in terms of L of the curves of complex structures starting at a given integrable almost-complex structure J.

A.M. Fino and A. Tomassini, in [FT10, Sects. 6 and 7], studied several examples of families of almost-complex structures constructed in such a way. We provide here some further examples, starting with a curve of almost-complex structures on the four-dimensional torus, see [AT11].

Example 4.11 ([AT11, pp. 420–422]). A curve of almost-complex structures through the standard Kähler structure on the four-dimensional torus.
Let (J_0, ω_0) be the standard Kähler structure on the four-dimensional torus $\mathbb{T}^4$ with coordinates $\{x^j\}_{j\in\{1,\dots,4\}}$, that is,

$$J_0 := \begin{pmatrix} & & -1 & \\ & & & -1 \\ 1 & & & \\ & 1 & & \end{pmatrix} \in \text{End}\left(T\mathbb{T}^4\right) \qquad \text{and}$$

$$\omega_0 := dx^1 \wedge dx^3 + dx^2 \wedge dx^4 \in \wedge^2 \mathbb{T}^4 .$$

Set

$$L := \begin{pmatrix} \ell & & & \\ & 0 & & \\ & & -\ell & \\ & & & 0 \end{pmatrix} \in \text{End}\left(T\mathbb{T}^4\right) ,$$

where $\ell \in \mathcal{C}^\infty(\mathbb{T}^4; \mathbb{R})$, that is, $\ell \in \mathcal{C}^\infty(\mathbb{R}^4; \mathbb{R})$ is a $\mathbb{Z}^4$-periodic function. For $t \in (-\varepsilon, \varepsilon)$ with $\varepsilon > 0$ small enough, define

$$J_{t,\ell} := (\text{id} - t\, L)\, J_0\, (\text{id} - t\, L)^{-1} = \begin{pmatrix} & & -\frac{1-t\ell}{1+t\ell} & \\ & & & -1 \\ \frac{1+t\ell}{1-t\ell} & & & \\ & 1 & & \end{pmatrix} \in \text{End}\left(T\mathbb{T}^4\right) ,$$

obtaining a curve of ω_0-compatible almost-complex structures on $\mathbb{T}^4$, see also Proposition 4.3. To simplify the notation, set

$$\alpha := \alpha(t, \ell) := \frac{1 - t\ell}{1 + t\ell} .$$

A co-frame for the holomorphic cotangent bundle of $\mathbb{T}^4$ with respect to $J_{t,\ell}$ is given by

$$\begin{cases} \varphi^1_{t,\ell} := d x^1 + i \, \alpha \, d x^3 \\ \varphi^2_{t,\ell} := d x^2 + i \, d x^4 \end{cases},$$

with respect to which we compute the structure equations

$$\begin{cases} d\varphi^1_{t,\ell} = i \, d\alpha \wedge d x^3 \\ d\varphi^2_{t,\ell} = 0 \end{cases}.$$

Note that, taking $\ell = \ell\left(x^1, x^3\right)$, the corresponding almost-complex structure $J_{t,\ell}$ is integrable, in fact, $(J_{t,\ell}, \omega_0)$ is a Kähler structure on $\mathbb{T}^4$. Recall that, $\mathbb{T}^4$ being four-dimensional, $J_{t,\ell}$ is C^∞-pure-and-full by [DLZ10, Theorem 2.3]. For the sake of simplicity, assume $\ell = \ell\left(x^2\right)$ depending just on x^2 and non-constant. Set

$$v_1 := d x^1 \wedge d x^2 - \alpha \, d x^3 \wedge d x^4 \,,$$

$$v_2 := d x^1 \wedge d x^4 - \alpha \, d x^2 \wedge d x^3 \,,$$

$$w_1 := \alpha \, d x^1 \wedge d x^3 \,,$$

$$w_2 := d x^2 \wedge d x^4 \,,$$

$$w_3 := d x^1 \wedge d x^2 + \alpha \, d x^3 \wedge d x^4 \,,$$

$$w_4 := d x^1 \wedge d x^4 + \alpha \, d x^2 \wedge d x^3 \,.$$

Using this notation, an arbitrary $J_{t,\ell}$-anti-invariant real 2-form $\psi = A\, v_1 + B\, v_2$, with $A, B \in C^\infty\left(\mathbb{T}^4; \mathbb{R}\right)$, is d-closed if and only if

$$\begin{cases} \dfrac{\partial A}{\partial x^3} - \dfrac{\partial B}{\partial x^1}\, \alpha = 0 \\[4pt] \dfrac{\partial A}{\partial x^4} - \dfrac{\partial B}{\partial x^2} = 0 \\[4pt] -\dfrac{\partial A}{\partial x^1}\, \alpha - \dfrac{\partial B}{\partial x^3} = 0 \\[4pt] -\dfrac{\partial B}{\partial x^4}\, \alpha - \dfrac{\partial A}{\partial x^2}\, \alpha - A\, \dfrac{\partial \alpha}{\partial x^2} = 0 \end{cases}. \qquad (4.1)$$

By solving (4.1), we obtain the solutions

$$\psi = \frac{A}{\alpha}\, v_1 + B\, v_2 \qquad \text{where} \qquad A, B \in \mathbb{R} \,.$$

Therefore, for $t \in (-\varepsilon, \varepsilon)$ with $\varepsilon > 0$ small enough, we have

$$\dim_{\mathbb{R}} H_{J_{t,\ell}}^{(2,0),(0,2)}\left(\mathbb{T}^4; \mathbb{R}\right) = 2 = \dim_{\mathbb{R}} H_{J_0}^{(2,0),(0,2)}\left(\mathbb{T}^4; \mathbb{R}\right) ,$$

and hence

$$\dim_{\mathbb{R}} H_{J_{t,\ell}}^{(1,1)}\left(\mathbb{T}^4; \mathbb{R}\right) = 4 = \dim_{\mathbb{R}} H_{J_0}^{(1,1)}\left(\mathbb{T}^4; \mathbb{R}\right) .$$

In particular, this agrees with the upper-semi-continuity, respectively lower-semi-continuity, property proven in [DLZ11, Theorem 2.6] for four-dimensional almost-complex manifolds.

Now, we turn our attention to the case of dimension greater than 4.

Example 4.12 ([AT11, pp. 422–423]). A curve of almost-complex structures through the standard Kähler structure on the six-dimensional torus.
Let (J_0, ω_0) be the standard Kähler structure on the six-dimensional torus $\mathbb{T}^6$ with coordinates $\{x^j\}_{j \in \{1,\dots,6\}}$, that is,

$$J_0 := \begin{pmatrix} & & & -1 & & \\ & & & & -1 & \\ & & & & & -1 \\ 1 & & & & & \\ & 1 & & & & \\ & & 1 & & & \end{pmatrix} \in \mathrm{End}\left(T\mathbb{T}^6\right) \qquad \text{and}$$

$$\omega_0 := d x^1 \wedge d x^4 + d x^2 \wedge d x^5 + d x^3 \wedge d x^6 .$$

Set

$$L = \begin{pmatrix} \ell & & & & & \\ & 0 & & & & \\ & & 0 & & & \\ & & & -\ell & & \\ & & & & 0 & \\ & & & & & 0 \end{pmatrix} \in \mathrm{End}\left(T\mathbb{T}^6\right) ,$$

where $\ell \in \mathcal{C}^\infty(\mathbb{T}^6; \mathbb{R})$, that is, $\ell \in \mathcal{C}^\infty(\mathbb{R}^6; \mathbb{R})$ is a $\mathbb{Z}^6$-periodic function. For $t \in (-\varepsilon, \varepsilon)$ with $\varepsilon > 0$ small enough, define

$$J_{t,\ell} := (\mathrm{id} - t\,L)\,J_0\,(\mathrm{id} - t\,L)^{-1}$$

$$= \begin{pmatrix} & \left| \begin{matrix} -\frac{1-t\ell}{1+t\ell} \\ & -1 \\ & & -1 \end{matrix} \right. \\ \hline \begin{matrix} \frac{1+t\ell}{1-t\ell} \\ & 1 \\ & & 1 \end{matrix} & \end{pmatrix} \in \mathrm{End}\left(T\mathbb{T}^6\right) ,$$

obtaining a curve of ω_0-compatible almost-complex structures on $\mathbb{T}^6$, see also Example 4.2. Setting

$$\alpha := \alpha(t,\ell) := \frac{1 - t\ell}{1 + t\ell} ,$$

a co-frame for the holomorphic cotangent bundle of $\mathbb{T}^6$ with respect to $J_{t,\ell}$ is given by

$$\begin{cases} \varphi_{t,\ell}^1 := \mathrm{d}x^1 + i\,\alpha\,\mathrm{d}x^4 \\ \varphi_{t,\ell}^2 := \mathrm{d}x^2 + i\,\mathrm{d}x^5 \\ \varphi_{t,\ell}^3 := \mathrm{d}x^3 + i\,\mathrm{d}x^6 \end{cases} ,$$

with respect to which the structure equations are

$$\begin{cases} \mathrm{d}\varphi_{t,\ell}^1 = i\,\mathrm{d}\alpha \wedge \mathrm{d}x^4 \\ \mathrm{d}\varphi_{t,\ell}^2 = 0 \\ \mathrm{d}\varphi_{t,\ell}^3 = 0 \end{cases} .$$

Note that if $\ell = \ell\left(x^1, x^4\right)$, then we get a curve of integrable almost-complex structures, in fact, of Kähler structures, on $\mathbb{T}^6$: in particular, in such a case, $J_{t,\ell}$ is C^∞-pure-and-full. Therefore, as an example, assume that $\ell = \ell\left(x^3\right)$ depends just on x^3 and is non-constant.

An arbitrary $J_{t,\ell}$-anti-invariant real 2-form

$$\psi = A\left(\mathrm{d}x^1 \wedge \mathrm{d}x^2 - \alpha\,\mathrm{d}x^4 \wedge \mathrm{d}x^5\right) + B\left(\mathrm{d}x^1 \wedge \mathrm{d}x^5 - \alpha\,\mathrm{d}x^2 \wedge \mathrm{d}x^4\right)$$

$$+ C\left(\mathrm{d}x^1 \wedge \mathrm{d}x^3 - \alpha\,\mathrm{d}x^4 \wedge \mathrm{d}x^6\right) + D\left(\mathrm{d}x^1 \wedge \mathrm{d}x^6 - \alpha\,\mathrm{d}x^3 \wedge \mathrm{d}x^4\right)$$

$$+ E\left(\mathrm{d}x^2 \wedge \mathrm{d}x^3 - \mathrm{d}x^5 \wedge \mathrm{d}x^6\right) + F\left(\mathrm{d}x^2 \wedge \mathrm{d}x^6 - \mathrm{d}x^3 \wedge \mathrm{d}x^5\right) ,$$

with $A, B, C, D, E, F \in \mathcal{C}^\infty \left(\mathbb{T}^6; \mathbb{R} \right)$, is d-closed if and only if

$$
\begin{cases}
\frac{\partial A}{\partial x^3} - \frac{\partial C}{\partial x^2} + \frac{\partial E}{\partial x^1} = 0 \\[4pt]
\frac{\partial A}{\partial x^4} - \frac{\partial B}{\partial x^1} \alpha = 0 \\[4pt]
\frac{\partial A}{\partial x^5} - \frac{\partial B}{\partial x^2} = 0 \\[4pt]
\frac{\partial A}{\partial x^6} - \frac{\partial D}{\partial x^2} + \frac{\partial F}{\partial x^1} = 0 \\[4pt]
\frac{\partial C}{\partial x^4} - \frac{\partial D}{\partial x^1} \alpha = 0 \\[4pt]
-\frac{\partial B}{\partial x^3} + \frac{\partial C}{\partial x^5} - \frac{\partial F}{\partial x^1} = 0 \\[4pt]
\frac{\partial C}{\partial x^6} - \frac{\partial D}{\partial x^3} = 0 \\[4pt]
-\frac{\partial A}{\partial x^1} \alpha - \frac{\partial B}{\partial x^4} = 0 \\[4pt]
-\frac{\partial C}{\partial x^1} \alpha - \frac{\partial D}{\partial x^4} = 0 \\[4pt]
\frac{\partial B}{\partial x^6} - \frac{\partial D}{\partial x^5} - \frac{\partial E}{\partial x^1} = 0 \\[4pt]
\frac{\partial (B \alpha)}{\partial x^3} - \frac{\partial D}{\partial x^2} \alpha + \frac{\partial E}{\partial x^4} = 0 \\[4pt]
\frac{\partial E}{\partial x^5} - \frac{\partial F}{\partial x^2} = 0 \\[4pt]
\frac{\partial E}{\partial x^6} - \frac{\partial F}{\partial x^3} = 0 \\[4pt]
-\frac{\partial A}{\partial x^2} \alpha - \frac{\partial B}{\partial x^5} \alpha = 0 \\[4pt]
-\frac{\partial B}{\partial x^6} \alpha - \frac{\partial C}{\partial x^2} \alpha - \frac{\partial F}{\partial x^4} = 0 \\[4pt]
-\frac{\partial E}{\partial x^2} - \frac{\partial F}{\partial x^5} = 0 \\[4pt]
-\frac{\partial (A \alpha)}{\partial x^3} - \frac{\partial D}{\partial x^5} \alpha + \frac{\partial F}{\partial x^4} = 0 \\[4pt]
-\frac{\partial (C \alpha)}{\partial x^3} - \frac{\partial D}{\partial x^6} \alpha = 0 \\[4pt]
-\frac{\partial E}{\partial x^3} - \frac{\partial F}{\partial x^6} = 0 \\[4pt]
-\frac{\partial A}{\partial x^6} \alpha + \frac{\partial C}{\partial x^5} \alpha - \frac{\partial E}{\partial x^4} = 0
\end{cases}
\tag{4.2}
$$

For $t \neq 0$ small enough, by solving (4.2), we obtain that the $J_{t,\ell}$-anti-invariant real d-closed 2-forms are

$$
\psi = \frac{C}{\alpha} \left(\mathrm{d} x^{13} - \alpha \, \mathrm{d} x^{46} \right) + D \left(\mathrm{d} x^{16} - \alpha \, \mathrm{d} x^{34} \right) + E \left(\mathrm{d} x^{23} - \mathrm{d} x^{56} \right)
$$
$$
+ F \left(\mathrm{d} x^{26} - \mathrm{d} x^{35} \right) ,
$$

where $C, D, E, F \in \mathbb{R}$.

For $t \neq 0$ small enough, we have

$$
\dim_{\mathbb{R}} H_{J_{t,\ell}}^{(2,0),(0,2)} \left(\mathbb{T}^6; \mathbb{R} \right) \leq 4 < 6 = \dim_{\mathbb{R}} H_{J_0}^{(2,0),(0,2)} \left(\mathbb{T}^6; \mathbb{R} \right) ,
$$

and hence the function $t \mapsto \dim_\mathbb{R} H_{J_{t,\ell}}^{(2,0),(0,2)}\left(\mathbb{T}^6; \mathbb{R}\right)$ is upper-semi-continuous at 0. On the other hand, the explicit computations for $H_{J_{t,\ell}}^{(1,1)}\left(\mathbb{T}^6; \mathbb{R}\right)$ are not so straightforward. In particular, it is not clear if $J_{t,\ell}$ remains still C^∞-full; note that $J_{t,\ell}$ is C^∞-pure by [DLZ10, Proposition 2.7] or [FT10, Proposition 3.2].

We recall here the construction of curves of almost-complex structures through an almost-complex structure J by means of a J-anti-invariant real 2-form, as introduced by J. Lee in [Lee04, Sect. 1], in the context of holomorphic curves on symplectic manifolds and Gromov and Witten invariants.

Let J be an almost-complex structure on a compact manifold X; let g be a J-Hermitian metric on X and fix $\gamma \in \left(\wedge^{2,0}X \oplus \wedge^{0,2}X\right) \cap \wedge^2 X$. Define $V_\gamma \in \mathrm{End}\,(TX)$ such that

$$\gamma\,(\cdot, \cdot\cdot) \;=\; g\left(V_\gamma\,\cdot, \cdot\cdot\right)\;; \tag{4.3}$$

a direct computation shows that $V_\gamma\,J + J\,V_\gamma = 0$. Therefore, setting

$$L_\gamma \;:=\; \frac{1}{2}\,V_\gamma\,J \;\in\; \mathrm{End}\,(TX)\;,$$

one gets that $L_\gamma\,J + J\,L_\gamma = 0$. For $t \in (-\varepsilon,\,\varepsilon)$ with $\varepsilon > 0$ small enough, define

$$J_{t,\gamma} \;:=\; \left(\mathrm{id} - t\,L_\gamma\right)\,J\,\left(\mathrm{id} - t\,L_\gamma\right)^{-1} \;\in\; \mathrm{End}\,(TX)\;,$$

obtaining a curve $\left\{J_{t,\gamma}\right\}_{t \in (-\varepsilon,\varepsilon)}$ of almost-complex structures associated with γ.

We give an example of a C^∞-pure-and-full structure on a non-Kähler manifold such that the stability property of the C^∞-pure-and-fullness holds along a curve obtained using the construction by J. Lee, see [AT11].

Example 4.13 ([AT11, pp. 423–425]). A curve of C^∞-pure-and-full almost-complex structures on the completely-solvable solvmanifold $N^6(c)$.
We recall that the manifold $N^6(c)$ is a compact six-dimensional completely-solvable solvmanifold defined, for suitable $c \in \mathbb{R}$, as the product

$$N^6(c) \;:=\; \left(\Gamma(c)\,\backslash\mathrm{Sol}(3)\right) \times \left(\Gamma(c)\,\backslash\mathrm{Sol}(3)\right)\;,$$

where $\mathrm{Sol}(3)$ is a completely-solvable Lie group and $\Gamma(c)$ is a co-compact discrete subgroup of $\mathrm{Sol}(3)$, [AGH63, Sect. 3], see Example 4.10. It has been studied in [BG90, Example 1] as an example of a cohomologically Kähler manifold, and in [FMS03, Example 3.4] by M. Fernández, V. Muñoz, and J.A. Santisteban, as an example of a formal manifold admitting a symplectic structure satisfying the Hard Lefschetz Condition and with no Kähler structure, [FMS03, Theorem 3.5]. A.M. Fino and A. Tomassini provided in [FT10, Sect. 6.3] a family of C^∞-pure-and-full structures on $N^6(c)$. We construct here a curve of C^∞-pure-and-full almost-complex structures on $N^6(c)$ using the construction by J. Lee, [Lee04, Sect. 1].

Let $\{e^i\}_{i\in\{1,\dots,6\}}$ be a co-frame for $N^6(c)$ such that the structure equations are

$$\begin{cases} de^1 = & c\,e^1 \wedge e^3 \\ de^2 = & -c\,e^2 \wedge e^3 \\ de^3 = & 0 \\ de^4 = & c\,e^4 \wedge e^6 \\ de^5 = & -c\,e^5 \wedge e^6 \\ de^6 = & 0 \end{cases}.$$

Take the almost-complex structure

$$J = \begin{pmatrix} -1 & & & \\ 1 & & & \\ & -1 & & \\ & 1 & & \\ & & -1 & \\ & & 1 & \end{pmatrix} \in \mathrm{End}\left(TN^6(c)\right) .$$

By A. Hattori's theorem [Hat60, Corollary 4.2], one computes

$$H^2_{dR}\left(N^6(c); \mathbb{R}\right) = \mathbb{R}\langle e^1 \wedge e^2,\ e^3 \wedge e^6 - e^4 \wedge e^5,\ e^3 \wedge e^6 + e^4 \wedge e^5\rangle ,$$

proving that $\left(N^6(c),\ J\right)$ is $\mathcal{C}^\infty$-pure-and-full and pure-and-full: indeed, the above harmonic representatives with respect to the $(\mathrm{Sol}(3) \times \mathrm{Sol}(3))$-left-invariant metric $g := \sum_{j=1}^6 e^j \odot e^j$ are of pure type with respect to J, and hence [FT10, Theorem 3.7] assures the $\mathcal{C}^\infty$-pure-and-fullness and the pure-and-fullness. Note that

$$H_J^{(2,0),(0,2)}\left(N^6(c); \mathbb{R}\right) = \mathbb{R}\langle e^3 \wedge e^6 + e^4 \wedge e^5\rangle ;$$

apply J. Lee's construction [Lee04, Sect. 1] to the real J-anti-invariant 2-form

$$\gamma := e^3 \wedge e^6 + e^4 \wedge e^5 :$$

the linear map $V \in \mathrm{End}(TX)$ representing γ as in (4.3) is

$$V = \begin{pmatrix} 0 & & & \\ & 0 & & \\ & & & -1 \\ & & -1 & \\ & & 1 & \\ & 1 & & \end{pmatrix} \in \mathrm{End}\left(TN^6(c)\right) ,$$

and then it is straightforward to compute

$$
L = \left(
\begin{array}{cc|cc}
0 & & & \\
 & 0 & & \\
\hline
 & & & -\frac{1}{2} \\
 & & \frac{1}{2} & \\
\hline
 & \frac{1}{2} & & \\
 & -\frac{1}{2} & &
\end{array}
\right) \in \mathrm{End}\left(TN^6(c)\right) ,
$$

and

$$
J_t := J_{t,\gamma} = \left(
\begin{array}{cc|cc|cc}
 & -1 & & & & \\
1 & & & & & \\
\hline
 & & -\frac{4-t^2}{4+t^2} & & -\frac{4t}{4+t^2} & \\
 & & \frac{4-t^2}{4+t^2} & & -\frac{4t}{4+t^2} & \\
\hline
 & & & \frac{4t}{4+t^2} & & -\frac{4-t^2}{4+t^2} \\
 & & \frac{4t}{4+t^2} & & \frac{4-t^2}{4+t^2} &
\end{array}
\right) \in \mathrm{End}\left(TN^6(c)\right) .
$$

To shorten the notation, set

$$
\alpha := \alpha(t) := \frac{4-t^2}{4+t^2} , \qquad \beta := \beta(t) := \frac{4t}{4+t^2} .
$$

A co-frame for the J_t-holomorphic cotangent bundle is given by

$$
\begin{cases}
\varphi_t^1 := e^1 + i\, e^2 \\
\varphi_t^2 := e^3 + i\left(\alpha\, e^4 + \beta\, e^6\right) \\
\varphi_t^3 := e^5 + i\left(-\beta\, e^4 + \alpha\, e^6\right)
\end{cases} .
$$

Since the real d-closed 2-forms

$$
\frac{1}{2i}\,\varphi_t^{1\bar{1}} , \qquad \frac{1}{2i}\,\varphi_t^{3\bar{3}} - \frac{\alpha}{c}\, d\,e^5 , \qquad \frac{1}{2i}\left(\beta\,\varphi_t^{2\bar{2}} + \alpha\left(\varphi_t^{2\bar{3}} - \varphi_t^{\bar{2}3}\right)\right) + \frac{1}{2i}\,\varphi_t^{3\bar{3}}
$$

generate three different cohomology classes, we get that, for $t \neq 0$ small enough,

$$
H_{dR}^2\left(N^6(c);\mathbb{R}\right) = H_{J_t}^{(1,1)}\left(N^6(c);\mathbb{R}\right) ,
$$

and so, in particular, J is $\mathcal{C}^\infty$-full and pure. A straightforward computation yields

$$H_{dR}^4\left(N^6(c);\mathbb{R}\right) = \mathbb{R}\left\langle *_g\left(\frac{1}{2\,\mathrm{i}}\,\varphi_t^{1\bar{1}}\right),\ *_g\left(\varphi_t^{3\bar{3}} - \frac{\alpha}{c}\,\mathrm{d}\,e^5\right) + \frac{\alpha}{c}\,\mathrm{d}\left(e^{125}\right),\right.$$

$$\left.\frac{\alpha}{4}\left(\varphi_t^{12\bar{1}\bar{3}} + \varphi_t^{\bar{1}2\bar{1}3}\right) + \frac{\beta}{4}\,\varphi_t^{12\bar{1}\bar{2}} + \frac{\alpha\,\beta}{c}\,\mathrm{d}\left(e^{125}\right)\right\rangle$$

$$= H_{J_t}^{(2,2)}\left(N^6(c);\mathbb{R}\right)\ ,$$

therefore $N^6(c)$ is also $\mathcal{C}^\infty$-full at the 4th stage and hence full and $\mathcal{C}^\infty$-pure.

We resume the content of the last example in the following theorem.

Theorem 4.9 ([AT11, Theorem 4.1]). *There exists a compact manifold $N^6(c)$ endowed with an almost-complex structure J and a J-Hermitian metric g such that:*

(i) *J is $\mathcal{C}^\infty$-pure-and-full;*
(ii) *each J-anti-invariant g-harmonic form gives rise to a curve $\{J_t\}_{t\in(-\varepsilon,\varepsilon)}$ of $\mathcal{C}^\infty$-pure-and-full almost-complex structures on $N^6(c)$, where $\varepsilon > 0$ is small enough, using J. Lee's construction;*
(iii) *furthermore, the function*

$$(-\varepsilon,\varepsilon)\ni t \mapsto \dim_\mathbb{R} H_{J_t}^{(2,0),(0,2)}\left(N^6(c);\mathbb{R}\right)\in\mathbb{N}$$

is upper-semi-continuous at 0.

4.3.2   The Semi-continuity Problem

Given a compact four-dimensional manifold X and a family $\{J_t\}_t$ of almost-complex structures on X, T. Drăghici, T.-J. Li, and W. Zhang studied in [DLZ11] the semi-continuity properties of the functions $t \mapsto \dim_\mathbb{R} H_{J_t}^+(X)$ and $t \mapsto \dim_\mathbb{R} H_{J_t}^-(X)$. They proved the following result.

Theorem 4.10 ([DLZ11, Theorem 2.6]). *Let X be a compact four-dimensional manifold and let $\{J_t\}_{t\in I\subseteq\mathbb{R}}$ be a family of ($\mathcal{C}^\infty$-pure-and-full) almost-complex structures on X, for $I \subseteq \mathbb{R}$ an interval. Then the function*

$$I \ni t \mapsto \dim_\mathbb{R} H_{J_t}^-(X)\in\mathbb{N}$$

is upper-semi-continuous, and therefore the function

$$I \ni t \mapsto \dim_\mathbb{R} H_{J_t}^+(X)\in\mathbb{N}$$

is lower-semi-continuous.

The previous result is closely related to the geometry of four-dimensional manifolds; more precisely, it follows from M. Lejmi's result in [Lej10, Lemma 4.1] that a certain operator is a self-adjoint strongly elliptic linear operator with kernel the harmonic J-anti-invariant 2-forms. In this section, we are concerned with establishing if a similar semi-continuity result could occur in dimension higher than 4, possibly assuming further hypotheses.

4.3.2.1 Counterexamples to Semi-continuity

First of all, we provide two examples showing that, in general, no semi-continuity property holds in dimension higher than 4.

The following result provides a counterexample to the upper-semi-continuity of $t \mapsto \dim_{\mathbb{R}} H_{J_t}^-$ in dimension greater than 4.

Proposition 4.9 ([AT12a, Proposition 4.1]). *The compact ten-dimensional manifold $\eta\beta_5$ is endowed with a C^∞-pure-and-full complex structure J and a curve $\{J_t\}_{t\in\Delta(0,\varepsilon)\subset\subset\mathbb{C}}$ of complex structures (which are non-C^∞-pure for $t \neq 0$), with $J_0 = J$, and $\varepsilon > 0$, such that the function*

$$\Delta(0,\varepsilon) \ni t \mapsto \dim_{\mathbb{R}} H_{J_t}^- (\eta\beta_5) \in \mathbb{N}$$

is not upper-semi-continuous.

Proof. The proof follows from the following example, [AT12a, Example 4.2].

Consider the nilmanifold $\eta\beta_5$ endowed with its natural complex structure J, as described in Example 4.5. We recall that, chosen a suitable co-frame $\{\varphi^j\}_{j\in\{1,\dots,5\}}$ of the holomorphic cotangent bundle, the complex structure equations are

$$\begin{cases} d\varphi^1 = d\varphi^2 = d\varphi^3 = d\varphi^4 = 0 \\ d\varphi^5 = -\varphi^{12} - \varphi^{34} \end{cases} .$$

By K. Nomizu's theorem [Nom54, Theorem 1], it is straightforward to compute

$$H_{dR}^2(\eta\beta_5;\mathbb{C}) = \mathbb{C}\langle\varphi^{13}, \varphi^{14}, \varphi^{23}, \varphi^{24}, \varphi^{\bar{1}3}, \varphi^{\bar{1}4}, \varphi^{\bar{2}3}, \varphi^{\bar{2}4}, \varphi^{12} - \varphi^{34}, \varphi^{\bar{1}\bar{2}} - \varphi^{\bar{3}\bar{4}}\rangle$$

$$\oplus \mathbb{C}\langle\varphi^{1\bar{1}}, \varphi^{1\bar{2}}, \varphi^{1\bar{3}}, \varphi^{1\bar{4}}, \varphi^{2\bar{1}}, \varphi^{2\bar{2}}, \varphi^{2\bar{3}}, \varphi^{2\bar{4}}, \varphi^{3\bar{1}}, \varphi^{3\bar{2}}, \varphi^{3\bar{3}}, \varphi^{3\bar{4}}, \varphi^{4\bar{1}}, \varphi^{4\bar{2}}, \varphi^{4\bar{3}}, \varphi^{4\bar{4}}\rangle$$

(where, as usually, we have listed the harmonic representatives with respect to the left-invariant Hermitian metric $\sum_{j=1}^5 \varphi^j \odot \bar{\varphi}^j$ instead of their classes, and we have shortened, e.g., $\varphi^{A\bar{B}} := \varphi^A \wedge \bar{\varphi}^B$). Hence the complex structure J is C^∞-pure-and-full by [FT10, Theorem 3.7], and

$$\dim_{\mathbb{R}} H_J^- (\eta\beta_5) = 10, \qquad\qquad \dim_{\mathbb{R}} H_J^+ (\eta\beta_5) = 16.$$

Now, for $\varepsilon > 0$ small enough, consider the curve $\{J_t\}_{t\in\Delta(0,\varepsilon)}$ of complex structures such that a co-frame for the J_t-holomorphic cotangent bundle is given by $\left\{\varphi_t^j\right\}_{j\in\{1,\ldots,5\}}$, where, for any $t \in \Delta(0,\varepsilon)$,

$$
\begin{cases}
\varphi_t^1 := \varphi^1 + t\,\bar\varphi^1 \\
\varphi_t^2 := \varphi^2 \\
\varphi_t^3 := \varphi^3 \\
\varphi_t^4 := \varphi^4 \\
\varphi_t^5 := \varphi^5
\end{cases},
$$

see Example 4.5. The structure equations with respect to $\left\{\varphi_t^j\right\}_{j\in\{1,\ldots,5\}}$ are

$$
\begin{cases}
d\varphi_t^1 = d\varphi_t^2 = d\varphi_t^3 = d\varphi_t^4 = 0 \\
d\varphi_t^5 = -\frac{1}{1-|t|^2}\,\varphi_t^{12} - \varphi_t^{34} - \frac{t}{1-|t|^2}\,\varphi_t^{2\bar1}
\end{cases}.
$$

When $\varepsilon > 0$ is small enough, for $t \in \Delta(0,\varepsilon)\setminus\{0\}$, the complex structure J_t is not $\mathcal{C}^\infty$-pure: indeed,

$$
H_{J_t}^{(1,1)}(\eta\beta_5;\mathbb{C}) \ni \left[\frac{t}{1-|t|^2}\,\varphi_t^{2\bar1} + d\varphi_t^5\right] = \left[-\frac{1}{1-|t|^2}\,\varphi_t^{12} - \varphi_t^{34}\right] \in H_{J_t}^{(2,0)}(\eta\beta_5;\mathbb{C}),
$$

where $\left[\frac{t}{1-|t|^2}\,\varphi_t^{2\bar1}\right] \in H_{dR}^2(\eta\beta_5;\mathbb{C})$ is a non-zero cohomology class by K. Nomizu's theorem [Nom54, Theorem 1]. Moreover, note that

$$
H_{J_t}^{(2,0),(0,2)}(\eta\beta_5;\mathbb{C}) \supseteq \mathbb{C}\left\langle \varphi_t^{13}, \varphi_t^{14}, \varphi_t^{23}, \varphi_t^{24}, \varphi_t^{\bar1\bar3}, \varphi_t^{\bar1\bar4}, \varphi_t^{\bar2\bar3}, \varphi_t^{\bar2\bar4}, \varphi_t^{12}, \varphi_t^{34}, \varphi_t^{\bar1\bar2}, \varphi_t^{\bar3\bar4}\right\rangle,
$$

hence, for every $t \in \Delta(0,\varepsilon)\setminus\{0\}$,

$$
\dim_{\mathbb{R}} H_{J_0}^-(\eta\beta_5) = 10 < 12 \le \dim_{\mathbb{R}} H_{J_t}^-(\eta\beta_5),
$$

and in particular $t \mapsto h_{J_t}^-$ is not upper-semi-continuous at 0.   □

The following result provides a counterexample to the lower-semi-continuity of $t \mapsto \dim_{\mathbb{R}} H_{J_t}^+$ in dimension greater than 4.

Proposition 4.10 ([AT12a, Proposition 4.3]). *The compact six-dimensional manifold $\mathbb{S}^3 \times \mathbb{T}^3$ is endowed with a $\mathcal{C}^\infty$-full (non-integrable) almost-complex structure J and a curve $\{J_t\}_{t\in\Delta(0,\varepsilon)\subset\mathbb{C}}$, where $\varepsilon > 0$, of (non-integrable) almost-complex structures (which are not $\mathcal{C}^\infty$-pure), with $J_0 = J$, such that*

$$
\Delta(0,\varepsilon) \ni t \mapsto \dim_{\mathbb{R}} H_{J_t}^+(\mathbb{S}^3 \times \mathbb{T}^3) \in \mathbb{N}
$$

is not lower-semi-continuous.

Proof. The proof follows from the following example, [AT12a, Example 4.4].

Consider the compact six-dimensional manifold $\mathbb{S}^3 \times \mathbb{T}^3$, and set a global co-frame $\{e^j\}_{j \in \{1,\dots,6\}}$ with respect to which the structure equations are

$$\left(23, \; -13, \; 12, \; 0^3\right) \; ;$$

consider the (non-integrable) almost-complex structure J defined requiring that

$$\begin{cases} \varphi^1 := e^1 + i \, e^4 \\ \varphi^2 := e^2 + i \, e^5 \\ \varphi^3 := e^3 + i \, e^6 \end{cases}$$

generate the $C^\infty\left(\mathbb{S}^3 \times \mathbb{T}^3; \mathbb{C}\right)$-module of $(1,0)$-forms on $\mathbb{S}^3 \times \mathbb{T}^3$. By the Künneth formula, one computes

$$H^2_{dR}(\mathbb{S}^3 \times \mathbb{T}^3; \mathbb{C}) = \mathbb{C}\left\langle e^{45}, \; e^{46}, \; e^{56} \right\rangle$$

$$= \left\langle \varphi^{12} + \varphi^{\bar{1}\bar{2}}, \; \varphi^{13} + \varphi^{\bar{1}\bar{3}}, \; \varphi^{23} + \varphi^{\bar{2}\bar{3}} \right\rangle = H^-_J\left(\mathbb{S}^3 \times \mathbb{T}^3\right)$$

$$= \left\langle \varphi^{1\bar{2}} - \varphi^{2\bar{1}}, \; \varphi^{1\bar{3}} - \varphi^{3\bar{1}}, \; \varphi^{2\bar{3}} - \varphi^{3\bar{2}} \right\rangle = H^+_J\left(\mathbb{S}^3 \times \mathbb{T}^3\right) \; .$$

For $\varepsilon > 0$ small enough, consider the curve $\{J_t\}_{t \in \Delta(0,\varepsilon) \subset \subset \mathbb{C}}$ of (non-integrable) almost-complex structures defined requiring that, for any $t \in \Delta(0,\varepsilon)$, the J_t-holomorphic cotangent bundle has co-frame

$$\begin{cases} \varphi^1_t := \varphi^1 + t \, \bar{\varphi}^1 \\ \varphi^2_t := \varphi^2 \\ \varphi^3_t := \varphi^3 \end{cases} \; .$$

By using the F.A. Belgun symmetrization trick, [Bel00, Theorem 7], we have that, for $t \in \Delta(0,\varepsilon) \setminus \mathbb{R}$,

$$\left[\varphi^{1\bar{2}} - \varphi^{2\bar{1}}\right] = \left[\frac{1}{1-|t|^2}\left(\varphi^{1\bar{2}}_t - \varphi^{2\bar{1}}_t\right) - \frac{1}{1-|t|^2}\left(\bar{t}\,\varphi^{12}_t + t\,\varphi^{\bar{1}\bar{2}}_t\right)\right] \notin H^+_{J_t}\left(\mathbb{S}^3 \times \mathbb{T}^3\right)$$

and

$$\left[\varphi^{1\bar{3}} - \varphi^{3\bar{1}}\right] = \left[\frac{1}{1-|t|^2}\left(\varphi^{1\bar{3}}_t - \varphi^{3\bar{1}}_t\right) - \frac{1}{1-|t|^2}\left(\bar{t}\,\varphi^{13}_t + t\,\varphi^{\bar{1}\bar{3}}_t\right)\right] \notin H^+_{J_t}\left(\mathbb{S}^3 \times \mathbb{T}^3\right);$$

indeed, the terms $\psi_1 := \bar{t}\,\varphi^{12}_t + t\,\varphi^{\bar{1}\bar{2}}_t$, respectively $\psi_2 := \bar{t}\,\varphi^{13}_t + t\,\varphi^{\bar{1}\bar{3}}_t$, cannot be written as the sum of a J_t-invariant form and a d-exact form: on the contrary, since

ψ_1, and ψ_2 are left-invariant, applying F.A. Belgun's symmetrization map, [Bel00, Theorem 7], we can suppose that the J_t-anti-invariant component of the d-exact term is actually the J_t-anti-invariant component of the differential of a left-invariant 1-form; but the image of the differential on the space of left-invariant 1-forms is

$$\mathrm{d} \wedge^1 \mathfrak{g}_{\mathbb{C}}^* = \mathbb{C}\left\langle \varphi_t^{23} + \varphi_t^{2\bar{3}} - \varphi_t^{3\bar{2}} + \varphi_t^{\bar{2}\bar{3}}, (1-\bar{t})\,\varphi_t^{13} + (1-\bar{t})\,\varphi^{1\bar{3}} - (1-t)\,\varphi^{3\bar{1}}\right.$$

$$\left. + (1-t)\,\varphi^{\bar{1}\bar{3}}, (1-\bar{t})\,\varphi_t^{12} + (1-\bar{t})\,\varphi^{1\bar{2}} - (1-t)\,\varphi^{2\bar{1}} + (1-t)\,\varphi^{\bar{1}\bar{2}}\right\rangle,$$

and hence one should have $t \in \mathbb{R}$. Hence, we have that, for $t \in \Delta(0, \varepsilon) \setminus \mathbb{R}$,

$$\dim_{\mathbb{R}} H_{J_t}^+\left(\mathbb{S}^3 \times \mathbb{T}^3\right) = 1 < 3 = \dim_{\mathbb{R}} H_{J_0}^+\left(\mathbb{S}^3 \times \mathbb{T}^3\right),$$

and consequently, in particular, $t \mapsto \dim_{\mathbb{R}} H_{J_t}^+\left(\mathbb{S}^3 \times \mathbb{T}^3\right)$ is not lower-semi-continuous at 0.   □

4.3.2.2 Semi-continuity in a Stronger Sense

Note that Propositions 4.9 and 4.10 force us to consider stronger conditions under which semi-continuity may occur, or to slightly modify the statement of the semi-continuity problem.

We turn our attention to the aim of giving a more precise statement of the semi-continuity problem. We notice that, for a compact four-dimensional manifold X endowed with a family $\{J_t\}_{t \in \Delta(0,\varepsilon)}$ of almost-complex structures, one does not have only the semi-continuity properties of $t \mapsto \dim_{\mathbb{R}} H_{J_t}^+(X)$ and $t \mapsto \dim_{\mathbb{R}} H_{J_t}^-(X)$, but one gets also that every J_0-invariant class admits a J_t-invariant class *close to it*. This is also a sufficient condition to assure that, if α is a J_0-compatible symplectic structure on X, then there is a J_t-compatible symplectic structure α_t on X for t small enough. Therefore, we are interested in the following problem.

Let X be a compact manifold endowed with an almost-complex structure J and with a curve $\{J_t\}_{t \in (-\varepsilon, \varepsilon) \subset \mathbb{R}}$ of almost-complex structures, where $\varepsilon > 0$ is small enough, such that $J_0 = J$. Suppose that

$$H_J^+(X) = \mathbb{C}\left\langle [\alpha^1], \dots, [\alpha^k] \right\rangle,$$

where $\alpha^1, \dots, \alpha^k$ are forms of type $(1, 1)$ with respect to J. We look for further hypotheses assuring that, for every $t \in (-\varepsilon, \varepsilon)$,

$$H_{J_t}^+(X) \supseteq \mathbb{C}\left\langle [\alpha_t^1], \dots, [\alpha_t^k] \right\rangle$$

with

$$\alpha_t^j = \alpha^j + o(1).$$

In this case, $(-\varepsilon, \varepsilon) \ni t \mapsto \dim_\mathbb{R} H_{J_t}^+(X) \in \mathbb{N}$ is a lower-semi-continuous function at 0.

Concerning this problem, we have the following result.

Proposition 4.11 ([AT12a, Proposition 4.5]). *Let X be a compact manifold endowed with an almost-complex structure J. Take $L \in \mathrm{End}(TX)$ and consider the curve $\{J_t\}_{t \in (-\varepsilon, \varepsilon) \subset \mathbb{R}}$ of almost-complex structures defined by*

$$J_t := (\mathrm{id} - t\, L)\, J\, (\mathrm{id} - t\, L)^{-1} \in \mathrm{End}(TX) \,,$$

where $\varepsilon > 0$ is small enough. For every $[\alpha] \in H_J^+(X)$ with $\alpha \in \wedge_J^{1,1}(X) \cap \wedge^2 X$, the following conditions are equivalent:

(i) there exists a family $\{\eta_t = \alpha + \mathrm{o}\,(1)\}_{t \in (-\varepsilon, \varepsilon)} \subseteq \wedge_{J_t}^{1,1}(X) \cap \wedge^2 X$ of real 2-forms, with $\varepsilon > 0$ small enough, depending real-analytically in t and such that $\mathrm{d}\, \eta_t = 0$, for every $t \in (-\varepsilon, \varepsilon)$;

(ii) there exists $\{\beta_j\}_{j \in \mathbb{N} \setminus \{0\}} \subseteq \wedge^2 X$ solution of the system

$$\mathrm{d}\Bigg(\beta_j + 2\alpha\,(L^j \cdot, \cdot\cdot) + 4 \sum_{k=1}^{j-1} \alpha\,(L^{j-k} \cdot, L^k \cdot\cdot) + 2\alpha\,(\cdot, L^j \cdot\cdot)$$

$$+ \sum_{h=1}^{j-1} \Bigg(2\beta_h\,(L^{j-h} \cdot, \cdot\cdot) + 4 \sum_{k=1}^{j-h-1} \alpha\,(L^{j-h-k} \cdot, L^k \cdot\cdot) + 2\alpha\,(\cdot, L^{j-h} \cdot\cdot)\Bigg)\Bigg)$$

$$= 0 \,, \tag{4.4}$$

varying $j \in \mathbb{N} \setminus \{0\}$, such that $\sum_{j \geq 1} t^j\, \beta_j$ converges.

In particular, the first order obstruction to the existence of η_t as in (i) reads: there exists $\beta_1 \in \wedge^2 X$ such that

$$\mathrm{d}\,(\beta_1 + 2\alpha(L \cdot, \cdot\cdot) + 2\alpha(\cdot, L \cdot\cdot)) = 0 \,. \tag{4.5}$$

Proof. Expanding J_t in power series with respect to t, one gets

$$J_t = J + \sum_{j \geq 1} 2 t^j\, J\, L^j \,,$$

and then, for every $\varphi \in \wedge^2 X$, one computes

$$J_t\, \varphi(\cdot, \cdot\cdot) = J\, \varphi(\cdot, \cdot\cdot) + 2t\, J\,(\varphi(L \cdot, \cdot\cdot) + \varphi(\cdot, L \cdot\cdot)) + \mathrm{o}\,(|t|)$$

and

$$\mathrm{d}_{J_t}^c\, \varphi = J_t^{-1}\, \mathrm{d}\, J_t\, \varphi = \mathrm{d}_J^c\, \varphi + 2t\, J_t\, \mathrm{d}\, J\,(\varphi(L \cdot, \cdot\cdot) + \varphi(\cdot, L \cdot\cdot)) + \mathrm{o}\,(|t|) \,.$$

Now, given $[\alpha] \in H_J^+(X)$ with $\alpha \in \wedge_J^{1,1}(X) \cap \wedge^2 X$, let $\{\beta_j\}_j$ be such that (4.4) holds and $\sum_{j \geq 1} t^j \beta_j$ converges, for $t \in (-\varepsilon, \varepsilon)$ with $\varepsilon > 0$ small enough; we define

$$\alpha_t := \alpha + \sum_{j \geq 1} t^j \beta_j \in \wedge^2 X$$

and

$$\eta_t := \frac{\alpha_t + J_t \alpha_t}{2} \in \wedge_{J_t}^{1,1} X \cap \wedge^2 X .$$

By construction, η_t is a J_t-invariant real 2-form, real-analytic in t, and such that $\eta_t = \alpha + \mathrm{o}(1)$. A straightforward computation yields

$$\mathrm{d}\,\eta_t = \sum_{j \geq 1} t^j \, \mathrm{d}\Bigg(\beta_j + 2\alpha\left(L^j \cdot, \cdot\cdot\right) + 4 \sum_{k=1}^{j-1} \alpha\left(L^{j-k}\cdot, L^k\cdot\cdot\right) + 2\alpha\left(\cdot, L^j\cdot\cdot\right)$$

$$+ \sum_{h=1}^{j-1} \Bigg(2\beta_h\left(L^{j-h}\cdot, \cdot\cdot\right) + 4 \sum_{k=1}^{j-h-1} \alpha\left(L^{j-h-k}\cdot, L^k\cdot\cdot\right) + 2\alpha\left(\cdot, L^{j-h}\cdot\cdot\right) \Bigg) \Bigg)$$

therefore $\mathrm{d}\,\eta_t = 0$.

Conversely, given $[\alpha] \in H_J^+(X)$ with $\alpha \in \wedge_J^{1,1}(X) \cap \wedge^2 X$, let $\eta_t \in \wedge_{J_t}^{1,1}(X) \cap \wedge^2 X$ be real-analytic in t and such that $\eta_t = \alpha + \mathrm{o}(1)$ and $\mathrm{d}\,\eta_t = 0$, for every $t \in (-\varepsilon, \varepsilon)$ with $\varepsilon > 0$ small enough. Defining $\beta_j \in \wedge^2 X$, for every $j \in \mathbb{N} \setminus \{0\}$, such that

$$\eta_t =: \alpha + \sum_{j \geq 1} t^j \beta_j ,$$

by the same computation we have that (4.4) holds, being $\mathrm{d}\,\eta_t = \mathrm{d}\left(\frac{\eta_t + J_t \eta_t}{2}\right) = 0$.

$\square$

Remark 4.13. We notice that, if $\mathrm{d}\,J_t = \pm J_t\,\mathrm{d}$ on $\wedge^2 M$ for any t, then one can simply let

$$\eta_t := \frac{\alpha + J_t \alpha}{2}$$

so that $\eta_t \in \wedge_{J_t}^{1,1}$ and $\mathrm{d}\,\eta_t = 0$. This is the case, for example, if any J_t is an Abelian complex structure; C. Maclaughlin, H. Pedersen, Y.S. Poon, and S. Salamon characterized in [MPPS06, Theorem 6] the 2-step nilmanifolds whose complex deformations are Abelian.

4.3.2.3   Counterexample to the Stronger Semi-continuity

In the following example, we provide an application of Proposition 4.11, showing a curve of almost-complex structures that does not have the semi-continuity property in the stronger sense described above, see [AT12a].

Example 4.14 ([AT12a, Example 4.8]). A curve of almost-complex structures that does not satisfy (4.5).
As in Example 4.10 and in Example 4.13, consider, for suitable $c \in \mathbb{R}$, [AGH63, Sect. 3], the solvmanifold

$$N^6(c) := (\Gamma(c) \backslash \mathrm{Sol}(3)) \times (\Gamma(c) \backslash \mathrm{Sol}(3)) ,$$

which has been studied in [FMS03, Example 3.4] as an example of a cohomologically Kähler manifold without Kähler structures, see also [BG90, Example 1]. In the following, we consider $N^6 := N^6(1)$. We recall that, with respect to a suitable co-frame, the structure equations of N^6 are

$$(12, \, 0, \, -36, \, 24, \, 56, \, 0) .$$

We look for a curve $\{J_t\}_{t \in (-\varepsilon, \varepsilon) \subset \mathbb{R}}$ of almost-complex structures on N^6, where $\varepsilon > 0$ is small enough, and for a J_0-invariant form α that do not satisfy the first-order obstruction (4.5) to the stronger semi-continuity problem stated above: therefore, there will not be a J_t-invariant class close to α, for any $t \in (-\varepsilon, \, \varepsilon)$.
   Consider the almost-complex structure represented by

$$J \;=\; \begin{pmatrix} & & & -1 & & \\ & & & & -1 & \\ & & & & & -1 \\ \hline 1 & & & & & \\ & 1 & & & & \\ & & 1 & & & \end{pmatrix} \;\in\; \mathrm{End}\left(TN^6\right) ,$$

and

$$L \;=\; \left(\begin{array}{c|c} \mathbf{A} & \mathbf{B} \\ \hline \mathbf{B} & -\mathbf{A} \end{array}\right) \;\in\; \mathrm{End}\left(TN^6\right) ,$$

where

$$\mathbf{A} \;=\; \left(a_i^j\right)_{i,j \in \{1,2,3\}} , \qquad \mathbf{B} \;=\; \left(b_i^j\right)_{i,j \in \{1,2,3\}}$$

are constant matrices; for

$$\alpha = e^{14}$$

we have

$$\mathrm{d}\left(\alpha(L\cdot,\cdot\cdot) + \alpha(\cdot,L\cdot\cdot)\right) = b_1^3\,e^{123} + a_1^2\,e^{125} - a_1^3\,e^{126} + b_1^3\,e^{136} - a_1^2\,e^{156}$$
$$+a_1^3\,e^{234} - b_1^2\,e^{245} - b_1^3\,e^{246} + a_1^3\,e^{346} + b_1^2\,e^{456}\,.$$

Then, choosing

$$L = \left(\begin{array}{c|c} & \begin{matrix} & b_1^3 \\ & 0 \\ & 0 \end{matrix} \\ \hline \begin{matrix} b_1^3 \\ 0 \\ 0 \end{matrix} & \end{array}\right) \in \mathrm{End}\left(TN^6\right)$$

with $b_1^3 \in \mathbb{R} \setminus \{0\}$, it is straightforward to check that there is no $(\mathrm{Sol}(3) \times \mathrm{Sol}(3))$-left-invariant $\beta \in \wedge^2 N^6$ such that

$$\mathrm{d}\beta = b_1^3\,e^{123} + b_1^3\,e^{136} - b_1^3\,e^{246}\,; \qquad (4.6)$$

hence, by applying the F.A. Belgun symmetrization trick, [Bel00, Theorem 7], there is no (possibly non-$(\mathrm{Sol}(3) \times \mathrm{Sol}(3))$-left-invariant) $\beta \in \wedge^2 N^6$ satisfying (4.6).

We resume the content of the last example in the following proposition.

Proposition 4.12 ([AT12a, Proposition 4.9]). *There exist a compact manifold* X *endowed with a* C^∞-*pure-and-full almost-complex structure* J_0, *and a curve* $\{J_t\}_{t\in(-\varepsilon,\varepsilon)}$ *of almost-complex structures on* X, *where* $\varepsilon > 0$ *is small enough, such that, for every* $t \in (-\varepsilon,\varepsilon)$, *there is no* J_t-*invariant class, real-analytic in* t, *close to any fixed* J_0-*invariant class.*

4.4 Cones of Metric Structures

In introducing and studying the subgroups $H_J^{(\bullet,\bullet)}(X;\mathbb{R})$ on a compact manifold X endowed with an almost-complex structure J, T.-J. Li and W. Zhang were mainly aimed by the problem of investigating the relations between the J-taming and the J-compatible symplectic cones on X. As follows by their theorem [LZ09, Theorem 1.1], whenever J is C^∞-full, then the subgroup $H_J^-(X)$ measures the difference between the J-taming cone and the J-compatible cone.

In this section, we discuss some results obtained in [AT12a], jointly with A. Tomassini, giving a counterpart of T.-J. Li and W. Zhang's theorem [LZ09, Theorem 1.1] in the semi-Kähler case, Theorem 4.19, and, in particular, comparing the cones of balanced metrics and of strongly-Gauduchon metrics on a compact complex manifold. Furthermore, concerning the search of a holomorphic-tamed non-Kähler example, [LZ09, p. 678], [ST10, Question 1.7], we show that no such example can exist among six-dimensional nilmanifolds endowed with left-invariant complex structures, Theorem 4.13, as proven in [AT11].

4.4.1 Sullivan's Results on Cone Structures

Firstly, we recall some results by D.P. Sullivan, [Sul76, Sect. I.1], concerning cone structures on a (differentiable) manifold X.

Fixed $p \in \mathbb{N}$, a *cone structure* of p-directions on X is a continuous field $C :=$ $\{C(x)\}_{x \in X}$, with $C(x)$ a compact convex cone in $\wedge^p(T_x X)$ for every $x \in X$.

A p-form ω on X is called *transverse* to a cone structure C if $\omega\lfloor_x(v) > 0$ for all $v \in C(x) \setminus \{0\}$ and for all $x \in X$; using the partitions of unity, a transverse form could be constructed for any given C, [Sul76, Proposition I.4].

Every cone structure C gives rise to a cone $\mathfrak{C}$ of *structure currents*, which are by definition the currents generated by the Dirac currents associated to the elements in $C(x)$, see [Sul76, Definition I.4]; the set $\mathfrak{C}$ is a compact convex cone in $\mathcal{D}_p X$.

The cone $\mathcal{Z}\mathfrak{C}$ of the *structure cycles* is defined as the sub-cone of $\mathfrak{C}$ consisting of d-closed currents; denote with $\mathcal{B}$ the set of d-exact currents.

Define the cone $H\mathfrak{C}$ in $H_p^{dR}(X; \mathbb{R})$ as the set of the classes of the structure cycles.

The dual cone of $H\mathfrak{C}$ in $H_{dR}^p(X; \mathbb{R})$ is denoted by $\check{H}\mathfrak{C}$ and is characterized by the relation

$$\left(\check{H}\mathfrak{C}, H\mathfrak{C} \right) \geq 0 ;$$

its interior is denoted by $\operatorname{int} \check{H}\mathfrak{C}$ and is characterized by the relation

$$\left(\operatorname{int} \check{H}\mathfrak{C}, H\mathfrak{C} \right) > 0 .$$

A cone structure of 2-directions is said to be *ample* if, for every $x \in X$, it satisfies that

$$C(x) \cap \operatorname{span}\{e \in S_\tau : \tau \text{ is a 2-plane}\} \neq \{0\} ,$$

where S_τ is the Schubert variety, given by the set of 2-planes intersecting τ in at least one line; by [Sul76, Theorem III.2], an ample cone structure admits non-trivial structure cycles.

When the $2n$-dimensional manifold X is endowed with an almost-complex structure J, the following cone structures turn out to be particularly interesting.

For a fixed $p \in \{0, \ldots, n\}$, let $C_{p,J}$ be the cone $\{C_{p,J}(x)\}_{x \in X}$, where, for every $x \in X$, the compact convex cone $C_{p,J}(x)$ is generated by the positive combinations of p-dimensional complex subspaces of $T_x X \otimes_{\mathbb{R}} \mathbb{C}$ belonging to $\wedge^{2p} (T_x X \otimes_{\mathbb{R}} \mathbb{C})$.

The cone $\mathfrak{C}_{p,J}$ of *complex currents* is defined as the compact convex cone, see [Sul76, Sect. III.4], of the structure currents.

The cone $\mathcal{Z}\mathfrak{C}_{p,J}$ of *complex cycles* is defined as the compact convex cone, see [Sul76, Sect. III.7], of the structure cycles.

The structure cone $C_{1,J}$ is ample, [Sul76, p. 249], therefore it admits non-trivial cycles.

We recall the following theorem by D.P. Sullivan, which follows by Hahn and Banach's theorem.

Theorem 4.11 ([Sul76, Theorem I.7]). *Let X be a compact differentiable manifold (with or without boundary) and let C be a cone structure of p-vectors defined on a compact subspace Y in the interior of X.*

(i) *There are always non-trivial structure cycles in Y or closed p-forms on X transversal to the cone structure.*

(ii) *If no closed transverse form exists, some non-trivial structure cycle in Y is homologous to zero in X.*

(iii) *If no non-trivial structure cycle exists, some transversal closed form is cohomologous to zero.*

(iv) *If there are both structure cycles and transversal closed forms, then*

  (a) *the natural map*

$$\{structure\ cycles\ on\ Y\} \to \{homology\ classes\ in\ X\}$$

  is proper and the image is a compact cone $C \subseteq H_p^{dR}(X;\mathbb{R})$, and

  (b) *the interior of the dual cone $\check{C} \subseteq H_{dR}^p(X;\mathbb{R})$ (that is, $\check{C}$ is the cone defined by the relation $\left(\check{C}, C\right) \geq 0$) consists precisely of the classes of closed forms transverse to C.*

4.4.2 The Cones of Compatible, and Tamed Symplectic Structures

Let X be a manifold endowed with an almost-complex structure J.

We recall that a symplectic form ω is said to *tame* J if it is positive on the J-lines, that is, if $\omega_x (v_x, J_x v_x) > 0$ for every $v_x \in T_x X \setminus \{0\}$ and for every $x \in X$, equivalently, if

$$\tilde{g}_J (\cdot, \cdot\cdot) := \frac{1}{2} (\omega (\cdot, J\cdot\cdot) - \omega (J\cdot, \cdot\cdot))$$

is a J-Hermitian metric on X with $\pi_{\wedge^{1,1}X}\omega$ as the associated $(1,1)$-form (the map $\pi_{\wedge^{1,1}X}\colon \wedge^\bullet X \to \wedge^{1,1}X$ being the natural projection onto $\wedge^{1,1}X$). A symplectic form ω is called *compatible* with J if it tames J and it is J-invariant, equivalently, if ω is the $(1,1)$-form associated to the J-Hermitian metric $g_J(\cdot,\cdots) := \omega(\cdot, J\cdots)$. In particular, an integrable almost-complex structure J is called *holomorphic-tamed* if it admits a taming symplectic form; on the other hand, the datum of an integrable almost-complex structure and a compatible symplectic form gives a Kähler structure.

4.4.2.1 Symplectic Cones and Donaldson's Question

Consider the J-*tamed cone* $\mathcal{K}^t_J$, which is defined as the set of the cohomology classes of the J-taming symplectic forms, namely,

$$\mathcal{K}^t_J := \left\{[\omega] \in H^2_{dR}(X;\mathbb{R}) \;:\; \omega \text{ is a } J\text{-taming symplectic form on } X\right\},$$

and the J-*compatible cone* $\mathcal{K}^c_J$, which is defined as the set of the cohomology classes of the J-compatible symplectic forms, namely,

$$\mathcal{K}^c_J := \left\{[\omega] \in H^2_{dR}(X;\mathbb{R}) \;:\; \omega \text{ is a } J\text{-compatible symplectic form on } X\right\}.$$

The set $\mathcal{K}^t_J$ is an open convex cone in $H^2_{dR}(X;\mathbb{R})$, and the set $\mathcal{K}^c_J$ is a convex sub-cone of $\mathcal{K}^t_J$ and it is contained in $H^{(1,1)}_J(X;\mathbb{R})$; moreover, both the sets are sub-cones of the *symplectic cone*

$$\mathcal{S} := \left\{[\omega] \in H^2_{dR}(X;\mathbb{R}) \;:\; \omega \text{ is a symplectic form on } X\right\}$$

in $H^2_{dR}(X;\mathbb{R})$.

T.-J. Li and W. Zhang proved the following result in [LZ09], concerning the relation between the J-tamed and the J-compatible cones.

Theorem 4.12 ([LZ09, Proposition 3.1, Theorem 1.1, Corollary 1.1]). *Let X be a compact manifold endowed with an almost-Kähler structure J (namely, J is an almost-complex structure on X such that $\mathcal{K}^c_J \neq \varnothing$). Then*

$$\mathcal{K}^t_J \cap H^{(1,1)}_J(X;\mathbb{R}) = \mathcal{K}^c_J \qquad \text{and} \qquad \mathcal{K}^c_J + H^{(2,0),(0,2)}_J(X;\mathbb{R}) \subseteq \mathcal{K}^t_J.$$

Moreover, if J is C^∞-full, then

$$\mathcal{K}^t_J = \mathcal{K}^c_J + H^{(2,0),(0,2)}_J(X;\mathbb{R}).$$

In particular, if $\dim X = 4$ and $b^+(X) = 1$, then $\mathcal{K}^t_J = \mathcal{K}^c_J$.

The proof is essentially based on [Sul76, Theorem I.7]. Note indeed that the closed forms transverse to the cone $C_{1,J}$ are exactly the J-taming symplectic forms. By [Sul76, Theorem I.7(iv)(b)], it follows that $\mathcal{K}_J^t$ is the interior of the dual cone $\check{H}\mathfrak{C}_{1,J} \subseteq H_{dR}^2(X;\mathbb{R})$ of $H\mathfrak{C}_{1,J} \subseteq H_2^{dR}(X;\mathbb{R})$, [LZ09, Theorem 3.2]. On the other hand, assumed that $\mathcal{K}_J^c$ is non-empty, by the Hahn and Banach separation theorem, $\mathcal{K}_J^c$ is the interior of the dual cone of $H\mathfrak{C}_{1,J} \subseteq H_{(1,1)}^J(X;\mathbb{R})$, [LZ09, Theorem 3.4]. Finally, when $\dim X = 4$, chosen a J-Hermitian metric g on X with associated $(1,1)$-form ω, one has $\wedge_g^+ X = \mathbb{R}\langle\omega\rangle \oplus \wedge_J^- X$, hence, in the almost-Kähler case, if $b^+(X) := \dim_\mathbb{R} H_g^+(X) = 1$, then $H_J^{(2,0),(0,2)}(X;\mathbb{R}) = \{0\}$, see [DLZ10, Proposition 3.1].

Whereas the previous theorem by T.-J. Li and W. Zhang could be intended as a "quantitative comparison" between the J-taming and the J-compatible symplectic cones on a compact manifold X endowed with an almost-complex structure J, one could ask what about their "qualitative comparison", namely, one could ask whether $\mathcal{K}_J^c$ being empty implies $\mathcal{K}_J^t$ being empty, too. The following question has been arisen by S.K. Donaldson in [Don06].

Question 4.3 ([Don06, Question 2]). Let X be a compact four-dimensional manifold endowed with an almost-complex structure J. If J is tamed by a symplectic form, is there a symplectic form compatible with J?

Remark 4.14. S.K. Donaldson's "tamed to compatible" question has a positive answer for $\mathbb{CP}^2$ by the works by M. Gromov [Gro85] and by C.H. Taubes, [Tau95]. When $b^+(X) = 1$ (where b^+ is the number of positive eigenvalues of the intersection pairing on $H_2(X;\mathbb{R})$), a possible positive answer to [Don06, Question 2], see also [TWY08, Conjecture 1.2], would be provided as a consequence of [Don06, Conjecture 1], see also [TWY08, Conjecture 1.1], concerning the study of the symplectic Calabi and Yau equation, which aims to generalize S.-T. Yau's theorem [Yau77, Yau78], solving the E. Calabi conjecture, [Cal57], to the non-integrable case. Some results concerning this problem have been obtained by several authors, see, e.g., [Wei07, TWY08, TW11a, Tau11, Zha13, LT12, FLSV11, BFV11], see also [TW11b]. More precisely, in [Wei07], all the estimates for the closedness argument of the continuity method applied to the symplectic Calabi and Yau equation, [Don06, Conjecture 1], are reduced to a C^0 a priori estimate of a scalar potential function, [Wei07, Theorem 1]; then, the existence of a solution of the symplectic Calabi and Yau equation is proven for compact four-dimensional manifolds X endowed with an almost-Kähler structure (J, ω, g) satisfying $\|\mathrm{Nij}_J\|_{L^1} < \varepsilon$, where $\varepsilon > 0$ depends just on the data, [Wei07, Theorem 2]. In [TWY08], it is shown that the C^∞ a priori estimates can be reduced to an integral estimate of a scalar function potential, [TWY08, Theorem 1.3]; furthermore, it is shown that [Don06, Conjecture 1] holds under a positive curvature assumption, [TWY08, Theorem 1.4]. In [TW11a], the symplectic Calabi and Yau equation is solved on the Kodaira and Thurston manifold $\mathbb{S}^1 \times (\mathbb{H}(3;\mathbb{Z})\backslash \mathbb{H}(3;\mathbb{R}))$ for any given left-invariant volume form, [TW11a, Theorem 1.1]; further results on the Calabi and Yau equation for torus-bundles over a two-dimensional torus have been provided in [FLSV11, BFV11]. In [Tau11], it is shown that, on a compact four-dimensional manifold with $b^+ = 1$ and endowed with a symplectic form ω, a generic ω-tamed almost-complex structure on X is compatible with a symplectic form on X, [Tau11, Theorem 1], which is defined by integrating over a space of currents that are defined by pseudo-holomorphic curves. The Taubes currents have been studied, both in dimension 4 and higher, also by W. Zhang in [Zha13]. In [LZ11], T.-J. Li and W. Zhang were concerned with studying Donaldson's "tamed to compatible" question for almost-complex structures on rational four-dimensional manifolds; they provided, in particular, an affirmative answer to [Don06, Question 2] for $\mathbb{S}^2 \times \mathbb{S}^2$ and for $\mathbb{CP}^2 \sharp \overline{\mathbb{CP}^2}$, see [DLZ12, Theorem 4.11]. In [LT12], a positive answer to S.K. Donaldson's question [Don06, Question 2] is provided in the Lie algebra setting, proving that, given a four-dimensional Lie algebra $\mathfrak{g}$ such that $B \wedge B = 0$ (where $B \subseteq \wedge^2\mathfrak{g}$ is the space of boundary 2-vectors), e.g., a four-dimensional unimodular Lie algebra, a linear (possibly non-integrable) complex structure is tamed by a linear symplectic form if and only if it is compatible with a linear symplectic form, [LT12, Theorem 0.2]. (For the non-unimodular case, see [DL13].)

In a sense, [LZ09, Corollary 1.1] provides evidences towards an affirmative answer for [Don06, Question 2], especially in the case $b^+ = 1$; confirmed in their opinion by the computations in [DLZ10] in the case $b^+ > 1$, T.-J. Li and W. Zhang speculated in [LZ09, p. 655] that the equality $\mathcal{K}^t_J = \mathcal{K}^c_J$ holds for a generic almost-complex structure J on a four-dimensional manifold.

The analogous of [Don06, Question 2] in dimension higher than 4 has a negative answer: counterexamples in the (non-integrable) almost-complex case can be found in [MT00] by M. Migliorini and A. Tomassini, and in [Tom02] by A. Tomassini; for another counterexample, see Theorem 4.6: more precisely, in Example 4.10, a curve $\{J_t\}_t$ of almost-complex structures on the six-dimensional completely-solvable solvmanifold $N^6(c)$ has been constructed, proving that J_0 admits an almost-Kähler structure and J_t, for $t \neq 0$, admits no almost-Kähler structure. Notwithstanding, since examples of non-Kähler holomorphic-tamed complex structures are not known, T.-J. Li and W. Zhang speculated a negative answer for the following question, also addressed by J. Streets and G. Tian in [ST10].

Question 4.4 ([LZ09, p. 678], [ST10, Question 1.7]). Do there exist non-Kähler holomorphic-tamed complex manifolds, of complex dimension greater than 2?

Remark 4.15 (Weiyi Zhang). As noticed[7] in [Zha13, Sect. 5], non-Kähler Moĭšezon manifolds are non-holomorphic-tamed, as proven in [Pet86]. In particular, they do satisfy the "tamed to compatible" question.

---

[7]I would like to thank Weiyi Zhang for having let me notice this fact.

4.4.2.2   Tameness Conjecture for Six-Dimensional Nilmanifolds

In view of the speculation in [LZ09, p. 678], and of [ST10, Question 1.7], one
could ask whether small deformations of the Iwasawa manifold, see Sect. 3.2.1, may
provide examples of non-Kähler holomorphic-tamed complex structures. In this
section, we prove that this is not the case: more precisely, we prove that no example
of left-invariant non-Kähler holomorphic-tamed complex structure can be found
on six-dimensional nilmanifolds. The same holds true, more in general, for higher
dimensional nilmanifolds, as proven by N. Enrietti, A.M. Fino, and L. Vezzoni,
[EFV12, Theorem 1.3].

We recall that a Hermitian metric g on a complex manifold X is called *pluriclosed* (or *strong Kähler with torsion*, shortly SKT), [Bis89], if the $(1,1)$-form ω
associated to g satisfies $\partial\bar{\partial}\omega = 0$.

By the following result, holomorphic-tamed manifolds admit pluri-closed
metrics.

Proposition 4.13 ([AT11, Proposition 3.1]). *Let X be a manifold endowed
with a symplectic structure ω and an ω-tamed complex structure J. Then the
$(1,1)$-form $\tilde{\omega} := \tilde{g}_J (J \cdot, \cdot\cdot)$ associated to the Hermitian metric $\tilde{g}_J (\cdot, \cdot\cdot) :=
\frac{1}{2} (\omega (\cdot, J\cdot\cdot) - \omega (J\cdot, \cdot\cdot))$ is $\partial\bar{\partial}$-closed, namely, $\tilde{g}$ is a pluri-closed metric on X.*

Proof. Decomposing ω in pure type components, set

$$\omega =: \omega^{2,0} + \omega^{1,1} + \overline{\omega^{2,0}}$$

where $\omega^{2,0} \in \wedge^{2,0}X$ and $\omega^{1,1} = \overline{\omega^{1,1}} \in \wedge^{1,1}X$. Since, by definition, $\tilde{\omega} =
\frac{1}{2} (\omega + J\omega)$, we have $\tilde{\omega} = \omega^{1,1}$. We get that

$$\mathrm{d}\omega = 0 \quad \Leftrightarrow \quad \begin{cases} \partial\omega^{2,0} = 0 \\ \partial\omega^{1,1} + \bar{\partial}\omega^{2,0} = 0 \end{cases},$$

and hence

$$\partial\bar{\partial}\tilde{\omega} = \partial\bar{\partial}\omega^{1,1} = -\bar{\partial}\partial\omega^{1,1} = \bar{\partial}^2\omega^{2,0} = 0 ,$$

proving that $\tilde{g}$ is a pluri-closed metric on X.   □

Now, we can prove the announced theorem.

Theorem 4.13 ([AT11, Theorem 3.3]). *Let $X = \Gamma \backslash G$ be a six-dimensional nilmanifold endowed with a G-left-invariant complex structure J. If X is not a torus,
then there is no symplectic structure ω on X taming J.*

Proof. Let ω be a (non-necessarily G-left-invariant) symplectic form on X taming
J. By F.A. Belgun's symmetrization trick, [Bel00, Theorem 7], setting

$$\mu(\omega) := \int_X \omega \lfloor_m \eta(m) ,$$

where η is a G-bi-invariant volume form on G such that $\int_X \eta = 1$, whose existence follows from J. Milnor's lemma [Mil76, Lemma 6.2], we get a G-left-invariant symplectic form on X taming J. Then, it suffices to prove that, on a non-torus six-dimensional nilmanifold, there is no left-invariant symplectic structure taming a left-invariant complex structure.

Hence, let ω be such a G-left-invariant symplectic structure. Then, by Proposition 4.13, X should admit a G-left-invariant pluri-closed Hermitian metric g. Hence, by [FPS04, Theorem 1.2], there exists a co-frame $\{\varphi^1, \varphi^2, \varphi^3\}$ for the J-holomorphic cotangent bundle such that

$$\begin{cases} d\varphi^1 = 0 \\ d\varphi^2 = 0 \\ d\varphi^3 = A\,\overline{\varphi}^1 \wedge \varphi^2 + B\,\overline{\varphi}^2 \wedge \varphi^2 + C\,\varphi^1 \wedge \overline{\varphi}^1 + D\,\varphi^1 \wedge \overline{\varphi}^2 + E\,\varphi^1 \wedge \varphi^2 \end{cases},$$

where $A, B, C, D, E \in \mathbb{C}$ are complex numbers such that $|A|^2 + |D|^2 + |E|^2 + 2\,\mathfrak{Re}\left(\bar{B}C\right) = 0$. Set

$$\omega =: \omega^{2,0} + \omega^{1,1} + \overline{\omega^{2,0}},$$

where

$$\omega^{2,0} = \sum_{i<j} a_{ij}\,\varphi^i \wedge \varphi^j, \qquad \omega^{1,1} = \frac{i}{2} \sum_{i,j=1}^{3} b_{i\bar{j}}\,\varphi^i \wedge \overline{\varphi}^j,$$

with $\{a_{ij}, b_{i\bar{j}}\}_{i,j} \subset \mathbb{C}$ such that $\omega^{1,1} = \overline{\omega^{1,1}}$. A straightforward computation yields

$$d\omega = 0 \quad \Leftrightarrow \quad \left(A = B = C = D = E = 0 \quad \text{or} \quad b_{3\bar{3}} = 0\right).$$

Since $b_{3\bar{3}} \neq 0$, we get $A = B = C = D = E = 0$, namely, X is a torus. $\qquad\square$

As a corollary, we get the following result concerning the speculation in [LZ09, p. 678], and [ST10, Question 1.7].

Theorem 4.14 ([AT11, Theorem 3.4]). *No small deformation of the complex structure of the Iwasawa manifold $\mathbb{I}_3 := \mathbb{H}(3;\mathbb{Z}[i])\backslash\mathbb{H}(3;\mathbb{C})$ can be tamed by any symplectic form.*

Actually, Theorem 4.13 holds also for higher dimensional nilmanifolds, as proved by N. Enrietti, A.M. Fino, and L. Vezzoni in [EFV12].

Theorem 4.15 ([EFV12, Theorem 1.3]). *Let $X = \Gamma\backslash G$ be a nilmanifold endowed with a G-left-invariant complex structure J. If X is not a torus, then there is no symplectic structure ω on X taming J.*

4.4.3   The Cones of Semi-Kähler, and Strongly-Gauduchon Metrics

Let X be a compact $2n$-dimensional manifold endowed with an almost-complex structure J. We recall that a non-degenerate 2-form ω on X is called *semi-Kähler*, [GH80, p. 40], if ω is the $(1, 1)$-form associated to a J-Hermitian metric on X (that is, $\omega(\cdot, J\cdot) > 0$ and $\omega(J\cdot, J\cdot) = \omega(\cdot, \cdot)$) and $d\left(\omega^{n-1}\right) = 0$; when J is integrable, a semi-Kähler structure is called *balanced*, [Mic82, Definition 1.4, Theorem 1.6].

We set

$$\mathcal{K}b_J^c := \{[\Omega] \in H_{dR}^{2n-2}(X; \mathbb{R}) \; : \; \Omega \in \wedge^{n-1,n-1}X \text{ is positive on the}$$

$$\text{complex } (n-1)\text{-subspaces of } T_x X \otimes_{\mathbb{R}} \mathbb{C}, \text{ for every } x \in X\} \; ,$$

and

$$\mathcal{K}b_J^t := \{[\Omega] \in H_{dR}^{2n-2}(X; \mathbb{R}) \; : \; \Omega \in \wedge^{2n-2}X \text{ is positive on the}$$

$$\text{complex } (n-1)\text{-subspaces of } T_x X \otimes_{\mathbb{R}} \mathbb{C}, \text{ for every } x \in X\} \; .$$

We note that $\mathcal{K}b_J^c$ and $\mathcal{K}b_J^t$ are convex cones in $H_{dR}^{2n-2}(X; \mathbb{R})$, and that $\mathcal{K}b_J^c$ is a sub-cone of $\mathcal{K}b_J^t$ and is contained in $H_J^{(n-1,n-1)}(X; \mathbb{R})$.

We recall the following trick by M.L. Michelsohn.

Lemma 4.2 ([Mic82, pp. 279–280]). *Let X be a compact $2n$-dimensional manifold endowed with an almost-complex structure J. Let Φ be a real $(n-1, n-1)$-form such that it is positive on the complex $(n-1)$-subspaces of $T_x X \otimes_{\mathbb{R}} \mathbb{C}$, for every $x \in X$. Then Φ can be written as $\Phi = \varphi^{n-1}$, where φ is a J-taming real $(1, 1)$-form. In particular, if Φ is d-closed, then φ is a semi-Kähler form.*

The previous Lemma allows us to confuse the cone $\mathcal{K}b_J^c$ with the cone generated by the $(n-1)$th powers of the semi-Kähler forms, namely,

$$\mathcal{K}b_J^c = \{[\omega^{n-1}] \; : \; \omega \text{ is a semi-Kähler form on } X\} \; .$$

In particular, if J is integrable, then the cone $\mathcal{K}b_J^c$ is just the cone of balanced structures on X. On the other hand, in the integrable case, $\mathcal{K}b_J^t$ is the cone of strongly-Gauduchon metrics on X. We recall that a *strongly-Gauduchon metric* on X, [Pop13], [Pop09, Definition 3.1], is a positive-definite $(1, 1)$-form γ on X such that the $(n, n-1)$-form $\partial\left(\gamma^{n-1}\right)$ is $\overline{\partial}$-exact. These metrics have been introduced by D. Popovici in [Pop13] in studying the limits of projective manifolds under deformations of the complex structure, and they turn out to be special cases of *Gauduchon metrics*, [Gau77], for which $\partial\left(\gamma^{n-1}\right)$ is just $\overline{\partial}$-closed; note that the notions of Gauduchon metric and of strongly-Gauduchon metric coincide if the $\partial\overline{\partial}$-Lemma holds, [Pop09, p. 15]. D. Popovici proved in [Pop09, Lemma 3.2] that a compact complex manifold X, of complex dimension n, carries a strongly-Gauduchon metric if and only if there exists a real d-closed $(2n-2)$-form Ω such that its component $\Omega^{(n-1,n-1)}$ of type $(n-1,n-1)$ satisfies $\Omega^{(n-1,n-1)} > 0$.

The aim of this section is to compare the cones $\mathcal{K}b_J^c$ and $\mathcal{K}b_J^t$, Theorem 4.19, in the same way as [LZ09, Theorem 1.1] does for $\mathcal{K}_J^c$ and $\mathcal{K}_J^t$ in the almost-Kähler case.

Note that $\mathcal{K}b_J^t$ can be identified with the set of the classes of d-closed $(2n-2)$-forms transverse to $\mathcal{C}_{n-1,J}$. On the other hand, we recall the following lemma.

Lemma 4.3 (see, e.g., [Sil96, Proposition I.1.3]). *Let X be a compact manifold endowed with an almost-complex structure J, and fix $p \in \mathbb{N}$. A structure current in $\mathcal{C}_{p,J}$ is a positive current of bi-dimension (p,p).*

As a direct consequence of [Sul76, Theorem I.7], we get the following result.

Theorem 4.16 ([AT12a, Theorem 2.6]). *Let X be a compact $2n$-dimensional manifold endowed with an almost-complex structure J. Then $\mathcal{K}b_J^t$ is non-empty if and only if there is no non-trivial d-closed positive currents of bi-dimension $(n-1,n-1)$ that is a boundary, i.e.,*

$$\mathcal{Z}\mathcal{C}_{n-1,J} \cap \mathcal{B} = \{0\}.$$

Furthermore, if we suppose that $0 \notin \mathcal{K}b_J^t$, then $\mathcal{K}b_J^t \subseteq H_{dR}^{2n-2}(X;\mathbb{R})$ is the interior of the dual cone $\check{H}\mathcal{C}_{n-1,J} \subseteq H_{dR}^{2n-2}(X;\mathbb{R})$ of $H\mathcal{C}_{n-1,J} \subseteq H_{2n-2}^{dR}(X;\mathbb{R})$.

Proof. Note that if $\omega \in \mathcal{K}b_J^t \neq \emptyset$, and if $\eta = d\xi$ is a non-trivial d-closed positive current of bi-dimension $(n-1,n-1)$ being a boundary, then

$$0 < (\eta,\ \pi_{\wedge^{n-1,n-1}X}\omega) = (\eta,\ \omega) = (d\xi,\ \omega) = (\xi,\ d\omega) = 0$$

(where $\pi_{\wedge^{n-1,n-1}X}\colon \wedge^\bullet X \to \wedge^{n-1,n-1}X$ is the natural projection onto $\wedge^{n-1,n-1}X$) yields an absurd.

To prove the converse, suppose that no non-trivial d-closed positive currents of bi-dimension $(n-1,n-1)$ is a boundary; then, by [Sul76, Theorem I.7(ii)], there exists a d-closed form that is transverse to $\mathcal{C}_{n-1,J}$, that is, $\mathcal{K}b_J^t$ is non-empty.

The last statement follows from [Sul76, Theorem I.7(iv)]: indeed, by the assumption $0 \notin \mathcal{K}b_J^t$, no d-closed transverse form is cohomologous to zero, therefore, by [Sul76, Theorem I.7(iii)], there exists a non-trivial structure cycle. □

We provide a similar characterization for $\mathcal{K}b_J^c$.

Theorem 4.17 ([AT12a, Theorem 2.7]). *Let X be a compact $2n$-dimensional manifold endowed with an almost-complex structure J. Suppose that $\mathcal{K}b_J^c \neq \emptyset$ and that $0 \notin \mathcal{K}b_J^c$. Then $\mathcal{K}b_J^c \subseteq H_J^{(n-1,n-1)}(X;\mathbb{R})$ is the interior of the dual cone $\check{H}\mathcal{C}_{n-1,J} \subseteq H_J^{(n-1,n-1)}(X;\mathbb{R})$ of $H\mathcal{C}_{n-1,J} \subseteq H_{(n-1,n-1)}^J(X;\mathbb{R})$.*

Proof. By the hypothesis $0 \notin \mathcal{K}_J^c$, we have that $\left(\mathcal{K}b_J^c, H\mathfrak{C}_{n-1,J}\right) > 0$, and therefore the inclusion $\mathcal{K}b_J^c \subseteq \text{int}\, \breve{H}\mathfrak{C}_{n-1,J}$ holds.

To prove the other inclusion, let $e \in H_J^{(n-1,n-1)}(X;\mathbb{R})$ be an element in the interior of the dual cone in $H_J^{(n-1,n-1)}$ of $H\mathfrak{C}_{n-1,J}$, i.e., e is such that $(e, H\mathfrak{C}_{n-1,J}) > 0$. Consider the isomorphism

$$\bar{\sigma}^{n-1,n-1}: H_J^{(n-1,n-1)}(X;\mathbb{R}) \xrightarrow{\simeq} \left(\frac{\overline{\pi_{\mathcal{D}_{n-1,n-1}X}\mathcal{Z}}}{\overline{\pi_{\mathcal{D}_{n-1,n-1}X}\mathcal{B}}}\right)^*,$$

[LZ09, Proposition 2.4] (where $\pi_{\mathcal{D}_{n-1,n-1}X}: \mathcal{D}_\bullet X \to \mathcal{D}_{n-1,n-1}X$ denotes the natural projection onto $\mathcal{D}_{n-1,n-1}X$): hence, $\bar{\sigma}^{n-1,n-1}(e)$ gives rise to a functional on $\frac{\overline{\pi_{\mathcal{D}_{n-1,n-1}X}\mathcal{Z}}}{\overline{\pi_{\mathcal{D}_{n-1,n-1}X}\mathcal{B}}}$, namely, to a functional on $\overline{\pi_{\mathcal{D}_{n-1,n-1}X}\mathcal{Z}}$ vanishing on $\overline{\pi_{\mathcal{D}_{n-1,n-1}X}\mathcal{B}}$; such a functional, in turn, gives rise to a hyperplane L in $\overline{\pi_{\mathcal{D}_{n-1,n-1}X}\mathcal{Z}}$ containing $\overline{\pi_{\mathcal{D}_{n-1,n-1}X}\mathcal{B}}$. Being a kernel hyperplane in a closed set, L is closed in $\mathcal{D}_{n-1,n-1}X \cap \mathcal{D}_{2n-2}X$; furthermore, L is disjoint from $\mathfrak{C}_{n-1,J} \setminus \{0\}$, by the choice made for e. Pick a J-Hermitian metric and let φ be its associated $(1, 1)$-form; consider

$$K := \left\{T \in \mathfrak{C}_{n-1,J} : T\left(\varphi^{n-1}\right) = 1\right\},$$

which is a compact set. Now, in the space $\mathcal{D}_{n-1,n-1}X \cap \mathcal{D}_{2n-2}X$, consider the closed set L, and the compact convex non-empty set K, which have empty intersection. By the Hahn and Banach separation theorem, there exists a hyperplane containing L, and then containing also $\overline{\pi_{\mathcal{D}_{n-1,n-1}X}\mathcal{B}}$, and disjoint from K. The functional on $\mathcal{D}_{n-1,n-1}X \cap \mathcal{D}_{2n-2}X$ associated to this hyperplane is a real $(n-1,n-1)$-form being d-closed, since it vanishes on $\overline{\pi_{\mathcal{D}_{n-1,n-1}X}\mathcal{B}}$, and positive on the complex $(n-1)$-subspaces of $T_xX \otimes_\mathbb{R} \mathbb{C}$, for every $x \in X$, that is, a J-compatible symplectic form. $\qquad\square$

The same argument as in [HL83, Proposition 12, Theorem 14] yields the following result, [AT12a, Theorem 2.8], which generalizes [HL83, Proposition 12, Theorem 14], [LZ09, p. 671], see also [Mic82, Theorem 4.7].

Theorem 4.18 ([HL83, Proposition 12, Theorem 14], [LZ09, p. 671], [AT12a, Theorem 2.8]). *Let X be a compact $2n$-dimensional manifold endowed with an almost-complex structure J, and denote by $\pi_{\mathcal{D}_{k,k}X}: \mathcal{D}_\bullet X \to \mathcal{D}_{k,k}X$ the natural projection onto $\mathcal{D}_{k,k}X$, for every $k \in \mathbb{N}$.*

(i) *If J is integrable, then there exists a Kähler metric if and only if $\mathfrak{C}_{1,J} \cap \pi_{\mathcal{D}_{1,1}X}\mathcal{B} = \{0\}$.*

(ii) *There exists an almost-Kähler metric if and only if $\mathfrak{C}_{1,J} \cap \overline{\pi_{\mathcal{D}_{1,1}X}\mathcal{B}} = \{0\}$.*

(iii) *There exists a semi-Kähler metric if and only if $\mathfrak{C}_{n-1,J} \cap \overline{\pi_{\mathcal{D}_{n-1,n-1}X}\mathcal{B}} = \{0\}$.*

Proof. Note that *(i)*, namely, [HL83, Proposition 12, Theorem 14], is a consequence of *(ii)*: indeed, if J is integrable, then J is closed, [HL83, Lemma 6], that is, $\pi_{\mathcal{D}_{1,1}X}\mathcal{B}$ is a closed set.

The proof of *(ii)*, namely, [LZ09, p. 671], being similar, we prove *(iii)*, following closely the proof of *(i)* in [HL83, Proposition 12, Theorem 14].

Firstly, note that if ω is a semi-Kähler form and

$$0 \neq \eta = \lim_{k \to +\infty} \pi_{\mathcal{D}_{n-1,n-1}X}(d\alpha_k) \in \mathfrak{C}_{n-1,J} \cap \overline{\pi_{\mathcal{D}_{n-1,n-1}X}\mathcal{B}} \neq \{0\} ,$$

where $\{\alpha_k\}_{k \in \mathbb{N}} \subset \mathcal{D}_{2n-1}X$, then

$$0 < \left(\eta, \omega^{n-1}\right) = \left(\lim_{k \to +\infty} \pi_{\mathcal{D}_{n-1,n-1}X}(d\alpha_k), \omega^{n-1}\right) = \lim_{k \to +\infty}\left(d\alpha_k, \omega^{n-1}\right)$$

$$= \lim_{k \to +\infty}\left(\alpha_k, d\omega^{n-1}\right) = 0 ,$$

yielding an absurd.

For the converse, fix a J-Hermitian metric and let φ be its associated $(1,1)$-form; the set

$$K := \left\{T \in \mathfrak{C}_{n-1,J} : T\left(\varphi^{n-1}\right) = 1\right\}$$

is a compact convex non-empty set in $\mathcal{D}_{n-1,n-1}X \cap \mathcal{D}_{2n-2}X$. By the Hahn and Banach separation theorem, there exists a hyperplane in $\mathcal{D}_{n-1,n-1}X \cap \mathcal{D}_{2n-2}X$ containing the closed subspace $\overline{\pi_{\mathcal{D}_{n-1,n-1}X}\mathcal{B}}$ and disjoint from K; hence, the real $(n-1,n-1)$-form associated to this hyperplane is a real d-closed $(n-1,n-1)$-form and is positive on the complex $(n-1)$-subspaces, namely, it is a semi-Kähler form. □

Now, we can prove the semi-Kähler counterpart of T.-J. Li and W. Zhang's [LZ09, Proposition 3.1, Theorem 1.1].

Theorem 4.19 ([AT12a, Theorem 2.9]). *Let X be a compact $2n$-dimensional manifold endowed with an almost-complex structure J. Assume that $\mathcal{K}b_J^c \neq \varnothing$ (that is, there exists a semi-Kähler structure on X) and that $0 \notin \mathcal{K}b_J^t$. Then*

$$\mathcal{K}b_J^t \cap H_J^{(n-1,n-1)}(X;\mathbb{R}) = \mathcal{K}b_J^c$$

and

$$\mathcal{K}b_J^c + H_J^{(n,n-2),(n-2,n)}(X;\mathbb{R}) \subseteq \mathcal{K}b_J^t .$$

Moreover, if J is $\mathcal{C}^\infty$-full at the $(2n-2)$th stage, then

$$\mathcal{K}b_J^c + H_J^{(n,n-2),(n-2,n)}(X;\mathbb{R}) = \mathcal{K}b_J^t .$$

Proof. By Theorem 4.16, $\mathcal{K}b_J^t \subseteq H_{dR}^{2n-2}(X;\mathbb{R})$ is the interior of the dual cone $\breve{H}\mathfrak{C}_{n-1,J} \subseteq H_{dR}^{2n-2}(X;\mathbb{R})$ of $H\mathfrak{C}_{n-1,J} \subseteq H_{2n-2}^{dR}(X;\mathbb{R})$, and, by Theorem 4.17, $\mathcal{K}b_J^c \subseteq H_J^{(n-1,n-1)}(X;\mathbb{R})$ is the interior of the dual cone $\breve{H}\mathfrak{C}_{n-1,J} \subseteq H_J^{(n-1,n-1)}(X;\mathbb{R})$ of $H\mathfrak{C}_{n-1,J} \subseteq H_{(n-1,n-1)}^J(X;\mathbb{R})$; therefore $\mathcal{K}b_J^t \cap H_J^{(n-1,n-1)}(X;\mathbb{R}) = \mathcal{K}b_J^c$.

The inclusion $\mathcal{K}b_J^c + H_J^{(n,n-2),(n-2,n)}(X;\mathbb{R}) \subseteq \mathcal{K}b_J^t$ follows straightforwardly noting that the sum of a semi-Kähler form and a J-anti-invariant $(2n-2)$-form is still d-closed and positive on the complex $(n-1)$-subspaces.

Finally, if J is C^∞-full at the $(2n-2)$th stage, then

$$\mathcal{K}b_J^t = \text{int } \breve{H}\mathfrak{C}_{n-1,J} = \text{int } \breve{H}\mathfrak{C}_{n-1,J} \cap H_{dR}^{2n-2}(X;\mathbb{R})$$

$$= \text{int } \breve{H}\mathfrak{C}_{n-1,J} \cap \left(H_J^{(n-1,n-1)}(X;\mathbb{R}) + H_J^{(n,n-2),(n-2,n)}(X;\mathbb{R}) \right)$$

$$\subseteq \mathcal{K}b_J^c + H_J^{(n,n-2),(n-2,n)}(X;\mathbb{R}) ,$$

and hence $\mathcal{K}b_J^c + H_J^{(n,n-2),(n-2,n)}(X;\mathbb{R}) = \mathcal{K}b_J^t$.   □

Remark 4.16. We note that, while the de Rham cohomology class of an almost-Kähler metric cannot be trivial, the hypothesis $0 \notin \mathcal{K}b_J^t$ in Theorem 4.19 is not trivial: J. Fu, J. Li, and S.-T. Yau proved in [FLY12, Corollary 1.3] that, for any $k \geq 2$, the connected sum $\left(\mathbb{S}^3 \times \mathbb{S}^3\right)^{\sharp k}$, endowed with the complex structure constructed from the conifold transitions, admits balanced metrics.

References

[AB90]   L. Alessandrini, G. Bassanelli, Small deformations of a class of compact non-Kähler manifolds. Proc. Am. Math. Soc. **109**(4), 1059–1062 (1990)

[AB91]   L. Alessandrini, G. Bassanelli, Compact p-Kähler manifolds. Geom. Dedicata **38**(2), 199–210 (1991)

[AB96]   L. Alessandrini, G. Bassanelli, The class of compact balanced manifolds is invariant under modifications, in *Complex Analysis and Geometry (Trento, 1993)*. Lecture Notes in Pure and Applied Mathematics, vol. 173 (Marcel Dekker, New York, 1996), pp. 1–17

[Aba88]   M. Abate, Annular bundles. Pac. J. Math. **134**(1), 1–26 (1988)

[ABDM11]   A. Andrada, M.L. Barberis, I.G. Dotti Miatello, Classification of abelian complex structures on 6-dimensional Lie algebras. J. Lond. Math. Soc. (2) **83**(1), 232–255 (2011)

[ABDMO05]   A. Andrada, M.L. Barberis, I.G. Dotti Miatello, G.P. Ovando, Product structures on four dimensional solvable Lie algebras. Homol. Homotopy Appl. **7**(1), 9–37 (2005)

[AC12]   D. Angella, S. Calamai, A vanishing result for strictly p-convex domains. Ann. Mat. Pura Appl., 16 pp. (2012). DOI:10.1007/s00222-012-0406-3, Online First, http://link.springer.com/article/10.1007/s10231-012-0315-5

[AC13]   D. Angella, S. Calamai, Bott–Chern cohomology and q-complete domains. C. R. Math. Acad. Sci. Paris **351**(9–10), 343–348 (2013)

[Aep65]   A. Aeppli, On the cohomology structure of Stein manifolds, in *Proceedings of a Conference on Complex Analysis*, Minneapolis, MN, 1964 (Springer, Berlin, 1965), pp. 58–70

[AFR12]   D. Angella, M.G. Franzini, F.A. Rossi, Degree of non-kählerianity for 6-dimensional nilmanifolds, arXiv:1210.0406 [math.DG], 2012

[AG62]   A. Andreotti, H. Grauert, Théorème de finitude pour la cohomologie des espaces complexes. Bull. Soc. Math. France **90**, 193–259 (1962)

[AGH63]   L. Auslander, L.W. Green, F.J. Hahn, *Flows on Homogeneous Spaces*, With the assistance of L. Markus and W. Massey, and an appendix by L. Greenberg. Annals of Mathematics Studies, vol. 53 (Princeton University Press, Princeton, 1963)

[AGS97]   E. Abbena, S. Garbiero, S. Salamon, Hermitian geometry on the Iwasawa manifold. Boll. Un. Mat. Ital. B (7) **11**(2, Suppl.), 231–249 (1997)

[AI01]   B. Alexandrov, S. Ivanov, Vanishing theorems on Hermitian manifolds. Differ. Geom. Appl. **14**(3), 251–265 (2001)

[AK12]   D. Angella, H. Kasuya, Bott-chern cohomology of solvmanifolds, arXiv:1212.5708v3 [math.DG], 2012

[AK13a]   D. Angella, H. Kasuya, Cohomologies of deformations of solvmanifolds and closedness of some properties, arXiv:1305.6709v1 [math.CV], 2013

[AK13b]   D. Angella, H. Kasuya, Symplectic Bott-Chern cohomology of solvmanifolds. arXiv:1308.4258v1 [math.SG] (2013)

[AL94]   M. Audin, J. Lafontaine (eds.), *Holomorphic Curves in Symplectic Geometry*. Progress in Mathematics, vol. 117 (Birkhäuser Verlag, Basel, 1994)

[Ale98]   L. Alessandrini, Correnti positive: uno strumento per l'analisi globale su varietà complesse. Rend. Sem. Mat. Fis. Milano **68**(1), 59–120 (1998)

[Ale10]   L. Alessandrini, Correnti positive e varietà complesse. Rend. Mat. Appl. (7) **30**(2), 145–181 (2010)

[Ale11]   L. Alessandrini, Classes of compact non-Kähler manifolds. C. R. Math. Acad. Sci. Paris **349**(19–20), 1089–1092 (2011)

[AMT09]   D.V. Alekseevskiĭ, K. Medori, A. Tomassini, Homogeneous para-Kählerian Einstein manifolds. Uspekhi Mat. Nauk **64**(1(385)), 3–50 (2009). Translation in Russ. Math. Surv. **64**(1), 1–43 (2009)

[AN54]   Y. Akizuki, S. Nakano, Note on Kodaira-Spencer's proof of Lefschetz theorems. Proc. Jpn. Acad. **30**(4), 266–272 (1954)

[AN71]   A. Andreotti, F. Norguet, Cycles of algebraic manifolds and $\partial\bar{\partial}$-cohomology. Ann. Scuola Norm. Sup. Pisa (3) **25**(1), 59–114 (1971)

[Ang11]   D. Angella, The cohomologies of the Iwasawa manifold and of its small deformations. J. Geom. Anal. **23**(3), 1355–1378 (2013)

[Ang13a]   D. Angella, Cohomologies of certain orbifolds. J. Geom. Phys. **171**, 117–126 (2013)

[Ang13b]   D. Angellla, Cohomological aspects of non-Kähler manifolds, Ph.D. Thesis, Università di Pisa (advisor A. Tomassini), 2013, arXiv:1302.0524v1 [math.DG], http://etd.adm.unipi.it/theses/available/etd-01072013-231626/

[AR12]   D. Angella, F.A. Rossi, Cohomology of **D**-complex manifolds. Differ. Geom. Appl. **30**(5), 530–547 (2012)

[AS05]   A. Andrada, S. Salamon, Complex product structures on Lie algebras. Forum Math. **17**(2), 261–295 (2005)

[AT11]   D. Angella, A. Tomassini, On cohomological decomposition of almost-complex manifolds and deformations. J. Symplectic Geom. **9**(3), 403–428 (2011)

[AT12a]   D. Angella, A. Tomassini, On the cohomology of almost-complex manifolds. Int. J. Math. **23**(2), 1250019, 25 (2012)

[AT12b]   D. Angella, A. Tomassini, Symplectic manifolds and cohomological decomposition. J. Symplectic Geom., arXiv:1211.2565v1 [math.SG] (2012 to appear)

[AT13a]   D. Angella, A. Tomassini, Inequalities à la Frölicher and cohomological decompositions, preprint, 2013

[AT13b]   D. Angella, A. Tomassini, On the $\partial\bar{\partial}$-Lemma and Bott-Chern cohomology. Invent. Math. **192**(1), 71–81 (2013)

[ATZ12]   D. Angella, A. Tomassini, W. Zhang, On cohomological decomposition of almost-Kähler structures. Proc. Am. Math. Soc., arXiv:1211.2928v1 [math.DG] (2012 to appear)

[Aus61]   L. Auslander, Discrete uniform subgroups of solvable Lie groups. Trans. Am. Math. Soc. **99**, 398–402 (1961)

[Aus73a]   L. Auslander, An exposition of the structure of solvmanifolds. I. Algebraic theory. Bull. Am. Math. Soc. **79**(2), 227–261 (1973)

[Aus73b]   L. Auslander, An exposition of the structure of solvmanifolds. II. G-induced flows. Bull. Am. Math. Soc. **79**(2), 262–285 (1973)

[AV65a]   A. Andreotti, E. Vesentini, Carleman estimates for the Laplace-Beltrami equation on complex manifolds. Inst. Hautes Études Sci. Publ. Math. **25**, 81–130 (1965)

[AV65b]   A. Andreotti, E. Vesentini, Erratum to: Carleman estimates for the Laplace-Beltrami equation on complex manifolds. Inst. Hautes Études Sci. Publ. Math. **27**, 153–155 (1965)

[Bai54]   W.L. Baily Jr., On the quotient of an analytic manifold by a group of analytic homeomorphisms. Proc. Natl. Acad. Sci. USA **40**(9), 804–808 (1954)

[Bai56]   W.L. Baily Jr., The decomposition theorem for V-manifolds. Am. J. Math. **78**(4), 862–888 (1956)

[Bal06]   W. Ballmann, *Lectures on Kähler Manifolds*. ESI Lectures in Mathematics and Physics (European Mathematical Society (EMS), Zürich, 2006)

[Bas99]   G. Bassanelli, Area-minimizing Riemann surfaces on the Iwasawa manifold. J. Geom. Anal. **9**(2), 179–201 (1999)

[BC65]   R. Bott, S.S. Chern, Hermitian vector bundles and the equidistribution of the zeroes of their holomorphic sections. Acta Math. **114**(1), 71–112 (1965)

[BDMM95]   M.L. Barberis, I.G. Dotti Miatello, R.J. Miatello, On certain locally homogeneous Clifford manifolds. Ann. Global Anal. Geom. **13**(3), 289–301 (1995)

[Bel00]   F.A. Belgun, On the metric structure of non-Kähler complex surfaces. Math. Ann. **317**(1), 1–40 (2000)

[BFV11]   E. Buzano, A. Fino, L. Vezzoni, The Calabi-Yau equation for T^2-bundles over $\mathbb{T}^2$: the non-Lagrangian case. Rend. Semin. Mat. Univ. Politec. Torino **69**(3), 281–298 (2011)

[BG88]   Ch. Benson, C.S. Gordon, Kähler and symplectic structures on nilmanifolds. Topology **27**(4), 513–518 (1988)

[BG90]   Ch. Benson, C.S. Gordon, Kähler structures on compact solvmanifolds. Proc. Am. Math. Soc. **108**(4), 971–980 (1990)

[BHPVdV04]   W.P. Barth, K. Hulek, C.A.M. Peters, A. Van de Ven, *Compact Complex Surfaces*, 2nd edn. Ergebnisse der Mathematik und ihrer Grenzgebiete. 3. Folge. A Series of Modern Surveys in Mathematics [Results in Mathematics and Related Areas. 3rd Series. A Series of Modern Surveys in Mathematics], vol. 4 (Springer, Berlin, 2004)

[Big69]   B. Bigolin, Gruppi di Aeppli. Ann. Sc. Norm. Super. Pisa (3) **23**(2), 259–287 (1969)

[Big70]   B. Bigolin, Osservazioni sulla coomologia del $\partial\bar{\partial}$. Ann. Sc. Norm. Super. Pisa (3) **24**(3), 571–583 (1970)

[Bis89]   J.-M. Bismut, A local index theorem for non-Kähler manifolds. Math. Ann. **284**(4), 681–699 (1989)

[Bis11a]   J.-M. Bismut, Hypoelliptic Laplacian and Bott-Chern cohomology, preprint (Orsay), 2011

[Bis11b]   J.-M. Bismut, Laplacien hypoelliptique et cohomologie de Bott-Chern. C. R. Math. Acad. Sci. Paris **349**(1–2), 75–80 (2011)

[Boc45]   S. Bochner, Compact groups of differentiable transformations. Ann. Math. (2) **46**(3), 372–381 (1945)

[Boc09]   C. Bock, On low-dimensional solvmanifolds, arXiv:0903.2926v4 [math.DG], 2009

[Bog78]   F.A. Bogomolov, Hamiltonian Kählerian manifolds. Dokl. Akad. Nauk SSSR **243**(5), 1101–1104 (1978)

[Bou04]   S. Boucksom, Divisorial Zariski decompositions on compact complex manifolds. Ann. Sci. École Norm. Sup. (4) **37**(1), 45–76 (2004)

[Bry88]   J.-L. Brylinski, A differential complex for Poisson manifolds. J. Differ. Geom. **28**(1), 93–114 (1988)

[Buc99]   N. Buchdahl, On compact Kähler surfaces. Ann. Inst. Fourier (Grenoble) **49**(1), 287–302 (1999)

[Buc08]   N. Buchdahl, Algebraic deformations of compact Kähler surfaces. II. Math. Z. **258**(3), 493–498 (2008)

[Cal57]   E. Calabi, On Kähler manifolds with vanishing canonical class, in *Algebraic Geometry and Topology. A Symposium in Honor of S. Lefschetz* (Princeton University Press, Princeton, 1957), pp. 78–89

[Cam91]   F. Campana, The class $\mathcal{C}$ is not stable by small deformations. Math. Ann. **290**(1), 19–30 (1991)

[Cao12]   J. Cao, Deformation of Kähler manifolds, `arXiv:1211.2058v1 [math.AG]`, `hal-00749923`, 2012

[Car30]   E. Cartan, La théorie des groupes finis et continus et l'analysis situs, 1930 (French)

[Cav05]   G.R. Cavalcanti, New aspects of the dd^c-lemma, Oxford University D. Phil Thesis, 2005, `arXiv:math/0501406 [math.DG]`

[Cav06]   G.R. Cavalcanti, The decomposition of forms and cohomology of generalized complex manifolds. J. Geom. Phys. **57**(1), 121–132 (2006)

[Cav07]   G.R. Cavalcanti, *Introduction to Generalized Complex Geometry*. Publicações Matemáticas do IMPA [IMPA Mathematical Publications] (Instituto Nacional de Matemática Pura e Aplicada (IMPA), Rio de Janeiro, 2007), 26o Colóquio Brasileiro de Matemática [26th Brazilian Mathematics Colloquium]

[CdS01]   A. Cannas da Silva, *Lectures on Symplectic Geometry*. Lecture Notes in Mathematics, vol. 1764 (Springer, Berlin, 2001)

[CE48]   C. Chevalley, S. Eilenberg, Cohomology theory of Lie groups and Lie algebras. Trans. Am. Math. Soc. **63**, 85–124 (1948)

[CF01]   S. Console, A. Fino, Dolbeault cohomology of compact nilmanifolds. Transform. Groups **6**(2), 111–124 (2001)

[CF11]   S. Console, A. Fino, On the de Rham cohomology of solvmanifolds. Ann. Sc. Norm. Super. Pisa Cl. Sci. (5) **X**(4), 801–818 (2011)

[CFAG96]   V. Cruceanu, P. Fortuny Ayuso, P.M. Gadea, A survey on paracomplex geometry. Rocky Mt. J. Math. **26**(1), 83–115 (1996)

[CFGU97]   L.A. Cordero, M. Fernández, A. Gray, L. Ugarte, Nilpotent complex structures on compact nilmanifolds, in *Proceedings of the Workshop on Differential Geometry and Topology*, Palermo, 1996, vol. 49 (1997), pp. 83–100

[CFGU00]   L.A. Cordero, M. Fernández, A. Gray, L. Ugarte, Compact nilmanifolds with nilpotent complex structures: Dolbeault cohomology. Trans. Am. Math. Soc. **352**(12), 5405–5433 (2000)

[CFK13]   S. Console, A.M. Fino, H. Kasuya, Modifications and cohomologies of solvmanifolds, `arXiv:1301.6042v1 [math.DG]`, 2013

[CFUG97]   L.A. Cordero, M. Fernández, L. Ugarte, A. Gray, A general description of the terms in the Frölicher spectral sequence. Differ. Geom. Appl. **7**(1), 75–84 (1997)

[CG04]   G.R. Cavalcanti, M. Gualtieri, Generalized complex structures on nilmanifolds. J. Symplectic Geom. **2**(3), 393–410 (2004)

[Cho49]   W.-L. Chow, On compact complex analytic varieties. Am. J. Math. **71**(4), 893–914 (1949)

[CM09]   V. Cortés, T. Mohaupt, Special geometry of Euclidean supersymmetry. III. The local r-map, instantons and black holes. J. High Energy Phys. **7**(066), 64 (2009)

[CM12]   S. Console, M. Macrì, Lattices, cohomology and models of six dimensional almost abelian solvmanifolds, `arXiv:1206.5977v1 [math.DG]`, 2012

[CMMS04]   V. Cortés, C. Mayer, T. Mohaupt, F. Saueressig, Special geometry of Euclidean supersymmetry. I. Vector multiplets. J. High Energy Phys. **3**(028), 73 pp. (2004) (electronic)

[CMMS05]   V. Cortés, C. Mayer, T. Mohaupt, F. Saueressig, Special geometry of Euclidean supersymmetry. II. Hypermultiplets and the c-map, J. High Energy Phys. **6**(025), 37 pp. (2005) (electronic)

[Con85]   A. Connes, Noncommutative differential geometry. Inst. Hautes Études Sci. Publ. Math. **62**, 257–360 (1985)

[Con06]   S. Console, Dolbeault cohomology and deformations of nilmanifolds. Rev. Un. Mat. Argentina **47**(1), 51–60 (2006)

[COUV11]   M. Ceballos, A. Otal, L. Ugarte, R. Villacampa, Classification of complex structures on 6-dimensional nilpotent Lie algebras, `arXiv:1111.5873v3 [math.DG]`, 2011

[CS05]   R. Campoamor-Stursberg, Some remarks concerning the invariants of rank one solvable real Lie algebras. Algebra Colloq. **12**(3), 497–518 (2005)

[dAFdLM92] L.C. de Andrés, M. Fernández, M. de León, J.J. Mencía, Some six-dimensional compact symplectic and complex solvmanifolds. Rend. Mat. Appl. (7) **12**(1), 59–67 (1992)

[Dar82]   G. Darboux, Sur le problème de Pfaff. Bull. Sci. Math. **6** 14–36, 49–68 (1882)

[dBM10]   P. de Bartolomeis, F. Meylan, Intrinsic deformation theory of CR structures. Ann. Sc. Norm. Super. Pisa Cl. Sci. (5) **9** (3), 459–494 (2010)

[dBT06]   P. de Bartolomeis, A. Tommasini, On solvable generalized Calabi-Yau manifolds. Ann. Inst. Fourier (Grenoble) **56** (5), 1281–1296 (2006)

[dBT12]   P. de Bartolomeis, A. Tommasini, Exotic deformations of Calabi-Yau manifolds (Déformations exotiques des variétés de Calabi-Yau). Ann. l'inst. Fourier **63**(2), 391–415 (2013). DOI:10.5802/aif.2764

[Dek00]   K. Dekimpe, Semi-simple splittings for solvable Lie groups and polynomial structures. Forum Math. **12**(1), 77–96 (2000)

[Dem86]   J.-P. Demailly, Sur l'identité de Bochner-Kodaira-Nakano en géométrie hermitienne, *Séminaire d'analyse*. P. Lelong-P. Dolbeault-H. Skoda, Années 1983/1984. Lecture Notes in Mathematics, vol. 1198 (Springer, Berlin, 1986), pp. 88–97

[Dem12]   J.-P. Demailly, Complex Analytic and Differential Geometry, http://www-fourier. ujf-grenoble.fr/~demailly/manuscripts/agbook.pdf, 2012, Version of Thursday June 21, 2012

[DF09]   A. Diatta, B. Foreman, Lattices in contact lie groups and 5-dimensional contact solvmanifolds, arXiv:0904.3113v1 [math.DG], 2009

[DGMS75]   P. Deligne, Ph.A. Griffiths, J. Morgan, D.P. Sullivan, Real homotopy theory of Kähler manifolds. Invent. Math. **29**(3), 245–274 (1975)

[DL13]   T. Drăghici, H. Leon, On 4-dimensional non-unimodular symplectic Lie algebras, preprint, 2013

[DLZ10]   T. Drăghici, T.-J. Li, W. Zhang, Symplectic forms and cohomology decomposition of almost complex four-manifolds. Int. Math. Res. Not. IMRN **2010**(1), 1–17 (2010)

[DLZ11]   T. Drăghici, T.-J. Li, W. Zhang, On the J-anti-invariant cohomology of almost complex 4-manifolds. Q. J. Math. **64**(1), 83–111 (2013)

[DLZ12]   T. Drăghici, T.-J. Li, W. Zhang, Geometry of tamed almost complex structures on 4-dimensional manifolds, in *Fifth International Congress of Chinese Mathematicians*. Part 1, 2, AMS/IP Studies in Advanced Mathematics, 51, pt. 1, vol. 2 (American Mathematical Society, Providence, 2012), pp. 233–251

[DM05]   S.G. Dani, M.G. Mainkar, Anosov automorphisms on compact nilmanifolds associated with graphs. Trans. Am. Math. Soc. **357**(6), 2235–2251 (2005)

[Dol53]   P. Dolbeault, Sur la cohomologie des variétés analytiques complexes. C. R. Acad. Sci. Paris **236**, 175–177 (1953)

[Don06]   S.K. Donaldson, Two-forms on four-manifolds and elliptic equations, Inspired by S.S. Chern, in *Nankai Tracts Mathematics*, vol. 11 (World Scientific Publishing, Hackensack, 2006), pp. 153–172

[DP04]   J.-P. Demailly, M. Păun, Numerical characterization of the Kähler cone of a compact Kähler manifold. Ann. Math. (2) **159**(3), 1247–1274 (2004)

[dR84]   G. de Rham, *Differentiable Manifolds*. Grundlehren der Mathematischen Wissenschaften [Fundamental Principles of Mathematical Sciences], vol. 266. (Springer, Berlin, 1984). Forms, currents, harmonic forms, Translated from the French by F.R. Smith, with an introduction by S.S. Chern

[DtER03]   N. Dungey, A.F.M. ter Elst, D.W. Robinson, *Analysis on Lie Groups with Polynomial Growth*. Progress in Mathematics, vol. 214 (Birkhäuser Boston, Boston, 2003)

[DZ12]   T. Drăghici, W. Zhang, A note on exact forms on almost complex manifolds. Math. Res. Lett. **19**(3), 691–697 (2012)

[EFV12]   N. Enrietti, A. Fino, L. Vezzoni, Tamed symplectic forms and strong Kahler with torsion metrics. J. Symplectic Geom. **10**(2), 203–224 (2012)

[Ehr47]   C. Ehresmann, Sur les espaces fibrés différentiables. C. R. Acad. Sci. Paris **224**, 1611–1612 (1947)

[Ehr49]   Ch. Ehresmann, *Sur la théorie des espaces fibrés*. Topologie algébrique, Colloques Internationaux du Centre National de la Recherche Scientifique, no. 12 (Centre de la Recherche Scientifique, Paris, 1949), pp. 3–15.

[ES93]   M.G. Eastwood, M.A. Singer, The Fröhlicher spectral sequence on a twistor space. J. Differ. Geom. **38**(3), 653–669 (1993)

[EVS80]   M.G. Eastwood, G. Vigna Suria, Cohomologically complete and pseudoconvex domains. Comment. Math. Helv. **55**(3), 413–426 (1980)

[FdLS96]   M. Fernández, M. de León, M. Saralegui, A six-dimensional compact symplectic solvmanifold without Kähler structures. Osaka J. Math. **33**(1), 19–35 (1996)

[Fed69]   H. Federer, *Geometric Measure Theory*. Die Grundlehren der mathematischen Wissenschaften, Band 153 (Springer, New York, 1969)

[FG86]   M. Fernández, A. Gray, The Iwasawa manifold, in *Differential Geometry, Peñíscola 1985*. Lecture Notes in Mathematics, vol. 1209 (Springer, Berlin, 1986), pp. 157–159

[FG04]   A. Fino, G. Grantcharov, Properties of manifolds with skew-symmetric torsion and special holonomy. Adv. Math. **189**(2), 439–450 (2004)

[FIdL98]   M. Fernández, R. Ibáñez, M. de León, The canonical spectral sequences for Poisson manifolds. Israel J. Math. **106**, 133–155 (1998)

[FLSV11]   A. Fino, Y.Y. Li, S. Salamon, L. Vezzoni, The Calabi-Yau equation on 4-manifolds over 2-tori. Trans. Am. Math. Soc. **365**(3), 1551–1575 (2013)

[FLY12]   J. Fu, J. Li, S.-T. Yau, Balanced metrics on non-Kähler Calabi-Yau threefolds. J. Differ. Geom. **90**(1), 81–129 (2012)

[FMS03]   M. Fernández, V. Muñoz, J.A. Santisteban, Cohomologically Kähler manifolds with no Kähler metrics. Int. J. Math. Math. Sci. **2003**(52), 3315–3325 (2003)

[FOT08]   Y. Félix, J. Oprea, D. Tanré, *Algebraic Models in Geometry*. Oxford Graduate Texts in Mathematics, vol. 17 (Oxford University Press, Oxford, 2008)

[FPS04]   A. Fino, M. Parton, S. Salamon, Families of strong KT structures in six dimensions. Comment. Math. Helv. **79**(2), 317–340 (2004)

[Fra11]   M.G. Franzini, *Deformazioni di varietà bilanciate e loro proprietà coomologiche*. Tesi di Laurea Magistrale in Matematica, Università degli Studi di Parma, 2011

[Frö55]   A. Frölicher, Relations between the cohomology groups of Dolbeault and topological invariants. Proc. Natl. Acad. Sci. USA **41**(9), 641–644 (1955)

[FT09]   A.M. Fino, A. Tomassini, Non-Kähler solvmanifolds with generalized Kähler structure. J. Symplectic Geom. **7**(2), 1–14 (2009)

[FT10]   A. Fino, A. Tomassini, On some cohomological properties of almost complex manifolds. J. Geom. Anal. **20**(1), 107–131 (2010)

[Fub04]   G. Fubini, Sulle metriche definite da una forma *Hermitiana*. Nota. Ven. Ist. Atti **63**((8)6) (1904), 501–513

[Fuj78]   A. Fujiki, On automorphism groups of compact Kähler manifolds. Invent. Math. **44**(3), 225–258 (1978)

[FY11]   J. Fu, S.-T. Yau, A note on small deformations of balanced manifolds. C. R. Math. Acad. Sci. Paris **349**(13–14), 793–796 (2011)

[Gau77]   P. Gauduchon, Le théorème de l'excentricité nulle. C. R. Acad. Sci. Paris Sér. A-B **285**(5), A387–A390 (1977)

[GH80]   A. Gray, L.M. Hervella, The sixteen classes of almost Hermitian manifolds and their linear invariants. Ann. Mat. Pura Appl. (4) **123**(1), 35–58 (1980)

[GH94]   Ph.A. Griffiths, J. Harris, *Principles of Algebraic Geometry*. Wiley Classics Library (Wiley, New York, 1994). Reprint of the 1978 original

[GK96]   M. Goze, Y. Khakimdjanov, *Nilpotent Lie Algebras*. Mathematics and its Applications, vol. 361 (Kluwer, Dordrecht, 1996)

[Gle52]   A.M. Gleason, Groups without small subgroups. Ann. Math. (2) **56**, 193–212 (1952)
[Gon98]   M.-P. Gong, Classification of nilpotent Lie algebras of dimension 7 (over algebraically closed fields and R), ProQuest LLC, Ann Arbor, MI, Thesis (Ph.D.), University of Waterloo, Canada, 1998
[Goo85]   Th.G. Goodwillie, Cyclic homology, derivations, and the free loopspace. Topology **24**(2), 187–215 (1985)
[GOV97]   V.V. Gorbatsevich, A.L. Onishchik, E.B. Vinberg, *Foundations of Lie Theory and Lie Transformation Groups* (Springer, Berlin, 1997). Translated from the Russian by A. Kozlowski, Reprint of the 1993 translation [ıt Lie groups and Lie algebras. I. Encyclopaedia of Mathematical Sciences, 20 (Springer, Berlin, 1993); MR1306737 (95f:22001)]
[Gri66]   Ph.A. Griffiths, The extension problem in complex analysis. II. Embeddings with positive normal bundle. Am. J. Math. **88**(2), 366–446 (1966)
[Gro85]   M. Gromov, Pseudoholomorphic curves in symplectic manifolds. Invent. Math. **82**(2), 307–347 (1985)
[Gua04a]   M. Gualtieri, Generalized complex geometry, Oxford University D. Phil Thesis, 2004, arXiv:math/0401221v1 [math.DG]
[Gua04b]   M. Gualtieri, Generalized geometry and the Hodge decomposition, arXiv:math/0409093v1 [math.DG], 2004
[Gua07]   D. Guan, Modification and the cohomology groups of compact solvmanifolds. Electron. Res. Announc. Am. Math. Soc. **13**, 74–81 (2007)
[Gua11]   M. Gualtieri, Generalized complex geometry. Ann. Math. (2) **174**(1), 75–123 (2011)
[Gui01]   V. Guillemin, Symplectic Hodge theory and the dδ-lemma, preprint (Massachusetts Insitute of Technology), 2001
[Han57]   J.-i. Hano, On Kaehlerian homogeneous spaces of unimodular Lie groups. Am. J. Math. **79**, 885–900 (1957)
[Har96]   R. Harshavardhan, Geometric structures of Lie type on 5-manifolds, Ph.D. Thesis, Cambridge University, 1996
[Has89]   K. Hasegawa, Minimal models of nilmanifolds. Proc. Am. Math. Soc. **106**(1), 65–71 (1989)
[Has06]   K. Hasegawa, A note on compact solvmanifolds with Kähler structures. Osaka J. Math. **43**(1), 131–135 (2006)
[Has10]   K. Hasegawa, Small deformations and non-left-invariant complex structures on six-dimensional compact solvmanifolds. Differ. Geom. Appl. **28**(2), 220–227 (2010)
[Hat60]   A. Hattori, Spectral sequence in the de Rham cohomology of fibre bundles. J. Fac. Sci. Univ. Tokyo Sect. I **8**(2), 289–331 (1960)
[Hir62]   H. Hironaka, An example of a non-Kählerian complex-analytic deformation of Kählerian complex structures. Ann. Math. (2) **75**(1), 190–208 (1962)
[Hir95]   F. Hirzebruch, *Topological Methods in Algebraic Geometry*. Classics in Mathematics (Springer, Berlin, 1995). Translated from the German and Appendix One by R.L.E. Schwarzenberger, With a preface to the third English edition by the author and Schwarzenberger, Appendix Two by A. Borel, Reprint of the 1978 edition
[Hit03]   N.J. Hitchin, Generalized Calabi-Yau manifolds. Q. J. Math. **54**(3), 281–308 (2003)
[Hit10]   N.J. Hitchin, Generalized geometry—an introduction, in *Handbook of Pseudo-Riemannian Geometry and Supersymmetry*. IRMA Lectures in Mathematics and Theoretical Physics, vol. 16 (European Mathematical Society, Zürich, 2010), pp. 185–208
[HL83]   F.R. Harvey, H.B. Lawson Jr., An intrinsic characterization of Kähler manifolds. Invent. Math. **74**(2), 169–198 (1983)
[HL11]   F.R. Harvey, H.B. Lawson Jr., The foundations of p-convexity and p-plurisubharmonicity in riemannian geometry, arXiv:1111.3895 [math.DG], 2011

[HL12]   F.R. Harvey, H.B. Lawson Jr., Geometric plurisubharmonicity and convexity: an introduction. Adv. Math. **230**(4–6), 2428–2456 (2012)

[HMT11]   R.K. Hind, C. Medori, A. Tomassini, On non-pure forms on almost complex manifolds. Proc. Am. Math. Soc. (2011 to appear)

[Hod35]   W.V.D. Hodge, Harmonic integrals associated with algebraic varieties. Proc. Lond. Math. Soc. (2) **39**, 249–271 (1935)

[Hod89]   W.V.D. Hodge, *The Theory and Applications of Harmonic Integrals*. Cambridge Mathematical Library (Cambridge University Press, Cambridge, 1989). Reprint of the 1941 original, With a foreword by Michael Atiyah

[Hör65]   L. Hörmander, L^2 estimates and existence theorems for the $\bar{\partial}$ operator. Acta Math. **113**(1), 89–152 (1965)

[Hör90]   L. Hörmander, *An Introduction to Complex Analysis in Several Variables*, 3rd edn. North-Holland Mathematical Library, vol. 7 (North-Holland Publishing, Amsterdam, 1990)

[Hum78]   J.E. Humphreys, *Introduction to Lie Algebras and Representation Theory*. Graduate Texts in Mathematics, vol. 9 (Springer, New York, 1978). Second printing, revised

[Huy05]   D. Huybrechts, *Complex Geometry*. Universitext (Springer, Berlin, 2005)

[Iit72]   S. Iitaka, Genus and classification of algebraic varieties. I. Sûgaku **24**(1), 14–27 (1972)

[IRTU03]   R. Ibáñez, Y. Rudyak, A. Tralle, L. Ugarte, On certain geometric and homotopy properties of closed symplectic manifolds, in *Proceedings of the Pacific Institute for the Mathematical Sciences Workshop "Invariants of Three-Manifolds"*, Calgary, AB, 1999, vol. 127 (2003), pp. 33–45

[Joy96a]   D.D. Joyce, Compact 8-manifolds with holonomy Spin(7). Invent. Math. **123**(3), 507–552 (1996)

[Joy96b]   D.D. Joyce, Compact Riemannian 7-manifolds with holonomy G_2. I, II. J. Differ. Geom. **43**(2), 291–328, 329–375 (1996)

[Joy99]   D.D. Joyce, A new construction of compact 8-manifolds with holonomy Spin(7). J. Differ. Geom. **53**(1), 89–130 (1999)

[Joy00]   D.D. Joyce, *Compact Manifolds with Special Holonomy*. Oxford Mathematical Monographs (Oxford University Press, Oxford, 2000)

[Joy07]   D.D. Joyce, *Riemannian Holonomy Groups and Calibrated Geometry*. Oxford Graduate Texts in Mathematics, vol. 12 (Oxford University Press, Oxford, 2007)

[JY93]   J. Jost, S.-T. Yau, A nonlinear elliptic system for maps from Hermitian to Riemannian manifolds and rigidity theorems in Hermitian geometry. Acta Math. **170**(2), 221–254 (1993)

[JY94]   J. Jost, S.-T. Yau, Correction to: "A nonlinear elliptic system for maps from Hermitian to Riemannian manifolds and rigidity theorems in Hermitian geometry" [Acta Math. **170**(2), 221–254 (1993); MR1226528 (94g:58053)]. Acta Math. **173**(2), 307 (1994)

[Käh33]   E. Kähler, Über eine bemerkenswerte Hermitesche Metrik. Abh. Math. Sem. Univ. Hamburg **9**(1), 176–186 (1933)

[Kas09]   H. Kasuya, Formality and hard Lefschetz property of aspherical manifolds. Osaka J. Math. **50**(2), 439–455 (2013)

[Kas11]   H. Kasuya, Hodge symmetry and decomposition on non-Kähler solvmanifolds, arXiv:1109.5929v4 [math.DG], 2011

[Kas12a]   H. Kasuya, De Rham and Dolbeault cohomology of solvmanifolds with local systems, arXiv:1207.3988v3 [math.DG], 2012

[Kas12b]   H. Kasuya, Degenerations of the Frölicher spectral sequences of solvmanifolds, arXiv:1210.2661v2 [math.DG], 2012

[Kas12c]   H. Kasuya, Differential Gerstenhaber algebras and generalized deformations of solvmanifolds, arXiv:1211.4188v2 [math.DG], 2012

[Kas12d]   H. Kasuya, Geometrical formality of solvmanifolds and solvable Lie type geometries, in *RIMS Kokyuroku Bessatsu B 39. Geometry of Transformation Groups and Combinatorics* (2012), pp. 21–34

[Kas13a]   H. Kasuya, Minimal models, formality, and hard Lefschetz properties of solvmani-
            folds with local systems. J. Differ. Geom. **93** (2), 269–297 (2013)
[Kas13b]   H. Kasuya, Techniques of computations of Dolbeault cohomology of solvmani-
            folds. Math. Z. **273**(1–2), 437–447 (2013)
[Kir08]   A. Kirillov Jr., *An Introduction to Lie Groups and Lie Algebras*. Cambridge Studies
            in Advanced Mathematics, vol. 113 (Cambridge University Press, Cambridge,
            2008)
[KMW10]   Y.-H. Kim, R.J. McCann, M. Warren, Pseudo-Riemannian geometry calibrates
            optimal transportation. Math. Res. Lett. **17** (6), 1183–1197 (2010)
[KN96]   S. Kobayashi, K. Nomizu, *Foundations of Differential Geometry. Vol. II*. Wiley
            Classics Library (Wiley, New York, 1996). Reprint of the 1969 original, A Wiley-
            Interscience Publication
[KNS58]   K. Kodaira, L. Nirenberg, D.C. Spencer, On the existence of deformations of
            complex analytic structures. Ann. Math. (2) **68**(2), 450–459 (1958)
[Kod54]   K. Kodaira, On Kähler varieties of restricted type (an intrinsic characterization of
            algebraic varieties). Ann. Math. (2) **60**, 28–48 (1954)
[Kod63]   K. Kodaira, On compact analytic surfaces. III. Ann. Math. (2) **78**(1), 1–40 (1963)
[Kod64]   K. Kodaira, On the structure of compact complex analytic surfaces. I. Am. J. Math.
            86, 751–798 (1964)
[Kod05]   K. Kodaira, *Complex Manifolds and Deformation of Complex Structures*. English
            edn. Classics in Mathematics (Springer, Berlin, 2005). Translated from the 1981
            Japanese original by Kazuo Akao
[Koo11]   R. Kooistra, Regulator currents on compact complex manifolds, ProQuest LLC,
            Ann Arbor, MI, Thesis (Ph.D.), University of Alberta, Canada, 2011
[Kos85]   J.-L. Koszul, *Crochet de Schouten-Nijenhuis et cohomologie*, Astérisque (1985),
            no. Numero Hors Serie. The mathematical heritage of Élie Cartan (Lyon, 1984),
            257–271
[Kra10]   M. Krahe, Para-pluriharmonic maps and twistor spaces, in *Handbook of Pseudo-
            Riemannian Geometry and Supersymmetry*. IRMA Lectures in Mathematics and
            Theoretical Physics, vol. 16 (European Mathematical Society, Zürich, 2010),
            pp. 497–557
[KS58]   K. Kodaira, D.C. Spencer, On deformations of complex analytic structures. I, II.
            Ann. Math. (2) **67**(2, 3), 328–401, 403–466 (1958)
[KS60]   K. Kodaira, D.C. Spencer, On deformations of complex analytic structures. III.
            Stability theorems for complex structures. Ann. Math. (2) **71**(1), 43–76 (1960)
[KS04]   G. Ketsetzis, S. Salamon, Complex structures on the Iwasawa manifold. Adv.
            Geom. **4**(2), 165–179 (2004)
[Kur62]   M. Kuranishi, On the locally complete families of complex analytic structures. Ann.
            Math. (2) **75**(3), 536–577 (1962)
[Kur65]   M. Kuranishi, New proof for the existence of locally complete families of complex
            structures, in *Proceedings of Conference on Complex Analysis*, Minneapolis, 1964
            (Springer, Berlin, 1965), pp. 142–154
[Lam99]   A. Lamari, Courants kählériens et surfaces compactes. Ann. Inst. Fourier (Greno-
            ble) **49**(1, vii, x), 263–285 (1999)
[Lee04]   J. Lee, Family Gromov-Witten invariants for Kähler surfaces. Duke Math. J. **123**(1),
            209–233 (2004)
[Lej10]   M. Lejmi, Stability under deformations of extremal almost-Kähler metrics in
            dimension 4. Math. Res. Lett. **17**(4), 601–612 (2010)
[Ler50a]   J. Leray, L'anneau spectral et l'anneau filtré d'homologie d'un espace localement
            compact et d'une application continue. J. Math. Pures Appl. (9) **29**, 1–80, 81–139
            (1950)
[Ler50b]   J. Leray, L'homologie d'un espace fibré dont la fibre est connexe. J. Math. Pures
            Appl. (9) **29**, 169–213 (1950)
[Lie80]   S. Lie, Theorie der Transformationsgruppen I. Math. Ann. **16**(4), 441–528 (1880)

[Lin13]   Y. Lin, Symplectic harmonic theory and the Federer-Fleming deformation theorem, arXiv:1112.2442v2 [math.SG], 2013

[LO94]   G. Lupton, J. Oprea, Symplectic manifolds and formality. J. Pure Appl. Algebra **91**(1–3), 193–207 (1994)

[LP92]   C. LeBrun, Y.S. Poon, Twistors, Kähler manifolds, and bimeromorphic geometry. II. J. Am. Math. Soc. **5**(2), 317–325 (1992)

[LT12]   T.-J. Li, A. Tomassini, Almost Kähler structures on four dimensional unimodular Lie algebras. J. Geom. Phys. **62** (7), 1714–1731 (2012)

[Lus10]   L. Lussardi, A Stampacchia-type inequality for a fourth-order elliptic operator on Kähler manifolds and applications. Atti Accad. Naz. Lincei Cl. Sci. Fis. Mat. Natur. Rend. Lincei (9) Mat. Appl. **21** (2), 159–173 (2010)

[LUV12]   A. Latorre, L. Ugarte, R. Villacampa, On the Bott-Chern cohomology and balanced Hermitian nilmanifolds, arXiv:1210.0395v1 [math.DG], 2012

[LZ09]   T.-J. Li, W. Zhang, Comparing tamed and compatible symplectic cones and cohomological properties of almost complex manifolds. Commun. Anal. Geom. **17**(4), 651–683 (2009)

[LZ11]   T.-J. Li, W. Zhang, J-symplectic cones of rational four manifolds, preprint, 2011

[Mac13]   M. Macrì, Cohomological properties of unimodular six dimensional solvable Lie algebras. Differ. Geom. Appl. **31**(1), 112–129 (2013)

[Mag86]   L. Magnin, Sur les algèbres de Lie nilpotentes de dimension ≤ 7. J. Geom. Phys. **3**(1), 119–144 (1986)

[Mal49]   A.I. Mal'tsev, On a class of homogeneous spaces. Izv. Akad. Nauk SSSR Ser. Mat. **13**, 9–32 (1949). Translation in Am. Math. Soc. Translation **1951**(39), 33 pp. (1951)

[Man04]   M. Manetti, Lectures on deformations of complex manifolds (deformations from differential graded viewpoint). Rend. Mat. Appl. (7) **24**(1), 1–183 (2004)

[Mat51]   Y. Matsushima, On the discrete subgroups and homogeneous spaces of nilpotent Lie groups. Nagoya Math. J. **2**, 95–110 (1951)

[Mat95]   O. Mathieu, Harmonic cohomology classes of symplectic manifolds. Comment. Math. Helv. **70**(1), 1–9 (1995)

[McC01]   J. McCleary, *A User's Guide to Spectral Sequences*, 2nd edn. Cambridge Studies in Advanced Mathematics, vol. 58 (Cambridge University Press, Cambridge, 2001)

[Mer98]   S.A. Merkulov, Formality of canonical symplectic complexes and Frobenius manifolds. Int. Math. Res. Not. **14**, 727–733 (1998)

[Mic82]   M.L. Michelsohn, On the existence of special metrics in complex geometry. Acta Math. **149**(3–4), 261–295 (1982)

[Mil76]   J. Milnor, Curvatures of left invariant metrics on Lie groups. Adv. Math. **21**(3), 293–329 (1976)

[Miy74]   Y. Miyaoka, Kähler metrics on elliptic surfaces. Proc. Jpn. Acad. **50**(8), 533–536 (1974)

[MK06]   J. Morrow, K. Kodaira, *Complex Manifolds* (AMS Chelsea Publishing, Providence, 2006). Reprint of the 1971 edition with errata

[Moĭ66]   B.G. Moĭšezon, On n-dimensional compact complex manifolds having n algebraically independent meromorphic functions. I, II, III. Izv. Akad. Nauk SSSR Ser. Mat. **30**(1, 2, 3), 133–174, 345–386, 621–656 (1966). Translation in Am. Math. Soc., Transl., II. Ser. **63**, 51–177 (1967)

[Mor58]   V.V. Morozov, Classification of nilpotent Lie algebras of sixth order. Izv. Vysš. Učebn. Zaved. Matematika **4**(5), 161–171 (1958)

[Mor07]   A. Moroianu, *Lectures on Kähler Geometry*. London Mathematical Society Student Texts, vol. 69 (Cambridge University Press, Cambridge, 2007)

[Mos54]   G.D. Mostow, Factor spaces of solvable groups. Ann. Math. (2) **60**(1), 1–27 (1954)

[Mos57]   G.D. Mostow, Errata, "Factor spaces of solvable groups." Ann. Math. (2) **66**(3), 590 (1957)

[MPPS06]   C. Maclaughlin, H. Pedersen, Y.S. Poon, S. Salamon, Deformation of 2-step nilmanifolds with abelian complex structures. J. Lond. Math. Soc. (2) **73**(1), 173–193 (2006)

[MT00]   M. Migliorini, A. Tomassini, Local calibrations of almost complex structures. Forum Math. **12**(6), 723–730 (2000)

[MT11]   C. Medori, A. Tomassini, On small deformations of paracomplex manifolds. J. Noncommut. Geom. **5**(4), 507–522 (2011)

[Mub63a]   G.M. Mubarakzjanov, Classification of real structures of Lie algebras of fifth order (classifkaziya vehestvennych structur algebrach lie pyatovo poryadko). Izv. Vysš. Učebn. Zaved. Matematika **34** (3), 99–106 (1963)

[Mub63b]   G.M. Mubarakzjanov, Classification of solvable Lie algebras of sixth order with a non-nilpotent basis element (classifkaziya rasreshimych structur algebrach lie shestovo poryadka a odnim nenilpoentnym bazisnym elementom). Izv. Vysš. Učebn. Zaved. Matematika **35**(4), 104–116 (1963)

[Mub63c]   G.M. Mubarakzjanov, O rasreshimych algebrach Lie (On solvable Lie algebras). Izv. Vysš. Učehn. Zaved. Matematika **32** (1), 114–123 (1963)

[MZ52]   D. Montgomery, L. Zippin, Small subgroups of finite-dimensional groups. Ann. Math. (2) **56**, 213–241 (1952)

[Nak75]   I. Nakamura, Complex parallelisable manifolds and their small deformations. J. Differ. Geom. **10**, 85–112 (1975)

[NN57]   A. Newlander, L. Nirenberg, Complex analytic coordinates in almost complex manifolds. Ann. Math. (2) **65**(3), 391–404 (1957)

[Nom54]   K. Nomizu, On the cohomology of compact homogeneous spaces of nilpotent Lie groups. Ann. Math. (2) **59**, 531–538 (1954)

[Ofm85a]   S. Ofman, Résultats sur les $d'd''$ et d''-cohomologies. Application à l'intégration sur les cycles analytiques. I. C. R. Acad. Sci. Paris Sér. I Math. **300**(2), 43–45 (1985)

[Ofm85b]   S. Ofman, Résultats sur les $d'd''$ et d''-cohomologies. Application à l'intégration sur les cycles analytiques. II. C. R. Acad. Sci. Paris Sér. I Math. **300**(5), 133–135 (1985)

[Ofm88]   S. Ofman, $d'd''$, d''-cohomologies et intégration sur les cycles analytiques. Invent. Math. **92**(2), 389–402 (1988)

[Pet86]   Th. Peternell, Algebraicity criteria for compact complex manifolds. Math. Ann. **275**(4), 653–672 (1986)

[Pop09]   D. Popovici, Limits of projective manifolds under holomorphic deformations, arXiv:0910.2032v1 [math.AG], 2009

[Pop10]   D. Popovici, Limits of Moishezon manifolds under holomorphic deformations, arXiv:1003.3605v1 [math.AG], 2010

[Pop11]   D. Popovici, Deformation openness and closedness of various classes of compact complex manifolds; examples. Ann. Sc. Norm. Super. Pisa Cl. Sci. DOI:10.2422/2036-2145.201110_008, arXiv:1102.1687v1 [math.AG] (2011 to appear)

[Pop13]   D. Popovici, Deformation limits of projective manifolds: Hodge numbers and strongly Gauduchon metrics. Invent. Math., 1–20 (2013) (English), Online First, http://dx.doi.org/10.1007/s00222-013-0449-0

[PT09]   H. Pouseele, P. Tirao, Compact symplectic nilmanifolds associated with graphs. J. Pure Appl. Algebra **213**(9), 1788–1794 (2009)

[Rag72]   M.S. Raghunathan, *Discrete Subgroups of Lie Groups* (Springer, New York, 1972). Ergebnisse der Mathematik und ihrer Grenzgebiete, Band 68

[Rai06]   J. Raissy, Normalizzazione di campi vettoriali olomorfi, Tesi di Laurea Specialistica in Matematica, Università di Pisa, 2006

[Rol07]   S. Rollenske, Nilmanifolds: complex structures, geometry and deformations, Ph.D, Thesis, Universität Bayreuth, 2007, http://opus.ub.uni-bayreuth.de/opus4-ubbayreuth/frontdoor/index/index/docId/280,

[Rol08]   S. Rollenske, The Frölicher spectral sequence can be arbitrarily non-degenerate. Math. Ann. **341**(3), 623–628 (2008)

[Rol09a]   S. Rollenske, Geometry of nilmanifolds with left-invariant complex structure and deformations in the large. Proc. Lond. Math. Soc. (3) **99**(2), 425–460 (2009)

[Rol09b]   S. Rollenske, Lie-algebra Dolbeault-cohomology and small deformations of nil-manifolds. J. Lond. Math. Soc. (2) **79**(2), 346–362 (2009)

[Rol11a]   S. Rollenske, Dolbeault cohomology of nilmanifolds with left-invariant complex structure, in *Complex and Differential Geometry*, ed. by W. Ebeling, K. Hulek, K. Smoczyk. Springer Proceedings in Mathematics, vol. 8 (Springer, Berlin, 2011), pp. 369–392

[Rol11b]   S. Rollenske, The Kuranishi space of complex parallelisable nilmanifolds. J. Eur. Math. Soc. (JEMS) **13**(3), 513–531 (2011)

[Ros12a]   F.A. Rossi, On deformations of **D**-manifolds and CR **D**-manifolds. J. Geom. Phys. **62**(2), 464–478 (2012)

[Ros12b]   F.A. Rossi, On Ricci-flat **D**-Kähler manifolds, preprint, 2012

[Ros13]   F.A. Rossi, D-complex structures on manifolds: cohomological properties and deformations, Ph.D. Thesis, Università di Milano Bicocca, 2013, http://hdl.handle.net/10281/41976

[Rot55]   W. Rothstein, Zur Theorie der analytischen Mannigfaltigkeiten im Raume von n komplexen Veränderlichen. Math. Ann. **129**(1), 96–138 (1955)

[Sak76]   Y. Sakane, On compact complex parallelisable solvmanifolds. Osaka J. Math. **13**(1), 187–212 (1976)

[Sal01]   S.M. Salamon, Complex structures on nilpotent Lie algebras. J. Pure Appl. Algebra **157**(2–3), 311–333 (2001)

[Sat56]   I. Satake, On a generalization of the notion of manifold. Proc. Natl. Acad. Sci. USA **42**(6), 359–363 (1956)

[Sch29]   J.A. Schouten, Über unitäre Geometrie. Proc. Amsterdam **32**, 457–465 (1929)

[Sch07]   M. Schweitzer, Autour de la cohomologie de Bott-Chern, arXiv:0709.3528 [math.AG], 2007

[Ser51]   J.-P. Serre, Homologie singulière des espaces fibrés. Applications. Ann. Math. (2) **54**, 425–505 (1951)

[Ser55]   J.-P. Serre, Un théorème de dualité. Comment. Math. Helv. **29**(9), 9–26 (1955)

[Ser56]   J.-P. Serre, Géométrie algébrique et géométrie analytique. Ann. Inst. Fourier (Grenoble) **6**, 1–42 (1955–1956)

[Sha86]   J.-P. Sha, p-convex Riemannian manifolds. Invent. Math. **83**(3), 437–447 (1986)

[Sil96]   A. Silva, $\partial\bar\partial$-closed positive currents and special metrics on compact complex manifolds, in *Complex Analysis and Geometry (Trento, 1993)*. Lecture Notes in Pure and Applied Mathematics, vol. 173 (Marcel Dekker, New York, 1996), pp. 377–441

[Siu83]   Y.T. Siu, Every $K3$ surface is Kähler. Invent. Math. **73**(1), 139–150 (1983)

[ST10]   J. Streets, G. Tian, A parabolic flow of pluriclosed metrics. Int. Math. Res. Not. IMRN **2010**(16), 3101–3133 (2010)

[Stu05]   E. Study, Kürzeste Wege im komplexen Gebiet. Math. Ann. **60**(3), 321–378 (1905)

[Sul76]   D.P. Sullivan, Cycles for the dynamical study of foliated manifolds and complex manifolds. Invent. Math. **36**(1), 225–255 (1976)

[Sul77]   D.P. Sullivan, Infinitesimal computations in topology. Inst. Hautes Études Sci. Publ. Math. **47**, 269–331 (1978)

[SvD30]   J.A. Schouten, D. van Dantzig, Über unitäre Geometrie. Math. Ann. **103**(1), 319–346 (1930)

[Tau95]   C.H. Taubes, The Seiberg-Witten and Gromov invariants. Math. Res. Lett. **2**(2), 221–238 (1995)

[Tau11]   C.H. Taubes, Tamed to compatible: symplectic forms via moduli space integration. J. Symplectic Geom. **9**(2), 161–250 (2011)

[Thu76]   W.P. Thurston, Some simple examples of symplectic manifolds. Proc. Am. Math. Soc. **55**(2), 467–468 (1976)

[Tia87]   G. Tian, Smoothness of the universal deformation space of compact Calabi-Yau manifolds and its Petersson-Weil metric, in *Mathematical Aspects of String Theory (San Diego, Calif., 1986)*. Advanced Series in Mathematical Physics, vol. 1 (World Scientific Publishing, Singapore, 1987), pp. 629–646

[Tia00]   G. Tian, *Canonical Metrics in Kähler Geometry*. Lectures in Mathematics ETH Zürich (Birkhäuser Verlag, Basel, 2000). Notes taken by Meike Akveld

[TK97]   A. Tralle, J. Kedra, Compact completely solvable Kähler solvmanifolds are tori. Int. Math. Res. Not. **15**, 727–732 (1997)

[TO97]   A. Tralle, J. Oprea, *Symplectic Manifolds with No Kähler Structure*. Lecture Notes in Mathematics, vol. 1661 (Springer, Berlin, 1997)

[Tod89]   A.N. Todorov, The Weil-Petersson geometry of the moduli space of SU($n \geq 3$) (Calabi-Yau) manifolds. I. Commun. Math. Phys. **126**(?), 325–346 (1989)

[Tom02]   A. Tomassini, Some examples of non calibrable almost complex structures. Forum Math. **14**(6), 869–876 (2002)

[Tom08]   A. Tomasiello, Reformulating supersymmetry with a generalized Dolbeault operator. J. High Energy Phys. **2**, 010, 25 (2008)

[Tur90]   P. Turkowski, Solvable Lie algebras of dimension six. J. Math. Phys. **31**(6), 1344–1350 (1990)

[TW11a]   V. Tosatti, B. Weinkove, The Calabi-Yau equation on the Kodaira-Thurston manifold. J. Inst. Math. Jussieu **10** (2), 437–447 (2011)

[TW11b]   V. Tosatti, B. Weinkove, The Calabi-Yau equation, symplectic forms and almost complex structures, in *Geometry and Analysis*. No. 1, Advanced Lectures in Mathematics (ALM), vol. 17 (International Press, Somerville, 2011), pp. 475–493

[TWY08]   V. Tosatti, B. Weinkove, S.-T. Yau, Taming symplectic forms and the Calabi-Yau equation. Proc. Lond. Math. Soc. (3) **97** (2), 401–424 (2008)

[TY11]   L.-S. Tseng, S.-T. Yau, Generalized cohomologies and supersymmetry, arXiv:1111.6968v1 [hep-th], 2011

[TY12a]   L.-S. Tseng, S.-T. Yau, Cohomology and Hodge theory on symplectic manifolds: I. J. Differ. Geom. **91**(3), 383–416 (2012)

[TY12b]   L.-S. Tseng, S.-T. Yau, Cohomology and Hodge theory on symplectic manifolds: II. J. Differ. Geom. **91**(3), 417–443 (2012)

[Uen75]   K. Ueno, *Classification Theory of Algebraic Varieties and Compact Complex Spaces*. Lecture Notes in Mathematics, vol. 439 (Springer, Berlin, 1975). Notes written in collaboration with P. Cherenack.

[Uga07]   L. Ugarte, Hermitian structures on six-dimensional nilmanifolds. Transform. Groups **12**(1), 175–202 (2007)

[UV09]   L. Ugarte, R. Villacampa, Non-nilpotent complex geometry of nilmanifolds and heterotic supersymmetry, arXiv:0912.5110v2 [math.DG], 2009

[Vai07]   I. Vaisman, Reduction and submanifolds of generalized complex manifolds. Differ. Geom. Appl. **25**(2), 147–166 (2007)

[Var86]   J. Varouchas, Propriétés cohomologiques d'une classe de variétés analytiques complexes compactes, *Séminaire d'analyse*. P. Lelong-P. Dolbeault-H. Skoda, Années 1983/1984. Lecture Notes in Mathematics, vol. 1198 (Springer, Berlin, 1986), pp. 233–243

[Ves67]   E. Vesentini, Lectures on Levi convexity of complex manifolds and cohomology vanishing theorems, Notes by M.S. Raghunathan. Tata Institute of Fundamental Research Lectures on Mathematics, No. 39 (Tata Institute of Fundamental Research, Bombay, 1967)

[Vin94]   È.B. Vinberg (ed.), *Lie Groups and Lie Algebras, III*. Encyclopaedia of Mathematical Sciences, vol. 41 (Springer, Berlin, 1994). Structure of Lie groups and Lie algebras, A translation of ıt Current problems in mathematics. Fundamental directions, vol. 41 (Russian), Akad. Nauk SSSR, Vsesoyuz. Inst. Nauchn. i Tekhn. Inform., Moscow, 1990 [MR1056485 (91b:22001)], Translation by V. Minachin [V.V. Minakhin], Translation edited by A.L. Onishchik and È.B. Vinberg

[Voi02]   C. Voisin, Théorie de Hodge et géométrie algébrique complexe, in *Cours Spécial-isés [Specialized Courses]*, vol. 10 (Société Mathématique de France, Paris, 2002)

[Voi04]   C. Voisin, On the homotopy types of compact Kähler and complex projective manifolds. Invent. Math. **157**(2), 329–343 (2004)

[Voi06]   C. Voisin, On the homotopy types of Kähler manifolds and the birational Kodaira problem. J. Differ. Geom. **72**(1), 43–71 (2006)

[Wan54]   H.-C. Wang, Complex parallisable manifolds. Proc. Am. Math. Soc. **5**(5), 771–776 (1954)

[War83]   F.W. Warner, *Foundations of Differentiable Manifolds and Lie Groups*. Graduate Texts in Mathematics, vol. 94 (Springer, New York, 1983). Corrected reprint of the 1971 edition

[Wei58]   A. Weil, *Introduction à l'étude des variétés kählériennes*, Publications de l'Institut de Mathématique de l'Université de Nancago, VI. Actualités Sci. Ind. no. 1267 (Hermann, Paris, 1958)

[Wei07]   B. Weinkove, The Calabi-Yau equation on almost-Kähler four-manifolds. J. Differ. Geom. **76**(2), 317–349 (2007)

[Wel74]   R.O. Wells Jr., Comparison of de Rham and Dolbeault cohomology for proper surjective mappings. Pac. J. Math. **53**(1), 281–300 (1974)

[Wel08]   R.O. Wells Jr., *Differential Analysis on Complex Manifolds*, 3rd edn. Graduate Texts in Mathematics, vol. 65 (Springer, New York, 2008). With a new appendix by Oscar Garcia-Prada

[Wey97]   H. Weyl, *The Classical Groups*. Princeton Landmarks in Mathematics (Princeton University Press, Princeton, 1997). Their invariants and representations, Fifteenth printing, Princeton Paperbacks

[Wu87]   H. Wu, Manifolds of partially positive curvature. Indiana Univ. Math. J. **36**(3), 525–548 (1987)

[Wu06]   C.-C. Wu, On the geometry of superstrings with torsion, ProQuest LLC, Ann Arbor, MI, Thesis (Ph.D.), Harvard University, 2006

[Yam05]   T. Yamada, A pseudo-Kähler structure on a nontoral compact complex paralleliz-able solvmanifold. Geom. Dedicata **112**, 115–122 (2005)

[Yan96]   D. Yan, Hodge structure on symplectic manifolds. Adv. Math. **120**(1), 143–154 (1996)

[Yau77]   S.-T. Yau, Calabi's conjecture and some new results in algebraic geometry. Proc. Natl. Acad. Sci. USA **74**(5), 1798–1799 (1977)

[Yau78]   S.-T. Yau, On the Ricci curvature of a compact Kähler manifold and the complex Monge-Ampère equation. I. Commun. Pure Appl. Math. **31**(3), 339–411 (1978)

[Ye08]   X. Ye, The jumping phenomenon of Hodge numbers. Pac. J. Math. **235**(2), 379–398 (2008)

[Zha13]   W. Zhang, From Taubes currents to almost Kähler forms. Math. Ann. **356**(3), 969–978

Index

$''F^q \left(\wedge^k X \otimes \mathbb{C} \right)$, 6
$'F^p \left(\wedge^k X \otimes \mathbb{C} \right)$, 6
A, 3
$A^{\bullet,\bullet}$, 79
B-transform, 26
$B^{\bullet,\bullet}$, 79
$C^{\bullet,\bullet}$, 79
$C_{p,J}$, 222
$D^{\bullet,\bullet}$, 79
$D_{\mathrm{d}+\mathrm{d}^\Lambda}$, 16
D_{dd^Λ}, 16
$E^{\bullet,\bullet}$, 79
$F^{\bullet,\bullet}$, 79
$GH^\bullet_{A_{\mathcal{J},H}}(X)$, 23
$GH^\bullet_{BC_{\mathcal{J},H}}(X)$, 23
$GH^\bullet_{\partial_{\mathcal{J},H}}(X)$, 23
$GH^\bullet_{\overline{\partial}_{\mathcal{J},H}}(X)$, 23
$GH_{dR_H}(X)$, 23
H, 10, 12, 17
H-spectrum, 10
H-twisted generalized complex structure, 21
$H\mathbb{C}$, 221
$H^+_J(X)$, xv, 152
$H^-_J(X)$, xv, 152
$H^S_J(X;\mathbb{C})$, 153
$H^S_J(X;\mathbb{R})$, 152
$H^\bullet_{\mathrm{d}+\mathrm{d}^\Lambda}(X;\mathbb{R})$, 91
$H^\bullet_{\mathrm{dd}^\Lambda}(X;\mathbb{R})$, 91
$H^\bullet_{\mathrm{d}+\mathrm{d}^\Lambda}(X;\mathbb{R})$, 16
$H^\bullet_{\mathrm{dd}^\Lambda}(X;\mathbb{R})$, 16
$H^\bullet_{(\delta_1,\delta_2;\delta_1\delta_2)}(A^\bullet)$, 88
$H^\bullet_{(\delta_1;\delta_1)}(A^\bullet)$, 88
$H^\bullet_{(\delta_1\delta_2;\delta_1,\delta_2)}(A^\bullet)$, 88

$H^\bullet_{(\delta_2;\delta_2)}(A^\bullet)$, 88
$H^{(p,q),(q,p)}_J(X;\mathbb{R})$, xiv
$H^{(p,q)}_J(X;\mathbb{C})$, xiv
   as generalization of Dolbeault cohomology, 165
$H^{(r,s)}_\omega(X;\mathbb{R})$, 19, 158
$H^S_J(\mathfrak{g};\mathbb{K})$, 168
$H^{\bullet,\bullet}_{\partial+}(X;\mathbb{R})$, 160
$H^{\bullet,\bullet}_A(X)$, 69
$H^{\bullet,\bullet}_{BC}(X)$, 66
$H^{\bullet,\bullet}_{\overline{\partial}}(\mathfrak{g}_{\mathbb{C}})$, $H^{\bullet,\bullet}_{\partial}(\mathfrak{g}_{\mathbb{C}})$, $H^{\bullet,\bullet}_{BC}(\mathfrak{g}_{\mathbb{C}})$, $H^{\bullet,\bullet}_A(\mathfrak{g}_{\mathbb{C}})$, 97
$H^\pm_g(X)$, 167
$H^J_S(X;\mathbb{R})$, 155
$H^\bullet_{dR}(X;\mathbb{R})$, 12
$H_{(\delta_1+\delta_2;\delta_1+\delta_2)}(\mathrm{Tot}\, A^\bullet)$, 88
$H^\bullet_{dR}(\mathfrak{g};\mathbb{K})$, 97
$H^\bullet_{dR}(\mathfrak{g};\mathbb{R})$, 168
J-anti-invariant subgroup of cohomology, 152
J-invariant subgroup of cohomology, 152
K-compatible symplectic form, 4
$K(\pi;n)$, 47
L, 10, 12, 17, 167
L^k_ω, 185
L_g, 45
$N^6(c)$, 197, 209, 219, 225
$T^{0,1}X$, 2
$T^{1,0}X$, 2
T_φ, 43
$U^k_{\mathcal{J}}$, 22
$[Z]$, 42
Δ^k, 131
$\mathrm{Doub}^{\bullet,\bullet}(A^\bullet)$, 90
Λ, 10, 12, 17, 166

D. Angella, *Cohomological Aspects in Complex Non-Kähler Geometry*,
Lecture Notes in Mathematics 2095, DOI 10.1007/978-3-319-02441-7,
© Springer International Publishing Switzerland 2014

Ω_X^k, 73
$\mathrm{Tot}^\bullet (A^{\bullet,\bullet})$, 89
$\bar{A}$, 3
$\check{H}\mathfrak{C}$, 221
d, 43
$\mathrm{d}\,\mathrm{d}^\Lambda$, 17
$\mathrm{d}\,\mathrm{d}^\Lambda$-Lemma, 19
$\mathrm{d}\,\mathrm{d}^{\mathcal{J}}$-Lemma, xi, xii
    stability, 30
$\mathrm{d}\,\mathrm{d}^c$-Lemma, *see* $\partial\bar\partial$-Lemma, 75
d^Λ, 14, 17
d^c, 34, 75
d_H, 21
$\mathrm{d}_H\,\mathrm{d}_H^{\mathcal{J}}$-Lemma, 23
    for generalized Kähler structures, 30
δ, 90
$\delta_1\delta_2$-Lemma
    cohomological characterization, 89, 91
    for graded bi-differential vector spaces,
        89
$\eta\beta_5$, 213
$\eta\beta_{2n+1}$, 181
int $\check{H}\mathfrak{C}$, 221
$(\mathrm{d}+\mathrm{d}^\Lambda)$-cohomology, 16
    and $(\mathrm{d}\,\mathrm{d}^\Lambda)$-cohomology, 17
    decomposition theorem, 18
    Hodge theory, 16
    primitive $(\mathrm{d}+\mathrm{d}^\Lambda)$-cohomology, 17
$(\mathrm{d}\,\mathrm{d}^\Lambda)$-cohomology, 16
    and $(\mathrm{d}+\mathrm{d}^\Lambda)$-cohomology, 17
    decomposition theorem, 18
    Hodge theory, 16
    primitive $(\mathrm{d}\,\mathrm{d}^\Lambda)$-cohomology, 18
$\left(\mathbb{S}^3\times\mathbb{S}^3\right)^{\sharp k}$, 232
$\left(\omega^{-1}\right)^k$, 14
$(\wedge^\bullet\mathfrak{g}^*, \mathrm{d})$, 96
$\left(\wedge_{\mathrm{inv}}^\bullet X, \mathrm{d}\lfloor_{\wedge_{\mathrm{inv}}^\bullet X}\right)$, 96
$(\wedge^{\bullet,\bullet}\mathfrak{g}_{\mathbb{C}}^*, \partial, \bar\partial)$, 96
$\left(\wedge_{\mathrm{inv}}^{\bullet,\bullet} X, \partial\lfloor_{\wedge_{\mathrm{inv}}^{\bullet,\bullet} X}, \bar\partial\lfloor_{\wedge_{\mathrm{inv}}^{\bullet,\bullet} X}\right)$, 96
$[Y]$, 12
$[\cdot, \cdot\cdot]$, 37
$\{X_t\}_{t\in B}$, 36
$\left\{\left({}''E_r^{\bullet,\bullet}, {}''\mathrm{d}_r\right)\right\}_{r\in\mathbb{N}}$, 7
$\left\{\left({}'E_r^{\bullet,\bullet}, {}'\mathrm{d}_r\right)\right\}_{r\in\mathbb{N}}$, 7
$\left\{\left(E_r^{\bullet,\bullet}, \mathrm{d}_r\right)\right\}_{r\in\mathbb{N}}$, 7
$\left\{\mathfrak{g}^{[n]}\right\}_{n\in\mathbb{N}}$, 46
$\left\{\mathfrak{g}^{\{n\}}\right\}_{n\in\mathbb{N}}$, 46
$\mathbb{H}(3;\mathbb{C})$, 111
$\mathbb{H}^{p+q-1}\left(X; \left(\mathcal{L}_{X\,p,q}^\bullet, \mathrm{d}_{\mathcal{L}_{X\,p,q}^\bullet}\right)\right)$, 72

$\mathbb{H}^{p+q-2}\left(X; \left(\mathcal{L}_{X\,p,q}^\bullet, \mathrm{d}_{\mathcal{L}_{X\,p,q}^\bullet}\right)\right)$, 72
$\mathbb{I}_3$, xiii, 112
D-complex structure, xiv, 3
    $\mathcal{C}^\infty$-pure-and-full, 159
    on the product of two equi-dimensional
        manifolds, 160
D-Dolbeault cohomology, 160
    non-finite-dimensional, 160
D-Hermitian metric, 4
D-holomorphic map, 3
D-Kähler
    and $\mathcal{C}^\infty$-pure-and-fullness, 160
    non-openness, 160
D-Kähler structure, 4
    non-openness, 196
$\mathbf{D}^n$, 3
$\mathcal{A}_X^k$, 43
$\mathcal{B}_{X\,p,q}^\bullet$, 72
$\mathcal{C}(\mathfrak{g})$, 96
$\mathcal{C}^\infty$-full, 153
    almost-Kähler
        non-$\mathcal{C}^\infty$-full on Iwasawa manifold, 193
    at the kth stage, 153
    complex-$\mathcal{C}^\infty$-full at the kth stage, 154
    linear-$\mathcal{C}^\infty$-full at the kth stage, 169
    linear-complex-$\mathcal{C}^\infty$-full at the kth stage,
        169
$\mathcal{C}^\infty$-pure, 153
    at the kth stage, 153
    complex-$\mathcal{C}^\infty$-pure at the kth stage, 154
    linear-$\mathcal{C}^\infty$-pure at the kth stage, 169
    linear-complex-$\mathcal{C}^\infty$-pure at the kth stage,
        169
$\mathcal{C}^\infty$-pure-and-full, xv, 153
    at the kth stage, 153
    complex-$\mathcal{C}^\infty$-pure-and-full at the kth stage,
        154
    example of $\mathcal{C}^\infty$-full non-$\mathcal{C}^\infty$-pure, 163
    example of $\mathcal{C}^\infty$-pure non-$\mathcal{C}^\infty$-full, 163
    example of a $\mathcal{C}^\infty$-pure-and-full and pure-
        and-full manifold (Fernández, de
        León, and Saralegui solvmanifold),
        171
    Fino and Tomassini non-$\mathcal{C}^\infty$-pure-and-full
        example, 179
    for (almost-)**D**-complex structure, 159
    for deformations of Iwasawa manifold, 200
    for solvmanifolds, 168
    linear-$\mathcal{C}^\infty$-pure-and-full at the kth stage,
        169
    linear-complex-$\mathcal{C}^\infty$-pure-and-full at the
        kth stage, 169

non-$\mathcal{C}^\infty$-pure-and-full non-compact
  example, 168
non-$\mathcal{C}^\infty$-pure example, 168
non-openness, xviii, 200
relations between $\mathcal{C}^\infty$-pure-and-fullness
  and pure-and-fullness, 161
$\mathcal{C}^\infty$-pure-and-fullness
  for symplectic structures, 158
$\mathcal{D}_X^k$, 43
$\mathcal{K}_J^c$, xv, 223
$\mathcal{K}_J^t$, xv, 223
$\mathcal{K}b_J^c$, xvi, 228
$\mathcal{K}b_J^t$, xvi, 228
$\mathcal{L}^\bullet_{X\,p,q}$, 72
$\mathcal{O}_X$, 73
$\mathcal{S}$, 223
$\mathcal{ZC}_{p,J}$, 222
$\overline{\mathcal{O}}_X$, 73
$\mathfrak{sl}(2;\mathbb{R})$-representation
  finite H-spectrum, 10
$\mathfrak{C}_{p,J}$, 222, 229
$\mathfrak{b}_m$, 116
$\mathfrak{h}_2$
  pluriclosed metrics, 137
$\mathfrak{h}_4$
  balanced metrics, 138
  pluriclosed metrics, 138
$\mathfrak{h}_7$, xiii, 101, 131, 187
$\mathfrak{sl}(2;\mathbb{R})$-module, 17
  commutation relations with d d$^\Lambda$, and d d$^\Lambda$,
    17
  on currents, 12
$\mathfrak{sl}(2;\mathbb{R})$-representation, 10
$\mathfrak{sl}(2;\mathbb{C})$ representation, 30
$PH^\bullet_{d+d^\Lambda}(X;\mathbb{R})$, 17
$PH^\bullet_{dd^\Lambda}(X;\mathbb{R})$, 18
$PH^s_d(X;\mathbb{R})$, 159
Sol(3), 197, 209
ad, 141
obs, 38
∇^C, 29
nilstep($\mathfrak{g}$), 46
ω-symplectically-harmonic form, 14
ω_{FS}, 28
$\square$, 5
$\square$-harmonic form, 6
$\overline{\Omega}^k_X$, 73
$\partial\overline{\partial}$-Lemma (for Kähler, Moǐšezon, class $\mathcal{C}$), 77
∂_+, 159
∂_-, 159
∂, 3
$\partial\overline{\partial}$-Lemma, x, xvii, 34, 67, 70, 75
  $(n-1,n)^{\mathrm{th}}$ weak $\partial\overline{\partial}$-Lemma, 196

and balanced metric in Wu theorem, 196
Bott-Chern characterization, 84
cohomological characterization, 92
completely-solvable Nakamura manifold in
  case (iii), 145
for complex manifolds, 75
for double complexes, 74
equivalent formulation on a complex
  manifold, 76
equivalent formulations (for double
  complexes), 74
formality, 35, 77
Hironaka non-Kähler manifold, 77
for Kähler manifolds, see theorem,
  $\partial\overline{\partial}$-Lemma for Kähler manifolds
openness, 87
LeBrun and Poon, and Campana non-class
  $\mathcal{C}$ manifold, 77
of solvmanifolds, 51, 77
for splitting-type solvmanifolds, 143
strongly-Gauduchon and Gauduchon, 228
$\partial_{\mathcal{J},H}$, 22
$\partial_{\mathcal{J},H}\overline{\partial}_{\mathcal{J},H}$-Lemma, 23
  cohomological characterization, 92
$\pi:\mathcal{X}\to B$, 36
τ, 3
$\widetilde{\Delta}_A$, 70
$\widetilde{\Delta}_{BC}$, 68
$\wedge^{p,q}_{+-}X$, 159
$\wedge^+_J X, \wedge^-_J X$, 152
$\wedge^\bullet_c X$, 42
$\wedge^\pm_g X$, 166
$\wedge^{p,q}_g X$, 2
$a^k, b^k, c^k, d^k, e^k, f^k$, 79
$a^{p,q}, b^{p,q}, c^{p,q}, d^{p,q}, e^{p,q}, f^{p,q}$, 79
b^+, 224
b_k, 78
e^{AB}, 48
$h^k_{\overline{\partial}}, h^k_\partial, h^k_{BC}, h^k_A$, 78
$h^{p,q}_{\overline{\partial}}, h^{p,q}_\partial, h^{p,q}_{BC}, h^{p,q}_A$, 78
p-Kähler, 180
$TX \oplus T^*X$, 20
$(T^+X)^*, (T^-X)^*$, 159
$\mathbb{CP}^n$, vii, 28
$\mathcal{D}_\bullet X$, 12
$\mathcal{D}_k X$, 42
$\mathcal{D}_{p,q}X$, 44
Nij$_K$, 3
$P\wedge^\bullet X$, 11, 166
$PD^\bullet X$, 12
$\overline{\partial}$, 3, 37
$\overline{\partial}^*$, 5
$\overline{\partial}_{\mathcal{J},H}$, 22

$(n-1,n)^{\text{th}}$ weak $\partial\bar{\partial}$-Lemma, 196
HLC, *see* Hard Lefschetz Condition
SKT, *see* strong Kähler with torsion metric, 226

Abate, M., ix, 65
Abelian complex structure, xiii, 102, 218
    on 6-dimensional nilmanifolds, 55
adjoint representation, 141
Aeppli cohomology, ix, 69
    Hodge theory, 70
    invariance for nilmanifolds, 108
    sheaf-theoretic interpretation, 71
    symplectic $\left(d\,d^{A}\right)$-cohomology, 16
Aeppli, A., ix, 65
Akizuki and Nakano identities, *see* theorem,
        Akizuki and Nakano identities
Akizuki, Y., 30
Alessandrini, L., ix, xvii, 65, 180, 200
algebraic manifold, vii
almost-**D**-complex structure, 3, 170
    $\mathcal{C}^{\infty}$-pure-and-full, 159
    integrability, 3
almost-Abelian, 63, 99
almost-c.p.s. structure, 3
almost-complex, 2
    $\mathcal{C}^{\infty}$-pure-and-full, 153
    integrable, 4
almost-complex structure, viii, xiv, 3
    compatible symplectic structure, 223
    curve of almost-complex structures, 203
    holomorphic-tamed, 223
    Lee curves, 209
    taming symplectic structure, 222
almost-Kähler metric, 157
almost-Kähler, 14, 184
    intrinsic characterization (Li and Zhang),
        230
    non-$\mathcal{C}^{\infty}$-full on Iwasawa manifold, 193
    non-openness, 200, 225
    openness for 4-dimensional manifolds,
        197
almost-Kähler metric, xv
almost-Kähler structure, xvii
almost-para-complex structure, 3
almost-product structure, 3
almost-subtangent structure, 3
analytic cycles, 66
Andrada, A., 55
Andreotti, A., ix, 8, 65
annular bundle, 65
aspherical, 47
associated $(1,1)$-form, 5

astheno-Kähler metric, vii, 70
Auslander, L., 47

Baily, W. L., 52
balanced, 228
balanced cone, 228
    characterization, 229
balanced metric, vii, xvi, xvii, 42, 66, 70, 131
    Fu and Yau example on $\#_{j=1}^{k}\left(\mathbb{S}^{3}\times\mathbb{S}^{3}\right)$
        with $k\geq 2$, 70
    Michelsohn characterization, 42
    non-closedness, 139
    on $\left(\mathbb{S}^{3}\times\mathbb{S}^{3}\right)^{\#k}$, 232
    on $\mathfrak{h}_{4}$, 138
    openness under additional conditions, 196
Barberis, M. L., 55
Bassanelli, G., ix, xvii, 65, 180, 200
Belgun symmetrization map, 50, 100, 170
Belgun symmetrization trick, *see* theorem,
        Belgun symmetrization trick, 226
Belgun symmetrization trick in the almost-
        complex setting, 170
Belgun, F. A., 50, 51, 170, 215, 216, 220, 226
Benson and Gordon conjecture, 51
Benson, Ch., 51
Betti number, 8, 63, 78
bi-graded bi-differential algebra
    of PD-type, 5
bi-graded bi-differential vector space
    associated to a graded bi-differential vector
        space, *see* canonical double complex
    cohomology, 88
    graded bi-differential vector space
        associated to a bi-graded bi-
        differential vector space, *see* total
        vector space
Bigolin, B., ix, 65
Bismut, J.-M., ix, 66
Bochner linearization theorem, 52
Bogomolov, F. A., 39
Bott, R., ix, 65
Bott-Chern characterization of $\partial\bar{\partial}$-Lemma,
        see theorem, Bott-Chern
        characterization of $\partial\bar{\partial}$-Lemma
Bott-Chern cohomology, ix, 66
    Hodge theory, 68
    invariance for nilmanifolds, 104, 107
    of holomorphically parallelizable
        solvmanifolds, 143
    of splitting-type solvmanifolds, 143
    sheaf-theoretic interpretation, 71
    symplectic $\left(d+d^{A}\right)$-cohomology, 16
Boucksom, S., ix, 66

bracket, 45
  derived bracket, 21
  twisted Courant bracket, 20
Brylinski
  Hodge theory for symplectic manifold, 13
Brylinski's conjecture, 14, 19
  for Kähler manifolds, 20
Brylinski, J.-L., xi, 12, 14
Buchdahl, N., 28

Calabi and Yau manifold, 39
Calabi conjecture, 224
Calabi, E., 27, 224
Campana, F., 28, 41, 77
canonical bundle, 22
canonical double complex, 90
canonical spectral sequence, 24
Cao, J., 29, 40
Cavalcanti, G. R., xi, 20, 55
Ceballos, M., xiv, 55, 87, 95, 130, 131, 134
Chern character, 66
Chern connection, 29
Chern, S. S., ix, 65
Chevalley and Eilenberg complex, 49
Chow, W. L., vii
class $\mathcal{C}$ of Fujiki, 35, 67, 70, 77
  conjecture on closedness, 41
  Demailly and Păun conjecture, 41
  Demailly and Paun characterization, 42
  non-openness, 77
Cliff, 21
Clifford
  action, 21
  algebra, 21
  annihilator, 21
closed property under deformations, see
      deformation, closedness
closedness
  non-closedness of balanced, 139
  non-closedness of strongly-Gauduchon,
      139
co-compact discrete subgroup, 46
co-isotropic, 12, 159
cohomological decomposition
  for 4-dimensional unimodular Lie algebras,
      173
cohomologically Kähler, 63
  completely-solvable Nakamura manifold,
      190
  $N^6(c)$, 197, 209, 219
cohomology
  D-Dolbeault, 160
  Aeppli, 69

Bott-Chern, 66
Dolbeault, 5
dot, 74
  generalized complex cohomologies, 23
  left-invariant cohomology on a
      solvmanifold, 49
  de Bartolomeis and Tomassini example,
      49, 99
  of graded bi-differential vector space, 88
  square, 75
  symplectic cohomologies, xi, 107
  symplectic cohomologies $((d + d^\Lambda)$-
      cohomology), 16
  symplectic cohomologies $((d\, d^\Lambda)$-
      cohomology), 16
  symplectic cohomologies (primitive
      $(d + d^\Lambda)$-cohomology), 17
  symplectic cohomologies (primitive
      $(d\, d^\Lambda)$-cohomology), 18
cohomology decomposition, x
  for Kähler manifolds, x
commutation relations
  between $d\, d^\Lambda$, $d\, d^\Lambda$, and L, Λ, H, 17
compatible, 223
compatible cone, xv, 223
complete family, see family, complete
completely-solvable
  Lie algebra, 46
completely-solvable solvmanifold, xii, 46
  Kähler, see theorem, Tralle and Kedra
      (characterization of Kähler
      completely-solvable solvmanifolds)
  Tralle and Kedra theorem, 51
complex
  almost-complex, 2
  as a generalized complex structure, 24
complex currents, 222
complex cycles, 222
complex Lie group, 45
complex manifold, viii
complex orbifold, 52
complex projective space, 28
complex structure, vii, 90
complex-$\mathcal{C}^\infty$-full at the kth stage, 154
complex-$\mathcal{C}^\infty$-pure-and-full at the kth stage,
      154
complex-$\mathcal{C}^\infty$-pure at the kth stage, 154
complex-analytic family of compact complex
      manifolds, xi
complex-analytic family of complex manifolds,
      see family, complex-analytic family
      of complex manifolds
complex-full at the kth stage, 157
complex-pure-and-full at the kth stage, 157

complex-pure at the kth stage, 157
cone structure, xvi, 221
    ample, 221
    of p-dimensional complex subspaces
        $(C_{p,J})$, 222
conjecture
    Brylinski, 19
    Calabi, 224
    complex holomorphic-tamed non-Kähler
        (Li and Zhang, and Streets and
        Tian), 225
    Dolbeault cohomology of nilmanifolds,
        103
    Donaldson, 157
    Donaldson (tamed to compatible), xvi
    Donaldson tamed to compatible question,
        224
    Drăghici, Li, and Zhang (generic vanishing
        of anti-invariant subgroup), 176
    Drăghici, Li, and Zhang (large anti-
        invariant subgroup implies
        integrability in dimension 4), 176
    invariance of cohomologies for
        nilmanifolds, 110
    large anti-invariant subgroup implies
        integrability, 177
    Li and Zhang, and Streets and Tian, xvi
Console, S., xiv, 97, 99, 100, 102, 103, 109,
        140
Cordero, L. A., 97
Courant bracket, 20
current, 12, 42
    dual, 12
    primitive, 12
curve of almost-complex structure, 203
    Lee construction, 209
cycles of algebraic manifolds, 65

Darboux theorem
    generalized Darboux theorem, 24
Darboux, G., 9
de Bartolomeis, P., 99, 203
de León, M., 90
de Rham cohomology, viii
    J-anti-invariant subgroup, 152
    J-invariant subgroup, 152
    for completely-solvable solvmanifolds,
        98
    for nilmanifolds, 98
    for solvmanifolds, 140
    of holomorphically parallelizable
        solvmanifolds, 142
    of Iwasawa manifold, 201

    of splitting-type solvmanifolds, 142
    primitive de Rham cohomology, 159
de Rham complex, 32
de Rham homology, 12, 43
    pure-and-full, 156
deformation, 36, 157
    and C^{∞}-pure-and-fullness for D-complex
        structures, 160
    closedness, 39
    complex-analytic family of complex
        manifolds, see family, complex-
        analytic family of complex
        manifolds
    differential graded Lie algebra, 37
    Hironaka example for non-closedness of
        Kähler, 41
    Kuranishi space, 39
        of holomorphically parallelizable
            nilmanifolds, 39
    Kuranishi theorem, 39
    LeBrun and Poon, and Campana example
        of non-openness for class $\mathcal{C}$, 41
    Maurer and Cartan equation, 37
    non-obstructed, 39
    obstruction, 38
    of D-complex structures, 160
    of Iwasawa manifold, xiii
    of nilmanifolds, 39
    of solvmanifolds, 39, 103
    openness, 39
Deligne, D., 74, 76
Deligne, P., x, 31, 35, 87
Demailly and Păun conjecture, 41
Demailly, J.-P., ix, 41, 42, 66
descending central series, 46
descending derived series, 46
dga, see differential graded algebra
differential
    decomposable, 32
differential graded algebra, 31
    associate to a Lie algebra, 49
    cohomology, 32
    de Rham complex, 32
    elementary extension, 32
    equivalence, 33
    formality, 33
        vanishing of the Massey products, 34
    graded-commutativity, 32
    graded-Leibniz rule, 32
    Massey product, 33
    minimal model, 33
    minimality, 32
    morphism, 32
    quasi-isomorphism, 32

Dirac current, 221
discrete co-compact subgroup
   Mal'tsev theorem, 48
Dolbeault and Grothendieck lemma,
   see theorem, Dolbeault and
   Grothendieck lemma
Dolbeault cohomology, viii, 5
   for nilmanifolds, 101
   vanishing for q-complete domains, 8
   vanishing for strictly pseudo-convex
      domains, 8
Donaldson conjecture
   for 4-dimensional unimodular Lie algebras,
      173
Donaldson tamed to compatible question, 185
Donaldson, S. K., xvi, 151, 157, 158, 173, 185,
      224, 225
dot, 74
Dotti, I. G., 55
double complex, 6
   dot, 74
   square, 75
double numbers, 3
Drăghici, T., xv, xviii, 157, 158, 165, 166, 168,
      175, 176, 184, 212
dual current, 12

Ehresmann, Ch., 36, 120
Eilenberg and MacLane space, 47
elliptic differential operator, 6, 16
   $\tilde{\Delta}_A$, 70
   Δ_{BC}, 68
Enrietti, N., 226, 227
equivalent differential graded algebras, 33
exhaustion function, 8
exterior differential
   and Lie bracket, 96

family
   complete, 36
   complex-analytic family of complex
      manifolds, 36
   differentiable family of complex manifolds,
      36
   Ehresmann theorem, 36
   pull-back, 36
   trivial, 36
   universal, 36
   versal, 36
      Kuranishi theorem, 39

Fernández, M., 90, 97
Fernández, de León, and Saralegui
   solvmanifold, 172
filtration
   natural filtrations on a complex manifold, 6
Fino, A. M., xiv, 30, 97, 99, 100, 102, 103,
      109, 140, 157, 162, 168, 170, 177,
      184, 185, 204, 209, 226, 227
formal, 189
   $N^6(c)$, 197, 209
   completely-solvable Nakamura manifold,
      191
   differentiable manifold, 33
   differential graded algebra, 33
   Hasegawa theorem for nilmanifolds, 51
formality, 63
   $\partial\bar{\partial}$-Lemma, 77
   of solvmanifolds, 51
Frölicher inequality, ix, 8
Frölicher spectral sequence, 165
   of compact complex surfaces, 166
Franzini, M. G., xiv, 130, 131
Frechet space, 42
Frölicher inequality, *see* theorem, Frölicher
   inequality
Fu, J., 196, 232
Fubini and Study metric, 28
Fujiki
   class $\mathcal{C}$, 35
full, 156
   at the kth stage, 156
   complex-full at the kth stage, 157
fundamental group
   of a nilmanifold, 47
   of a solvmanifold, 47

GAGA, vii
Gauduchon
   and $\partial\bar{\partial}$-Lemma, 228
Gauduchon metric, vii, 70, 228
generalized complex, 21
   B-transform, 26
   defined by the canonical bundle, 22
   defined by the maximal isotropic
      sub-bundle, 21
   generalized Darboux theorem, 24
   generalized Kähler, 30
   of type 0 (symplectic), 25
   regular point, 22
   type, 22
   type n (complex), 24
generalized complex cohomologies, 23

generalized complex geometry, 66
generalized complex structure, xi, 92
  cohomologies, 23
generalized Kähler, 30
  Fino and Tomassini example, 30
geometrically formal, 191
  completely-solvable Nakamura manifold, 191
Gleason, A. M., 45
Gordon, C. S., 51
graded bi-differential vector space, 88
  $\delta_1\delta_2$-Lemma, 89
  associated to a bi-graded bi-differential vector space, *see* total vector space
  bi-graded bi-differential vector space associated to a graded bi-differential vector space, *see* canonical double complex
  cohomologies, 88
Grauert, H., 8
Gray, A., 97
Griffiths, Ph. A., x, 31, 35, 74, 76, 87
Gromov and Witten invariants, 209
Gromov, M., 224
group-object, 45
Gualtieri, M., xi, 20, 24, 55
Guan, D., xiv, 140
Guillemin, V., 15

Hörmander, L., 8
Hamiltonian isotopy, 17
Hard Lefschetz Condition, xi, xii, 15, 19, 51, 177, 185
  $N^6(c)$, 197, 209
  cohomological characterization, 91
  completely-solvable Nakamura manifold, 191
  for 6-dimensional solvmanifolds, 63
  for Kähler manifolds, 19
  for solvmanifolds, 63
  of solvmanifolds, 51
  pure-and-fullness, 162
harmonic
  ω-symplectically-harmonic form, 14
  $\overline{\square}$-harmonic form, 6
Harvey, F. R., 42, 170
Hasegawa, K., 47, 51, 187, 191
Hattori, A., xiv, 49, 51, 97, 98, 140, 145, 168, 172, 183, 192, 210
Heisenberg group, 111
Hermitian metric, vii, 5
  associated (1, 1)-form, 5
  Hodge-$*$-operator, 5
Hilbert fifth problem, 45
Hind, R. K., 158
Hironaka manifold, 41, 77
Hironaka, H., 41, 77
Hitchin, N. J., xi, 20
HLC
  for 4-dimensional solvmanifolds, 54
Hodge and Frölicher spectral sequences, xi, 7, 76, 93
  degeneration at the first step ($E_1 \simeq E_\infty$), 76, 86, 87
  for Poisson manifolds, 90
  for symplectic manifolds, 90
  for symplectic structures, 91
Hodge decomposition, 143, 165
  for compact complex surfaces, 166
  of solvmanifolds, 51, 77
Hodge decomposition theorem, xi
Hodge number, 8, 41
Hodge structure, 87
Hodge structure of weight k, 76
Hodge symmetry, 143
Hodge theory, viii, ix
  $d_H\, d^{\mathcal{J}}_H$-Lemma, 23
  $\partial\overline{\partial}$-Lemma, 34
  $\partial_{\mathcal{J},H}\overline{\partial}_{\mathcal{J},H}$-Lemma, 23
  Brylinski's conjecture, 14
  for Aeppli cohomology, 70
  for Bott-Chern cohomology, 68
  for complex manifolds, 5
  for symplectic manifolds, xi, 13
  Hard Lefschetz Condition, 15
Hodge to de Rham spectral sequences, 7
Hodge, W., 28
Hodge-$*$-operator, 5
  self-dual and anti-self-dual decomposition, 166
holomorphic p-form, 5
holomorphic curves, 209
holomorphic-tamed, 223
  and pluri-closed metric on nilmanifolds, 226
  for 6-dimensional nilmanifolds, 226
  for nilmanifolds, 227
  non-Kähler (Li and Zhang, and Streets and Tian question), 225
holomorphically parallelizable, xiii
  Bott-Chern cohomology, 143
  Dolbeault cohomology, 142
holomorphically parallelizable complex structure, 102
homogeneous space, 46

homomorphism
   of Lie algebras, 45
   of Lie groups, 45
hypercohomology of complexes of sheaves,
   72

Ibáñez, R., 90
inequality à la Frölicher, 90
   for generalized complex structures, 92
   for symplectic structures, 91
inequality à la Frölicher for Bott-Chern
   cohomology, see theorem,
   inequality à la Frölicher
inequality à la Frölicher, x
integrability
   Newlander and Nirenberg theorem, 4
   of almost-D-complex structures, 3
   of almost-complex structures, 4
invariant object
   see left-invariant object, 45
isotropy group, 46
Iwasawa
   C^∞-pure-and-full of deformations, 200
Iwasawa manifold, xiii, xvii, 91,
   112
   Aeppli cohomology, 114
   almost-Kähler non-C^∞-full, 193
   almost-Kähler structure, 91
   as $\eta\beta_3$, 181
   balanced metric, 112, 178
   Bott-Chern cohomology, 114
   Bott-Chern cohomology of deformations,
   126
   classes of deformations, 116
   cohomologies of deformations, 130
   cohomology of deformations, 82, 83, 86
   conjugate Dolbeault cohomology, 114
   de Rham cohomology, 120, 201
   Dolbeault cohomology, 114
   Dolbeault cohomology of deformations,
   122
   double complex of left-invariant forms, 112
   holomorphic-tamed structures on
   deformations, 227
   Kuranishi space, 113
   structure equations of deformations, 118,
   200
   symplectic cohomologies, 91

Jacobi identity, 45
Jordan decomposition, 141
Joyce, D. D., 52

Kähler, E., 27
Kähler
   completely-solvable solvmanifolds,
   see theorem, Tralle and Kedra
   (characterization of Kähler
   completely-solvable solvmanifolds)
   generalized Kähler, 30
   Harvey and Lawson characterization, 42
   Hironaka example for non-closedness, 41
   intrinsic characterization (Harvey and
   Lawson), 230
   nilmanifolds, see theorem, Benson
   and Gordon (characterization of
   Kähler nilmanifolds), see theorem,
   Hasegawa (characterization of
   Kähler nilmanifolds)
   openness, 196
   solvmanifolds, see theorem, Hasegawa
   (characterization of Kähler
   solvmanifolds)
   stability, 68
   stability under deformations, see theorem,
   Kodaira and Spencer
Kähler current, 42
Kähler identities, x, see theorem, Kähler
   identities
Kähler manifold, vii
Kähler metric, 28
   defined by an osculating Hermitian metric,
   29
   defined by Chern connection, 29
Kähler structure, 223
   Lefschetz decomposition theorem, 15, 19
Künneth formula, 215
Kasuya, H., xiv, 39, 51, 77, 103, 140, 142,
   143
Kedra, J., 51
Kodaira and Thurston manifold, 224
Kodaira embedding theorem, vii
Kodaira problem, 28
Kodaira, K., xvii, 28, 36, 196, 200
Kooistra, R., ix, 66
Koszul, J.-L., xi, 12, 90
Kuperberg, G., 113
Kuranishi space, xiii, 39
   of Iwasawa manifold, 113
Kuranishi, M., 36, 39

Latorre, A., 131, 140
Lawson, H. B., 42, 170
LeBrun, C., 41, 77
Lee curves of almost-complex structures, 209
Lee, J., xviii, 209, 210

Lefschetz decomposition, 10, 158
  for Kähler manifolds, 31
Lefschetz decomposition theorem, xi, 15, 19
Lefschetz-type operator, 185
Lefschetz-type property, xvii, 158, *see* Zhang's
      Lefschetz-type property
left-invariant, 48, 96
  complex structure, 96
left-invariant form, xii
left-invariant object, 45
left-translation on a Lie group, 45
Lejmi, M., 213
Leon, H., 158
Levi form, 8
Li, J., 232
Li, T.-J., xiv–xvi, xviii, 151, 152, 157, 158,
      165, 166, 168, 173, 175, 176, 184,
      212, 220, 221, 223–225, 231
Lie algebra, 45
  bracket, 45
  cohomological properties of 4-dimensional
      unimodular Lie algebra, 173
  completely-solvable, 46
  descending central series, 46
  descending derived series, 46
  homomorphism, 45
  Jacobi identity, 45
  naturally associated to a Lie group, 45
  nilpotent, 46
  of rigid type, 46
  solvable, 46
  step of nilpotency, 46
Lie group, 45
  complex Lie group, 45
  homomorphism, 45
  left-invariant object, 45
  left-translation, 45
  Lie algebra naturally associated to a Lie
      group, 45
  nilpotent, 46
  of class C^k, 45
  of class C^k, 45
  solvable, 46
  topological Lie group, 45
  translation, *see* Lie group, left-translation
  unimodular, 48
Lin, Y., 13, 159
linear-C^∞-full at the kth stage, 169
linear-C^∞-pure-and-full at the kth stage, 169
  relation between linear-C^∞-pure-and-
      fullness at the kth stage and
      C^∞-pure-and-fullness at the kth
      stage, 171
linear-C^∞-pure at the kth stage, 169

linear-complex-C^∞-full at the kth stage, 169
linear-complex-C^∞-pure-and-full at the kth
      stage, 169
linear-complex-C^∞-pure at the kth stage, 169
Lupton, G., 51
Lussardi, L., ix, 66

Maclaughlin, C., 218
Macrì, M., 63, 91, 107
Magnússon, G. Þ., 40
Mal'tsev, A. I., 47, 48, 187
Massey product, 189
  m^{th} order Massey product, 34
  triple Massey product, 33
Mathieu, O., xi, 13, 15
Matsushima, Y., 49
Maurer and Cartan equation, *see* deformation,
      Maurer and Cartan equation
MC, *see* deformation, Maurer and Cartan
      equation
Medori, C., 158
Merkulov, S. A., 15
metric cones
  balanced cone, 228
  quantitative comparison, 231
  semi-Kähler cone, 228
  strongly-Gauduchon cone, 228
metric related, 176
metric structure, vii
Meylan, F., 203
Michelsohn, M. L., vii, 42, 228
Migliorini, M., 225
Milnor, J., 48, 50, 100, 170, 227
minimal differential graded algebra, 32
minimal model, 33
Moïšezon manifold, 35, 40, 41, 77
  Hironaka manifold, 77
  tamed to compatible, 225
modification, 77
Montgomery, D., 45
Morgan, J., x, 31, 35, 74, 76, 87
Morozov, V. V., 55
Mostow bundle, 49
Mostow condition, 50, 63, 99
Mostow structure theorem, 49
Mostow, G. D., xiv, 47, 99, 140
Mubarakzjanov, G. M., 55, 63

Nakamura manifold, xiv, 99
  Bott-Chern cohomology of the completely-
      solvable Nakamura manifold, 145,
      150

completely-solvable, 109, 144, 145, 190, 193
de Rham cohomology of the completely-solvable Nakamura manifold, 145
de Rham cohomology of the holomorphically parallelizable Nakamura manifold, 150
Dolbeault cohomology of the completely-solvable Nakamura manifold, 145
Dolbeault cohomology of the holomorphically parallelizable Nakamura manifold, 150
holomorphically parallelizable, 109, 147, 149
Nakamura, I., xiii, 111, 113, 122, 144, 147, 190
Nakano, S., 30
natural pairing on $TX \oplus T^*X$, 20
Newlander, A., 4
Nijenhuis tensor, 4
for almost-**D**-complex structures, 3
nilmanifold, xii, 46
6-dimensional, 55
6-dimensional and complex, 130
Benson and Gordon theorem, 51
Bott-Chern cohomology of 6-dimensional nilmanifolds, 131
deformations, 39
fundamental group, 47
Hasegawa theorem, 51
holomorphic-tamed, 226, 227
invariance of cohomologies, 104, 107
Kähler, see theorem, Benson and Gordon theorem (characterization of Kähler nilmanifolds), see theorem, Hasegawa (characterization of Kähler nilmanifolds)
left-invariant object, 48
Mal'tsev rigidity theorem, 47
parallelizable, 47
pluriclosed metrics on 6-dimensional nilmanifolds, 137
rational structure, 101
Salamon notation, 48
nilpotent
Lie algebra, 46
Lie group, 46
step of nilpotency, 46
nilpotent complex structure, xiii, 102
on 6-dimensional nilmanifolds, 55
nilpotent operator, 141
nilradical, 140

Nirenberg, L., 4, 36
Nomizu, K., 49, 51, 97, 121, 168, 180, 181, 188, 193, 201, 213, 214
non-obstructed, 39
Norguet, F., ix, 65

obstruction, see deformation, obstruction
Ofman, S., ix, 66
open property under deformations, see deformation, openness
openness
for Kähler, xvii
non-openness of **D**-Kähler, 160, 197
non-openness of $\mathcal{C}^\infty$-pure-and-full, 200
of $\partial\bar{\partial}$-Lemma, xi
of invariance of cohomology for nilmanifolds, 109
of left-invariance of cohomology, 102
Oprea, J., 51
orbifold, 52
of global-quotient type, 52
osculate, x, 29
Otal, A., xiv, 55, 87, 95, 131, 134

Păun, M., 41, 42
para-complex structure, 3
parallelizable, 47
nilmanifolds and solvmanifolds, 47
PD-type, 5
Pedersen, H., 218
pluriclosed metric, vii, 70, 131, 226
and holomorphic-tamed on nilmanifolds, 226
$\mathfrak{h}_2$, 137
on 6-dimensional nilmanifolds, 137
on $\mathfrak{h}_4$, 138
Poincaré lemma, 72
Poincaré lemma for currents, see theorem, Poincaré lemma for currents
Poisson bi-vector, 14
Poisson manifold, 90
Poon, Y. S., 41, 77, 218
Popovici, D., 41, 77, 228
positive current, 229
positivity on the J-line, 222
potential function, vii
primitive $(d + d^\Lambda)$-cohomology, 17, 158
primitive $(d\,d^\Lambda)$-cohomology, 18
primitive cohomology, xi
and $H_\omega^{(r,s)}(X; \mathbb{R})$, 158
primitive de Rham cohomology $(PH_d^s(X; \mathbb{R}))$, 159

primitive current, 12
primitive de Rham cohomology, 159
primitive form, 11, 166
principal torus-bundle series, 102
projective
    limit, 228
projective manifold, vii
    limit, 41
pseudo-convex
    strictly, 8
pseudo-holomorphic curve, 225
pure, 156
    at the kth stage, 156
    complex-pure at the kth stage, 157
pure-and-full, 156
    at the kth stage, 156
    complex-pure-and-full at the kth stage, 157

q-complete, 8
qis, *see* quasi-isomorphism
quantum inner state manifold, 39
quasi-isomorphism, 32

rational complex structure, xiii, 101
rational manifold, 225
rational structure, 101
real homotopy type of complex manifolds, x
regular point, 22
regularization process, 43
rigid type, 46
Rollenske, S., 39, 97
Rossi, F. A., xiv, 130, 196

Sakane, Y., 97
Salamon, S., 218
Salamon, S. M., 55
Satake, I., 52
Schouten, J. A., 27
Schubert variety, 221
Schweitzer, M., ix, 66, 72, 104
semi-continuity, xviii
    Drăghici, Li, and Zhang theorem in
        dimension 4, 212
    non-lower-semi-continuity of $\dim_{\mathbb{R}} H_{J_t}^+$,
        214
    non-upper-semi-continuity of $\dim_{\mathbb{R}} H_{J_t}^-$,
        213
    of subgroups of cohomology for **D**-complex
        manifolds, 160
    stronger semi-continuity, 216
semi-Kähler, xvi, 177, 228

intrinsic characterization, 230
semi-Kähler cone, 228
semi-Kähler structure, 157
semi-simple operator, 141
Serre duality, 6
sheaf-theoretic, 65
solvable
    completely-solvable, 46
    Lie algebra, 46
    Lie group, 46
    rigid type, 46
    type (I), 46
    type (R), 46
solvmanifold, xii, 46
    1-dimensional, 52
    2-dimensional, 52
    3-dimensional, 52
    4-dimensional, 52
    5-dimensional, 55
    6-dimensional, 63
    $\partial\bar{\partial}$-Lemma, 51, 77
    Bott-Chern cohomology, 143
    completely-solvable, 46
    de Rham cohomology, 140
    deformations, 39, 103
    Dolbeault cohomology, 142
    formality, 51
    fundamental group, 47
    Hard Lefschetz Condition, 51, 63
    Hasegawa theorem, 51
    HLC for 4-dimensional, 54
    Hodge decomposition, 51, 77
    Kähler, *see* theorem, Hasegawa
        (characterization of Kähler
        solvmanifolds)
    left-invariant object, 48
    left-invariant symplectic cohomologies, 91
    Mal'tsev rigidity theorem, 47
    Mostow theory on cohomology of
        solvmanifolds, 99
    of splitting type, 141
    parallelizable, 47
    Salamon notation, 48
    three-dimensional holomorphically
        parallelizable, 111, 147
spectral sequence
    canonical spectral sequence, 24
    Hodge and Frölicher spectral sequence, 7
    Hodge to de Rham spectral sequences, 7
Spencer, D. C., xvii, 36, 196, 200
splitting type
    solvmanifold, 141
splitting-type solvmanifold
    $\partial\bar{\partial}$-Lemma, 143

Bott-Chern cohomology, 143
Dolbeault cohomology, 142
square, 75
stable
under small deformations of the
almost-complex structure, 196
under small deformations of the complex
structure, 196
stable (principal) torus-bundle series, 102
stable property under deformations, *see*
deformation, openness
step of nilpotency, 46
Streets, J., xvi, 225
strictly pseudo-convex, 8
string theory, 66
strong Kähler with torsion metric, 131, 226
stronger semi-continuity, 216
counterexample, 219
strongly-Gauduchon
and $\partial\bar{\partial}$-Lemma, 228
non-closedness, 139
strongly-Gauduchon cone, 228
characterization, 229
strongly-Gauduchon metric, vii, xvi, 41, 131,
228
structure current, 221, 229
complex currents ($\mathfrak{C}_{p,J}$), 222
structure cycle, 221
complex cycles ($\mathcal{Z}\mathfrak{C}_{p,J}$), 222
Sullivan, D. P., x, xvi, 27, 31, 35, 74, 76, 87,
221, 222
symplectic, 9
as a generalized complex structure, 25
symplectic Calabi and Yau equation, 224
symplectic co-differential operator, 14
symplectic cohomologies, 91, 107
symplectic cohomology
$d\,d^{\Lambda}$-Lemma, 19
$\left(d + d^{\Lambda}\right)$-cohomology, 16
$\left(d\,d^{\Lambda}\right)$-cohomology, 16
Hodge theory, 16
primitive $\left(d + d^{\Lambda}\right)$-cohomology, 17
primitive $\left(d\,d^{\Lambda}\right)$-cohomology, 18
primitive de Rham cohomology, 159
symplectic cones, xv, 157, 223
compatible cone, xv, 223
for Lie algebras, 158
linear symplectic cones, 173
qualitative comparison (Donaldson tamed
to compatible question), 224
quantitative comparison (Li and Zhang
theorem), 223
tamed cone, xv, 223

symplectic structure, vii, xi, 90, 170
canonical orientation, 13
compatible, 223
taming, 222
symplectic-$\star$-operator, 14
symplectically-harmonic form, xi
for Kähler manifolds, 20
symplectomorphism, 17

tamed cone, xv, 223
tamed to compatible, xvi, 158, 224
almost-complex counterexample in
dimension higher than 4, 225
complex holomorphic-tamed non-Kähler
(Li and Zhang, and Streets and Tian
question), 225
for Lie algebras, 158, 225
for Moĭšezon manifolds, 225
state of art, 224
taming symplectic structure, 222
Taubes current, 185, 225
Taubes, C. H., 185, 224
theorem
$\mathcal{C}^{\infty}$-pure-and-fullness for left-invariant
D-complex 4-dimensional
nilmanifolds, 160
$\partial\bar{\partial}$-Lemma for Kähler, Moĭšezon, class $\mathcal{C}$,
77
$\partial\bar{\partial}$-Lemma for Kähler manifolds, 34
Akizuki and Nakano identities, 30
Alessandrini and Bassanelli (non-openness
of balanced), 200
Andreotti and Grauert, 8
Belgun symmetrization trick, 50
Belgun symmetrization trick in the
almost-complex setting, 170
Benson and Gordon, 51
Bochner linearization theorem, 52
Bock (HLC for 4-dimensional
solvmanifolds), 54
Bogomolov, *see* theorem, Tian and Todorov
Bott-Chern characterization of $\partial\bar{\partial}$-Lemma,
84
Bott-Chern cohomology of 6-dimensional
nilmanifolds, 131
Bott-Chern cohomology of
holomorphically parallelizable
solvmanifolds, 143
Bott-Chern cohomology of splitting-type
solvmanifolds, 143
Brylinski, and Fernández, Ibáñez, and
de León (degeneration of Hodge

and Frölicher spectral sequence for
symplectic structures), 91

Buchdahl, 41

Cartan and Lie, *see* theorem, Lie and
Cartan

Cavalcanti ($d\,d^\Lambda$-Lemma for symplectic
manifolds), 19

Cavalcanti ($d_H\,d_H^{\mathcal{J}}$-Lemma and
decomposition), 23

Cavalcanti (characterization of $d_H\,d_H^{\mathcal{J}}$-
Lemma), 24

Ceballos, Otal, Ugarte, and Villacampa
(non-closedness of strongly-
Gauduchon, and balanced),
139

Chow, vii

cohomological characterization of $\delta_1\delta_2$-
Lemma (for graded bi-differential
vector spaces), 89

cohomological characterization of
$\partial_{\mathcal{J},H}\bar{\partial}_{\mathcal{J},H}$-Lemma for generalized
complex structures, 92

cohomologies of deformations of Iwasawa
manifold, 130

comparison of metric cones, 231

Console and Fino (Dolbeault cohomology
for nilmanifolds), 101

Console and Fino (inclusion of left-
invariant Dolbeault cohomology),
100

Console and Fino (openness of left-
invariance of cohomology),
102

Cordero, Fernández, Gray, and Ugarte, 101

Darboux, 9

decomposition of $H_{dR}^2(X;\mathbb{R})$ for
symplectic manifolds, 159

Deligne, Griffiths, Morgan, and Sullivan,
67

Deligne, Griffiths, Morgan, and Sullivan
($\partial\bar{\partial}$-Lemma and formality), 35, 77

Deligne, Griffiths, Morgan, and Sullivan
(behaviour under modification), 76

Deligne, Griffiths, Morgan, and Sullivan
(characterization of $\partial\bar{\partial}$-Lemma
for double complexes), 74

Deligne, Griffiths, Morgan, and Sullivan
(characterization of $\partial\bar{\partial}$-Lemma
for complex manifolds), 76

Deligne, Griffiths, Morgan, and Sullivan
(equivalent formulations of
$\partial\bar{\partial}$-Lemma), 74, 76

Demailly and Păun, 42

Dolbeault (sheaf-theoretic interpretation of
Dolbeault cohomology), 5

Dolbeault and Grothendieck lemma, 5, 72

Drăghici, Li, and Zhang ($\mathcal{C}^\infty$-pure-
and-fullness and $\partial\bar{\partial}$-Lemma),
165

Drăghici, Li, and Zhang ($\mathcal{C}^\infty$-pure-and-
fullness for complex manifolds
satisfying $\partial\bar{\partial}$-Lemma), 166

Drăghici, Li, and Zhang ($\mathcal{C}^\infty$-pure-and-
fullness for complex surfaces),
166

Drăghici, Li, and Zhang ($\mathcal{C}^\infty$-pure-and-
fullness for Kähler manifolds),
166

Drăghici, Li, and Zhang ($\mathcal{C}^\infty$-pure-
and-fullness in dimension 4),
166

Drăghici, Li, and Zhang (semi-continuity
property in dimension 4), 212

Eastwood and Vigna Suria, 9

Ehresmann, 36

Fino and Tomassini ($\mathcal{C}^\infty$-pure-and-
fullness for pure-type harmonic
representatives), 162

Fino and Tomassini (pure-and-fullness for
Hard Lefschetz Condition), 162, 177

Fino and Tomassini, and Drăghici, Li,
and Zhang ($\mathcal{C}^\infty$-pureness of
almost-Kähler), 177, 184

Fino, Parton, and Salamon (6-dimensional
pluri-closed nilmanifolds), 227

Frölicher inequality, ix, 8, 81

Fu and Yau (openness of balanced under
additional conditions), 196

GAGA, vii

Gualtieri (generalized Darboux), 24

Guillemin, 19

Hörmander, 8

Hahn and Banach separation theorem, 230

Hano, 51

Harvey and Lawson, 42

Harvey and Lawson (intrinsic
characterization of Kähler),
230

Hasegawa (characterization of Kähler
nilmanifolds), 51

Hasegawa (characterization of Kähler
solvmanifolds), 51

Hattori, xii, 50, 98

Hodge, 6

Hodge decomposition theorem for Kähler
manifolds, 31

holomorphic tamed for 6-dimensional
    nilmanifolds, 226
holomorphic tamed for nilmanifolds, 227
inequality *à la* Frölicher
    for complex structures, 92
inequality *à la* Frölicher (for graded
    bi-differential vector spaces), 89
inequality *à la* Frölicher, 82
inequality *à la* Frölicher for generalized
    complex structures, 92
inequality *à la* Frölicher for symplectic
    structures, 91
intrinsic characterization of semi-Kähler,
    230
invariance of cohomologies, 104
invariance of cohomologies for
    nilmanifolds, 107
invariance of cohomology for nilmanifolds
    under deformations, 109
Kähler identities, 30
Kasuya (de Rham cohomology of
    solvmanifolds), 140
Kasuya (Dolbeault cohomology of
    holomorphically parallelizable
    solvmanifolds), 142
Kasuya (Dolbeault cohomology of
    splitting-type solvmanifolds), 142
Kasuya (Hodge symmetry and
    decomposition for splitting-
    type solvmanifolds), 142
Kodaira and Spencer, 40
Kodaira and Spencer (stability of
    Kählerness, 196
Kodaira embedding, vii, 28
Kodaira, Miyaoka, and Siu, 41
Kuranishi, 39
Kuranishi space of Iwasawa manifold, 113
Lamari, 41
LeBrun and Poon, and Campana, 41
Lefschetz decomposition theorem for
    Kähler manifolds, 31
Li and Tomassini (cohomological
    decomposition for 4-dimensional
    unimodular Lie algebras), 173
Li and Tomassini (comparison of linear
    symplectic cones), 173
Li and Tomassini (Donaldson conjecture
    for 4-dimensional unimodular Lie
    algebras), 173
Li and Zhang (comparison between
    symplectic cones), 223
Li and Zhang (intrinsic characterization of
    almost-Kähler), 230
Lie and Cartan, 45

Lin ($\mathfrak{sl}(2; \mathbb{R})$-module on currents), 12
Lupton and Oprea, 51
Macrì (left-invariant symplectic
    cohomologies), 91
Mal'tsev, 48
Mal'tsev (fundamental group of a
    nilmanifold), 47
Mal'tsev (rigidity for nilmanifolds), 47
Mathieu, 19
Merkulov, 19
Michelsohn, 42
Michelsohn trick, 228
Milnor lemma, 48
Mostow (cohomology of solvmanifolds),
    99
Mostow (rigidity for solvmanifolds), 47
Mostow structure theorem, 49
Newlander and Nirenberg, viii, 4
Nomizu, xii, 49, 98
non-lower-semi-continuity of $\dim_{\mathbb{R}} H_{J_t}^+$,
    214
non-openness of **D**-Kähler, 160, 197
non-upper-semi-continuity of $\dim_{\mathbb{R}} H_{J_t}^-$,
    213
openness of $\partial\bar{\partial}$-Lemma, 87
pluriclosed metrics on 6-dimensional
    nilmanifolds, 137
Poincaré lemma, 72
Poincaré lemma for currents, 43
Popovici (strongly-Gauduchon metric),
    228
relation between linear-$\mathcal{C}^\infty$-pure-and-
    fullness at the kth stage and
    $\mathcal{C}^\infty$-pure-and-fullness at the kth
    stage, 171
relations between $\mathcal{C}^\infty$-pure-and-fullness
    and pure-and-fullness, 161
result *à la* Nomizu for $H_J^S(X; \mathbb{K})$, 170
Rollenske, 39
Rollenske (deformations of nilmanifolds),
    103
Rollenske (Dolbeault cohomology for
    nilmanifolds), 101
Rollenske (Kuranishi space of
    holomorphically parallelizable
    nilmanifolds), 116
Sakane, 101
Schweitzer (finite-dimensionality of
    Bott-Chern cohomology), 68
Schweitzer (Hodge theory for Bott-Chern
    cohomology), 68
Schweitzer (isomorphism between Bott-
    Chern and Aeppli cohomologies),
    71

Schweitzer (semi-continuity of the
    dimension of Bott-Chern
    cohomology), 68
Schweitzer lemma, 72
Serre duality, 6
Sullivan, 222
theorem à la Nomizu for Bott-Chern
    cohomology, xii
Tian and Todorov, 39
Tomasiello (stability of $d\,d^{\mathcal{J}}$-Lemma), 30
Tralle and Kedra, 51
Tseng and Yau (decomposition for
    $\left(d+d^{\Lambda}\right)$-cohomology), 18
Tseng and Yau (decomposition for
    $\left(d\,d^{\Lambda}\right)$-cohomology), 18
Tseng and Yau (Hodge theory for
    symplectic $\left(d+d^{\Lambda}\right)$-cohomology),
    16
Tseng and Yau (Hodge theory for
    symplectic $\left(d\,d^{\Lambda}\right)$-cohomology), 16
Tseng and Yau (invariance of
    symplectic cohomologies
    under symplectomorphisms and
    Hamiltonian isotopies), 17
Weyl identity, 167
Wu (openness of balanced under additional
    conditions), 196
Yan ($\mathfrak{sl}(2;\mathbb{R})$-representation), 10
Yan (Lefschetz decomposition), 10
Yan (solution to the Brylinski conjecture),
    19
Zhang's Lefschetz type property and
    $\mathcal{C}^{\infty}$-pure-and-fullness, 186
Thurston, W., 51
Tian, G., xvi, 39, 225
Tomasiello, A., 30
Tomassini, A., 30, 88, 99, 157, 158, 162, 168,
    170, 173, 177, 184, 185, 204, 209,
    221, 225
topological Lie group, 45
torus-bundle series, xiii, 102
    principal, 102
    stable, 102
total vector space, 89
Tralle, A., 51
transform
    B-transform, 26
translation on a Lie group, see Lie group,
    left-translation
transverse form, 221

Tseng and Yau (isomorphism between
    $\left(d+d^{\Lambda}\right)$-cohomology and
    $\left(d\,d^{\Lambda}\right)$-cohomology), 17
Tseng, L.-S., ix, xi, 13, 16, 66, 91, 158
Turkowski, P., 63
twisted differential, 21
twistor space, 41, 77
type (I), 46
type (R), 46
type of a generalized complex structure, 22
    type 0 (symplectic), 25
    type n (complex), 24

Ugarte, A., 55, 87
Ugarte, L., xiv, 40, 55, 95, 97, 131, 134, 140
universal family, see family, universal

V-manifold, 52
van Dantzig, D., 27
Varouchas exact sequences, 80
Varouchas, J., ix, x, 65, 79
versal family, see family, versal
Vezzoni, L., 226, 227
Villacampa, R., xiv, 55, 87, 95, 131, 134, 140
Voisin, C., 28

weak-$*$ topology, 42
Weil, A., 27
Weinstein and Thurston problem, 51
Weinstein, A., 51
Weyl identity, 167, 185
Weyl, H., 20
Wu, C. C., 196

Yan, D., xi, 13, 15
Yau, S.-T., ix, xi, 13, 16, 27, 28, 66, 91, 158,
    196, 224, 232

Zariski decomposition, 66
Zhang's Lefschetz type property
    and $\mathcal{C}^{\infty}$-pure-and-fullness, 186
Zhang's Lefschetz-type property, 185
Zhang, W., xiv–xviii, 151–153, 155, 157, 158,
    165, 166, 168, 175, 176, 184, 185,
    212, 220, 221, 223–225, 231
Zippin, L., 45

LECTURE NOTES IN MATHEMATICS  Springer

Edited by J.-M. Morel, B. Teissier; P.K. Maini

Editorial Policy (for the publication of monographs)

1. Lecture Notes aim to report new developments in all areas of mathematics and their applications - quickly, informally and at a high level. Mathematical texts analysing new developments in modelling and numerical simulation are welcome.

   Monograph manuscripts should be reasonably self-contained and rounded off. Thus they may, and often will, present not only results of the author but also related work by other people. They may be based on specialised lecture courses. Furthermore, the manuscripts should provide sufficient motivation, examples and applications. This clearly distinguishes Lecture Notes from journal articles or technical reports which normally are very concise. Articles intended for a journal but too long to be accepted by most journals, usually do not have this "lecture notes" character. For similar reasons it is unusual for doctoral theses to be accepted for the Lecture Notes series, though habilitation theses may be appropriate.

2. Manuscripts should be submitted either online at www.editorialmanager.com/lnm to Springer's mathematics editorial in Heidelberg, or to one of the series editors. In general, manuscripts will be sent out to 2 external referees for evaluation. If a decision cannot yet be reached on the basis of the first 2 reports, further referees may be contacted: The author will be informed of this. A final decision to publish can be made only on the basis of the complete manuscript, however a refereeing process leading to a preliminary decision can be based on a pre-final or incomplete manuscript. The strict minimum amount of material that will be considered should include a detailed outline describing the planned contents of each chapter, a bibliography and several sample chapters.

   Authors should be aware that incomplete or insufficiently close to final manuscripts almost always result in longer refereeing times and nevertheless unclear referees' recommendations, making further refereeing of a final draft necessary.

   Authors should also be aware that parallel submission of their manuscript to another publisher while under consideration for LNM will in general lead to immediate rejection.

3. Manuscripts should in general be submitted in English. Final manuscripts should contain at least 100 pages of mathematical text and should always include

   - a table of contents;
   - an informative introduction, with adequate motivation and perhaps some historical remarks: it should be accessible to a reader not intimately familiar with the topic treated;
   - a subject index: as a rule this is genuinely helpful for the reader.

   For evaluation purposes, manuscripts may be submitted in print or electronic form (print form is still preferred by most referees), in the latter case preferably as pdf- or zipped ps-files. Lecture Notes volumes are, as a rule, printed digitally from the authors' files. To ensure best results, authors are asked to use the LaTeX2e style files available from Springer's web-server at:

   ftp://ftp.springer.de/pub/tex/latex/svmonot1/ (for monographs) and
   ftp://ftp.springer.de/pub/tex/latex/svmultt1/ (for summer schools/tutorials).

Additional technical instructions, if necessary, are available on request from lnm@springer.com.

4. Careful preparation of the manuscripts will help keep production time short besides ensuring satisfactory appearance of the finished book in print and online. After acceptance of the manuscript authors will be asked to prepare the final LaTeX source files and also the corresponding dvi-, pdf- or zipped ps-file. The LaTeX source files are essential for producing the full-text online version of the book (see http://www.springerlink.com/openurl.asp?genre=journal&issn=0075-8434 for the existing online volumes of LNM). The actual production of a Lecture Notes volume takes approximately 12 weeks.

5. Authors receive a total of 50 free copies of their volume, but no royalties. They are entitled to a discount of 33.3 % on the price of Springer books purchased for their personal use, if ordering directly from Springer.

6. Commitment to publish is made by letter of intent rather than by signing a formal contract. Springer-Verlag secures the copyright for each volume. Authors are free to reuse material contained in their LNM volumes in later publications: a brief written (or e-mail) request for formal permission is sufficient.

Addresses:

Professor J.-M. Morel, CMLA,
École Normale Supérieure de Cachan,
61 Avenue du Président Wilson, 94235 Cachan Cedex, France
E-mail: morel@cmla.ens-cachan.fr

Professor B. Teissier, Institut Mathématique de Jussieu,
UMR 7586 du CNRS, Équipe "Géométrie et Dynamique",
175 rue du Chevaleret
75013 Paris, France
E-mail: teissier@math.jussieu.fr

For the "Mathematical Biosciences Subseries" of LNM:

Professor P. K. Maini, Center for Mathematical Biology,
Mathematical Institute, 24-29 St Giles,
Oxford OX1 3LP, UK
E-mail: maini@maths.ox.ac.uk

Springer, Mathematics Editorial, Tiergartenstr. 17,
69121 Heidelberg, Germany,
Tel.: +49 (6221) 4876-8259

Fax: +49 (6221) 4876-8259
E-mail: lnm@springer.com